LIBRARY
LEWIS-CLARK STATE COLLEGE
LEWISTON, IDAHO

D0758249

Our Chemical Selves

Gender, Toxics, and Environmental Health

Edited by Dayna Nadine Scott

UBCPress · Vancouver · Toronto

23 22 21 20 19 18 17 16 15   5 4 3 2 1

Printed in Canada on FSC-certified ancient-forest-free paper (100% post-consumer recycled) that is processed chlorine- and acid-free.

Library and Archives Canada Cataloguing in Publication

Our chemical selves : gender, toxics, and environmental health / edited by Dayna Nadine Scott.

Includes bibliographical references and index.
Issued in print and electronic formats.
ISBN 978-0-7748-2833-8 (bound). – ISBN 978-0-7748-2834-5 (pbk.). – ISBN 978-0-7748-2835-2 (pdf). – ISBN 978-0-7748-2836-9 (epub)

1. Environmental toxicology – Canada 2. Women – Health and hygiene – Canada 3. Pollution – Environmental aspects – Canada 4. Pollution – Political aspects – Canada 5. Pollution – Social aspects – Canada 6. Feminism – Canada I. Scott, Dayna Nadine, author, editor

RA566.5.C3O97 2015   362.1969'800971   C2014-906369-5
C2014-906370-9

Canadä

UBC Press gratefully acknowledges the financial support for our publishing program of the Government of Canada (through the Canada Book Fund), the Canada Council for the Arts, and the British Columbia Arts Council.

This book has been published with the help of a grant from the Canadian Federation for the Humanities and Social Sciences, through the Awards to Scholarly Publications Program, using funds provided by the Social Sciences and Humanities Research Council of Canada.

UBC Press
The University of British Columbia
2029 West Mall
Vancouver, BC V6T 1Z2
www.ubcpress.ca

Contents

Figures and Tables

Figures

Tables

Abbreviations

ABS	acrylonitrile-butadiene-styrene
BBP	butyl benzyl phthalate
BPA	bisphenol A
CAW	Canadian Auto Workers
CBCRI	Canadian Breast Cancer Research Initiative
CCNM	Canadian College of Naturopathic Medicine
CDW	Federal-Provincial-Territorial Committee on Drinking Water
CEDAW	Convention on the Elimination of All Forms of Discrimination against Women
CEPA	Canadian Environmental Protection Act, 1999
CFIA	Canadian Food Inspection Agency
CIC	Citizenship and Immigration Canada
CIHR	Canadian Institutes of Health Research
CMP	Chemicals Management Plan
CRTC	Canadian Radio-television Telecommunications Commission
DBP	dibutyl phthalate
DDT	dichlorodiphenyltrichloroethane
DEHP	di(2-ethylhexyl) phthalate or bis(2-ethyhexyl) phthalate

DES	diethylstilbestrol
DiBP	diisobutyl phthalate
DiIDP	diisodecyl phthalate
DiNP	diisononyl phthalate
DNOP	di-*n*-octyl phthalate
DSL	Domestic Substances List
EDCs	endocrine-disrupting chemicals
ENGOs	environmental non-governmental organizations
EPC	European Patent Convention
EPR	extended producer responsibility principles
FASD	Fetal Alcohol Spectrum Disorder
GMOs	genetically modified organisms
HRSDC	Human Resources and Skills Development Canada
HWS	hazardous waste sites
IARC	International Agency for Research on Cancer
MAC	maximum acceptable concentration
MACs	maximum acceptable concentrations
MEHHP	mono-(2-ethlyl-5-hydroxyhexyl) phthalate
MEHP	mono-(2-ethylhexyl) phthalate
MEK	methyl ethyl ketone
MEOHP	mono-(2-ethly-5-oxohexyl) phthalate
MEOP	mono-(2-ethyl-5-oxohexyl) phthalate
MEP	mono-ethyl phthalate
MiBP	mono-isobutyl phthalate
MIREC	Maternal-Infant Research on Environmental Chemicals study
MMP	monomethyl phthalate
MSDS	material safety data sheet
MSW	municipal solid waste
NHANES	National Health and Nutrition Examination Survey
NiCd	nickel-cadmium
NIOSH	National Institute for Occupational Safety and Health
NP	4-nonylphenol
NPRI	National Pollutant Release Inventory

NSLC	Nova Scotia Liquor Corporation
OECD	Organization for Economic Co-operation and Development
OHCOW	Occupational Health Clinics for Ontario Workers
OHSA	Occupational Health and Safety Act (Ontario)
PAH	polycyclic aromatic hydrocarbon
PBB	polybrominated biphenyls
PBDE	polybrominated diphenyl ether
PCB	polychlorinated biphenyl
POP	persistent organic pollutant
PTSD	post-traumatic stress disorder
PVC	polyvinyl chloride
rBGH	recombinant bovine growth hormone
SAWP	Seasonal Agricultural Workers Program
SNAc	Significant New Activity Notices
TCDD	2,3,7,8-tetrachlorodibenzo-*p*-dioxin
TCE	trichloroethylene
TCEP	(tris(2-chloroethyl) phosphate
TEXB	total effective xenoestrogen burden
TFWP	Temporary Foreign Workers Program
WOHIS	Windsor Occupational Health Information Service
WOSH	Windsor Occupational Safety and Health

Foreword Water Is Life

Josephine Mandamin

I have the habit of introducing myself the way I've been taught to speak to the Creator and to the Spirits. I acknowledge them as they stand behind me, the protectors of our nations as the Eagle clans and all the spirits that have been offered tobacco to be with us in the work that we do as Anishinabe people. I introduce myself with my Anishinabe name, Beedawsigaye of the Awassisi, which means "the one who comes with the light." My Clan is the Fish Clan, and therefore a relative to the bullhead and catfish. They are fighters and really powerful fish. I am part of that clan which I have watched in action since I was a kid. I know what they are like. They are very powerful fish. I am also a fourth-degree Midewewin person. I have entered the Midewewin lodge four times. I was stood up – meaning I was presented, acknowledged, and recognized – as an Ogichidaw in 2004 at Pipestone sundance. And so I have also been holding the three maidens' pipe for the Ogitchidaw in Pipestone.

I also want to express how privileged we are to be where we are today in terms of understanding the impacts of pollution and the poison that is around us, which is very detrimental to our health – not only our health but that of generations to come. I want to talk about the role that is given to us as Anishinabe people, and confess that I also ask 'Why me?' even though that was something that was never, never, never put to us to ask when something is given to us.

The year 2000 was when messages about the water – what's going to happen to the water – began to be talked about. And I was at a sundance when I heard an elder talk about the water – how in thirty years from now, that water was going to cost as much as an ounce of gold. And standing there, I could not believe how it could be, because our water is so beautiful when I go to the reserve. When I go to Thunder Bay, I go to Kakabeka Falls and get my fresh water, spring water. And so I could not fathom how an ounce of water would cost as much as an ounce of gold. And I don't know today how much an ounce of gold is – maybe $500, $600. I cannot imagine how much it is going to be in thirty years. When the elder finished talking, he looked at the crowd and asked, "What are you going to do about it?" And that question hit me right in the heart. I felt it was directed at me. And so for two and a half to three years I went around talking to people about it, talking to other women about it. One thing he said was that women have to start picking up their bundles, doing their work. And that was another thing that I had to find out. What does a bundle mean? What is my work as Anishinabe? And I spent that time talking, encouraging women to start picking up their work. And then listening to the elder grandmothers.

Finally, in the winter of 2002, we were sitting around the house with tea, coffee, and bannock – five of us women talking about water. And I said, there has been so much happening – many people have said, "Yes, we'll do something about it," but no one seems to be doing anything about it. And one woman said, "Lake Superior is here" – you could see it from where we were – "we can walk around Lake Superior." And we laughed about it – how you talk about something, and you laugh about it, like it's a big joke. To see us women walking around Lake Superior. And then we talked about something else. And then came back to it. And Christine said: "A pail of water – walk around the lake with a pail of water." And so it came to be that that's what we planned. And I know that was Easter time. And we chose Easter Sunday as the day we would start at the south shore of Lake Superior. And so we started walking. And at that time there was a mark on the ground where we started walking. There was a bunch of men and women walking behind me. Then after twenty minutes I looked back, and there was nobody there. So I was walking alone. I continued alone for the rest of the day. And we had my pickup truck, an old Chevy with an orange light on top. And a lady was driving my truck. I was walking with my staff and water pail. I asked a man where the others were, and the man said they had other things to do, they had work to do – they had positions, school, work they had to go to. So I was alone.

So the next day, we were at one of these places when an old man came and said, "We want to do the travelling song for you." So we stopped at this plaque by the side of the road. He brought his son and grandson, and the three did a travelling song for us. And I was standing there and wondering what would become of us. Because we were standing there, and there was Lake Superior, and we were alone, and we were wondering how this was going to happen. And before the drumming finished, a car pulled up. And this old man got out of his vehicle and stood there and waited until the drumming stopped. When it finished, he came up to us and said he saw us and had to turn back and come and say that his father had said to him when he was young that there would be a time when the women would walk around the Great Lakes. And he said he was very moved by what we were doing. So he gave us an eagle feather and I put it on the pail and he left. I looked at the person with me and asked, "So does this mean we have to walk around the five Great Lakes?" Because originally it was only Lake Superior. And he said, "I don't know." He saw how hard it was for me to walk with the staff and pail at the same time. When we started walking with the water, we had to keep moving with it. It keeps going, like the river. And it's like the songs. The water songs are never-ending. And we have to honour the songs as we walk.

On the third day, my friend from Saskatchewan came to help. And that helped a bit and we reached Duluth. And our Grand Chief, Eddie Benton, came to help. And there was a powwow in Duluth and we received some donations. We started with $85 – a donation from Easter Monday. That's all we had. We didn't ask. We just did it. And that's something that we have always done as Anishinabe people. We just do it, without question, without funding. Then people come out and help. All we did was ask for permission. We asked Annie Wilson and Al Hunter for their blessing. They said what we were doing was good but that we should take extra socks. That was the advice we got and that is what we did. We took stuff that we needed, the bare essentials.

Then we had Minnesota. In Minneapolis-St. Paul, a group of young people came to help us push along. Then we got to Thunder Bay. And Lake Superior – when I think of it, I think of it as very, very strong water. And it reminds me of a woman, because it is very unpredictable. Sometimes it is very gentle and kind, sometimes very powerful and overwhelming, and you do not know what to expect from one moment to another. I remember I was waiting at Old Woman Bay. The water was beautiful and calm. I didn't see this wave coming and it almost swept me off my feet. So I always think of Lake Superior as an old woman – very powerful, very strong, unpredictable. And it has also taken many lives – you may have heard the history of Lake Superior. In 2003, we finished in thirty-six days – over a month it took us to walk around. During that time, we didn't ask for funds – we only relied on people's goodness to fill up our tank. If we had money, we'd stay at a hotel. It was a very spiritual walk. There are many things that I could tell you. How did it happen? Sometimes you question.

In 2004, we did the upper part of Lake Michigan. That Michigan walk was very telling – that was the time that we understood how our grandmothers, grandfathers – our ancestors – had left us a legacy that we are to remember. We saw many signs as we walked around the upper peninsula – we saw signs that our grandfathers, grandmothers were there. We saw the shapes, the drawings on the rocks. We saw trees showing that the elders must have been there. And it was very powerful, very beautiful. And we visited this place in Mount Pleasant. We walked that mountain. There's no mountain – I don't know why it's called mountain. And we visited a museum there, and that is where we really understood that we were really destined to do what we were doing, to be there. Because it showed us that our people were very strong, even though they had nothing. They were resilient. And then we looked at the other society. You know, what are the white people leaving for the children? We saw the destruction, the roads, the mining, the trees that are being cut – bald-headed mother. And we thought of our mother too. How she is being destroyed by the money-changers that are making money off her – the mining, for progress, for money. And so we understood that we would never do that to our children, our grandchildren that are yet to come. We would leave the message for them that we did something – that we are doing something. And the white people are leaving behind the destruction – the construction, as they call it – but it is really destruction. The mining that is gouging our mother. Her hair that is being cut. And we see women who are going through the same thing because of the radiation and the cancer that is going through our bodies. Everything that Mother Earth is going through, our women are going through – the hysterectomies, their bodies being cut out like Mother Earth. So we see that destruction is not really our way – that the white people are doing the destruction to the environment. So Lake Michigan was very telling. We saw that our work is something that is to be continued. And it was also the place where we heard that Nestlé was selling water from Lake Michigan as spring water around the world. And we had demonstrations. We came through demonstrations where people were fighting the Nestlé bottling works. And we saw trucks that looked like milk trucks that said "Spring Water from Michigan." And they were selling the water all over the world. And they probably still are.

And so in 2004-5, it was Lake Huron – that was when we understood how the men have to start to walk with the women – in balance. And it was in Sudbury, or in Whitefish, that we were invited to speak to a group of Chiefs and Councillors. And I spoke to them about the water walk and I did it all in the language because I knew many of them are fluent speakers. So I spoke to them about the balance of the fire and the water – how the mother, father, and son earth walk in balance. When I finished talking, no one said anything. And then the elder Gordon Waindubeins stood up and said that everything I said was true. And when we finished, we had a standing ovation from the Chief and Councillors when we were leaving the building.

The next day, we had men coming out to the highway waiting for us, to walk with us. That was the year when the men really started coming out. And yesterday I remembered the walk that we did from Sarnia to Port Huron. I remember they wouldn't let us walk to get on the other side. And Karl and Lincoln were the two who were walking with me that time and who were crossing the bridge with me with the water. But they wouldn't let us cross the bridge because the governor or someone had to be told that we were going to walk. We waited half an hour, and then out of frustration I just started walking and the two men started walking behind me. The two guards couldn't do anything. I dared them to stop me. And they wouldn't dare do anything. It took us about fifteen minutes to walk across. It didn't take long – I just sped along, with Lincoln and Karl right behind me. We got to the other side and there was a drum waiting for us – the Americans were waiting on the other side. So Lake Huron was a reminder of how men can also be good supports for us as women who care for the water because they are also life givers for us. Because they take care of us and they bring the warmth in our communities and our lives. And so we respect the men too and they come and join us in ceremonies.

When I think about how our men, our young men, boys, and young women have to carry that life within their bodies and they need to start thinking about whether they are going to pass it to their children and grandchildren – how is that going to be passed on? And the only way is through the fire and water – they have to be water and fire keepers. And that is what we are doing here today – bringing the message to people that we must do our work. In 2006, we walked Lake Ontario. You know, a lot of nuclear waste goes through that water. We couldn't touch the water because we were afraid. We saw many deaths – many fish on the shore. When we crossed the border from Kingston into New York, one of the girls said this water is so heavy. And it was heavy – we were really tired from carrying this water. Our shoulders hurt at the end of the day. So we kept changing the water to see if it was the water. And it was the water. A year later, someone sent me an article written by scientists – they tested the water in Lake Ontario and it was labelled heavy water. We knew that the year before. It was heavy because of the mercury, the chemicals – all that stuff in the water that makes it heavy. And we did experience that. So when we think of Ontario we think of the correlation between science and the Anishinabe way of life. That we know things that it takes the white people a while to understand. They have to do a scientific study to prove what we already know. So we know we are very smart people.

In 2007, we walked Lake Erie. To be stoic, you have to be patient. And Lake Erie was where we had people come and make fun of us. They would do the war whoops when we would walk by them. Trucks would drive by and they would yell, "Get a job!" And you wanted to yell back at them but I kept telling the young women, "Don't listen to them." And we walked through a town and there would be young men walking behind the young women who were walking with the water making really sexual remarks to them. So it was really hard. Then we got to Windsor – we got to Windsor after we crossed the bridge in Detroit. Detroit was a hard walk too. We started off really early in the morning – it was still dark. We wanted to get through Detroit and to the Canadian border as early as possible. When we got there, my nephew let out a sigh of relief and said, "It's good to be home." It was really hard for them because of the racism and the work we were doing – people making fun of us. People would stop us and ask us if we were crazy – they would say, "You crazy Indians." And we were crazy. Anyone in their right mind wouldn't be doing what we were doing. So we would say, "Yes, we are crazy – crazy for the water." And they would tell us about what they were doing for the lakes – dredging. And we'd say, "Yes, you think you are doing a good thing, but we know what the dredging does to the water." And we'd talk about the dredging and how it kills all the plants under the water when they dredge. So they think they are doing a good thing, but they don't really know what they are doing. So we were met by many people on the other side. We were met by the Anglicans, different churches. They really helped us on our way until we got to our destination again.

The next year, we walked the bottom part of Lake Michigan because we did the upper part in 2004. And that is something – I listen to my elders – when we walked Lake Michigan in 2004, my elder said, "I had a dream. I don't want to discourage you from doing all of Lake Michigan but I had a dream that something was going to happen to the water walkers if they go through Chicago." He said, "I don't want to discourage you. I'll let you decide." I went to my sister and said this is what our grandchief said, and she said he has always had good dreams. We decided not to do all of Lake Michigan that year. And now that I look back, I can understand. Because there were a lot of political things happening in Chicago in 2004 because of Nestlé selling the bottled water.

In 2008 when we went through Chicago, we were welcomed. Even Toyota came and lent us their hybrid vehicle because it was very hard walking through the city. In Wisconsin, Milwaukee, they also gave us a big welcome. And the young people there really gave us an overwhelming welcome to their area. They were just in awe because of what we were doing. And they promised to take care of Lake Michigan. They promised to do it for the rest of their lives – to take care of the water and protect it. And that was encouraging for us – to know that there would be young people when we're old and gray and can't do it anymore. That the next generations would be there for us.

When we finished walking Lake Michigan, we also acknowledged the young people, acknowledging the young people because they are the now generation – I don't call them the next generation. They are the ones now sitting around listening to what is happening, watching what is happening. They are very concerned – a little angry – but very concerned about what is happening to the environment, to Mother Earth, wondering what they can do about it. And that is what I ask them. I ask them what they can do about it. So they go to their own communities and think of what they can do. And I tell them how they can fast for the water. Fasting is a very powerful, spiritual way of tapping into your dreams, into Mother Earth when you sit on her, in her lap. She comes to you and you can feel her presence. When you are without food or water for four days and four nights, you can appreciate that first taste of water, first little bit of food. If all of you sitting here took your children and families to fast, can you imagine how much less garbage there would be out there, how much less wasted food, how much less water you would use? And I usually ask people how much water did you use today? How much water did you waste today? And that makes them think of how they are using the water. How they can use the water in a good way. So they don't leave the water running – when they're brushing their teeth or showering. So there are many ways that we can look at how we are doing in protecting the water.

When we finished walking the five Great Lakes, we sat back and said we'd done it. I was at a meeting in Akwesasne near Cornwall, and Henry Lickers came up to me and said, "You are not done yet. The Great Lakes water flows down to the St. Lawrence River and to the ocean." I told my sister what he said and she said, "Why doesn't Henry Lickers walk the St. Lawrence River?" So we started in 2009. We started in Kingston and we walked the St. Lawrence River. The river reminds me of the ocean. It just gets bigger and bigger and bigger. When we stopped on the rivers, we took our time and really acknowledged the St. Lawrence River. We were told to stop at this place called Rivière-la-Madeleine" – we were told that is the place where the salt water meets the fresh water. And we ended our walk there. We'd stop every little while and taste the water but it still tasted fresh. By the time we did get there, it started to taste salty. We know how the salt water works with the fresh water. We know what salt does. It is a cleanser. I use salt when I feel really, really tired. I put Mediterranean sea salt in the bathtub for my back. You can feel the energy of the salt working on your body. So when we finished the St. Lawrence River, that was it. We took a whole year off. Then one of our eastern walkers talked about the salt water – how we need to bring it together. And so the four directions came into being. So this spring, we're going to be walking with the salt water, taking it from the East, the South, the North, the West, and the Gulf of Mexico, where all the oil has been spilled on the water. So that salt water from the four directions will be carried to the place where we started in 2003. And there will be a big celebration where we intermingle the salt water with the fresh water. Every time I think of the salt water and I look at the water, I have a feeling that it is waiting, waiting, waiting – waiting for us to go and get that water and walk with it and bring it to the centre of Lake Superior. So that is what we are going to do on the 12th of June. We are going to do a ceremony on the water to intermingle the spring water and salt water, so that it will be united. And so when people ask what is going to happen in 2012, I always listen to my teachers. And I say, where are you going to be in 2013 – you are going to be one year older. Nothing is going to happen. It is just a new beginning. When I think about how we can – not just as Anishinabe people, as Aboriginal people, people from all walks of life – we can all do something. What can we do? We can fast. Go out there and sit on Mother Earth. Listen to her. Be with her. Be without food and water for a few days. And know what it is to be without. And I sincerely hope that politicians, the rich people, the 5 percent who are rich people, can do that someday. To be without food and water – just to know what it is like to be without for four days. And have that little drink after four days, to know that first taste of food, how precious it is. To never waste food again. And to think of the animals also. I know that they are without food. Polar bears are finding no place to rest when they're swimming in the waters. They rest on icebergs and the icebergs are almost all gone. Their food is going – the seals are going somewhere else and they cannot find the food. So we have to think of our relatives. They came to us when we needed them. That's why we have our clans – our clans are very precious to us to understand. I know what my fish clan means. I know that it is to work for the water on behalf of all my clans. I know my relatives – the bear clans – how they have responsibility for the medicines. How they go poking around Mother Earth for medicines. And so our clans are very important.

Young people have a responsibility also. They will be mothers, fathers, aunts, uncles, great-grandparents soon. And I am sure that they want what it is we want also for the next generations. To bring that spring water to our young people so they can taste the water, how beautiful it is. To see the animals, the deer running around. The moose, the rabbits, that they will not be extinct. That is a dream that I have – that they will always be around for us. At a time when we needed them, they were there for us. To take care of them, like they took care of us. Migwech.

Acknowledgment

This text is adapted from remarks given to the Community Forum on Pollution and Action, Sarnia/Aamjiwnaang, February 2011.

Acknowledgments

This volume began as a project of the National Network on Environments and Women's Health (NNEWH) in 2008-9. Many of the chapters were initially written at that time by participants in a public symposium on "Consuming Chemicals" hosted by York University and a policy forum held in Ottawa with representatives from Health Canada and other relevant audiences. We are indebted to all of the students, policy makers, and academics who attended these events. The NNEWH was funded at that time through the Women's Health Contribution Program of Health Canada. Although the initial policy workshop that led to this volume was thus made possible through a financial contribution from Health Canada, the views expressed here do not necessarily represent those of Health Canada.

York University students have been engaged throughout. Lauren Rakowski provided comprehensive editing assistance at an early stage; Adrian Roomes contributed expert research assistance; Sarah Lewis spearheaded the process by which the chapters were updated and brought into renewed conversation with each other in 2012, and Vanessa Scanga ably brought us to the final stretches. Ellen Sweeney and Sarah Wiebe were doctoral students I worked with during this period – Ellen's work ethic motivated me, and Sarah's creativity inspired me. Jyoti Phartiyal has been the soul of NNEWH since I've known it, and she is an absolute wonder at what she does.

My collaboration with the Health and Environment Committee of Aamjiwnaang First Nation continued throughout the work on this volume, and my friends and collaborators in Aamjiwnaang have undoubtedly shaped this book in a myriad of ways. The work that Ada Lockridge, Ron Plain, Wilson Plain Sr. and Wilson Plain Jr., Ted White, Mike Plain, Vanessa Gray, and countless others have been undertaking to focus attention on the issue of chronic pollution exposures and their environmental health costs has benefited generations of activists, Indigenous and non-Indigenous, across the country and beyond. We owe them a debt of gratitude. The young people of the Aamjiwnaang Green Teens, especially Lindsay Gray and Mckay Swanson, were a strong source of inspiration to me as they overcame fears and challenges and developed into strong leaders.

I am also greatly indebted to my colleagues Pat Armstrong and Anne Rochon Ford, who were tremendous supports and sources of strength over the past several years as I attempted to balance the work of directing NNEWH with the joys and terrors of parenting, the delightful distractions that come with teaching and writing and trying to get tenure, and the reality that, as my great friend Liora Salter says, "life intervenes."

I lost my father during the course of this project. He was devastated by cancer over four short months. I continue to be stopped daily by the magnitude of the gaping hole this has left in my heart and in my family. My father shaped my work in a very indirect and diffuse way: he was not an easy person to persuade, and he was very smart. I had to work hard if I wanted him on my side. I've learned that I will be a better advocate if I imagine my opponents to be like him; if I respect them and try to persuade them anyway.

To Adrian, Rohan, and Tatiana,
For daily lessons in the art of resistance

Our Chemical Selves

Introduction The Production of Pollution and Consumption of Chemicals in Canada

Dayna Nadine Scott, Lauren Rakowski, Laila Zahra Harris, and Troy Dixon

This volume explores both the processes that *produce* chemicals (and chemical pollution) and the paths of exposure to chemicals that come about through our everyday *consuming*. In other words, we take the position that the consumption of chemicals in Canada is inseparable from the generation of pollution in this country. Exposure to chemicals occurs throughout our everyday lives – in the food we eat, the air we breathe, and the products we consume; on the street, in our schools and workplaces, and in our homes. These exposures, even though they are pervasive and widespread, are also decidedly uneven in their distribution. In fact, their distribution tends to vary along familiar social gradients, with disproportionate burdens falling on low-income, racialized, and Indigenous communities. At the same time, the effects of the exposures (and the burden of managing them) appear to rest disproportionately on the shoulders of women.

The "consumption" of chemicals takes on multiple meanings in this volume. Some authors interpret it literally, focusing on the chemical inputs we consume in our food, in our drinking water, in alcoholic beverages, and in the air we breathe. Other authors talk about the consumption of chemicals through the use of everyday consumer goods that indirectly expose us to chemicals, such as bisphenol A (BPA) in plastic packaging or canned food. These authors also tackle the implications of individual "precautionary consumption," noting the difficult choices we make daily about what to buy, which chemicals to "consume," and which to avoid, knowing that women vary dramatically in their capacity to engage in these practices. Some authors focus on workplace exposures, highlighting health risks to specific groups of workers and emphasizing the importance of making connections between occupational health and the environmental health movement. For still others, the "consumption" of chemicals is about the chemicals we are exposed to simply through living and breathing in our everyday living and work environments – an extension of the fact that in modern society humans rely on a variety of products, such as petroleum and plastic, the production and consumption of which inevitably result in chemical inputs to our air, water, and soil. In this chapter, we try to piece together a conceptual framework that captures each of these meanings and their interconnections.

A central aim of this volume is to expand our collective understanding of the links between social inequity, environmental risks, and the gendered division of health burdens in Canada. That the interconnections are crucial is made obvious in Josephine Mandamin's foreword. Mandamin, the Anishinabe grandmother best known for leading the Mother Earth Water Walk, demonstrates with her words and actions – between 2002 and 2009 she led annual walks around the five Great Lakes to raise awareness of the political, ecological, social, and spiritual significance of water – that social justice and healthy ecosystems are inextricably linked. She inspires in us a collaborative spirit and a determination to formulate a general theoretical foundation that can propel us towards greater understanding of what we need to do next. Thus we attempt here to articulate a framework capable of grounding the questions and issues that bind the various and diverse contributions of the authors. By establishing an analytical basis for subsequent chapters, this introductory chapter aims to guide readers along the overall trajectory of the volume, and influence the direction of future research, thinking, and policy on gender, chemicals, and environmental health justice.

New research is steadily emerging that links exposures to certain chemicals, at very low doses and at key times, to various environmental health harms (Batt 2008-9; Cooper and Vanderlinden 2009; Eyles et al. 2011; Gray, Nudelman, and Engel 2010; Krupp 2000; Chapter 10, this volume). These key times, known as *critical windows of vulnerability* – such as during early development, puberty, and pregnancy – can have a distinctly biological, developmental, and thus gendered nature. Emerging understandings of these diffuse, nuanced, chronic effects of chemicals on our bodies directly challenge the competing notion of "thresholds" that anchors toxicology and forms the bedrock of our regulatory approaches to toxic chemicals (Scott 2009b). The prevailing paradigm in toxicology, that "the dose makes the poison," is starting to unravel. At least for whole categories of key chemicals, we are finding that in many cases where we thought there were thresholds for health effects, there actually may not be. The implications of this for women are enormous and are the focus of this volume.

A Feminist Political Economy of Pollution

The theoretical foundation that we seek to develop can be framed as a "feminist political economy of pollution." In our conception, political economy is closely related to an environmental justice framework, and offers a strong basis from which to launch the remainder of the collection. At its most fundamental level, the theory interrogates systemic issues of power and ownership relating to the question of who profits from and exerts exploitative control over ecological resources, economic capital, and social labour (Gosine and Teelucksingh 2008). But we also widen the scope to allow consideration of how exploitative relationships between industrial actors and marginalized workers extend into peoples' everyday physical realities (Gosine and Teelucksingh 2008). As Verchick (1996) explains, we are working to "shatter the walls between health, occupational, and environmental issues and re-imagine the environment in ways that directly affect [our] everyday lives" (47). In this way, the political economy of pollution focuses attention on the inseparable links between profit incentives, the unsustainable production of waste, exploitative labour practices, racialization, and differential exposure to pollutants. Besides those who experience economic disparities, the "exploited" also comprise all of those who, directly or indirectly, at times unknowingly, take on additional health risks as a result of their place in the hierarchies and contours of capitalism. Women, accordingly, are at the forefront (Rahder 2009).

Feminist activists in the environmental justice movement are increasingly turning their attention to environmental harms derived not only from air, water, or soil contamination but also from toxic workplaces, urban planning and transit decisions, and conditions in public housing, among others (Adamson, Evans, and Stein 2002). The lineage of social and environmental injustice cannot be overlooked as we traverse the territory of environments and women's health, and the risks that have come to signify the post-industrial era. Political economists Jennifer Clapp and Peter Dauvergne outline the roots of the "political economy of pollution," pointing to recent waves of postwar industrialization and domination as an extension of slavery, patriarchy, colonialism, and imperialism (Clapp and Dauvergne 2005). It is becoming increasingly clear in Canada that racialized communities (Teelucksingh 2011; Nelson 2002) and Indigenous communities on reserves (Hoover et al. 2012) bear much more than their "fair share" of our environmental burdens, and that within these communities, women are disproportionately harmed (Agyeman et al. 2009; Rahder 2009; see also Wiebe 2013).

Our approach also has much in common with work in political ecology. The precepts of ecology, in fact, give a strong indication of what the overall model represents. Ecology refers to "a bounded system of dynamic interdependent relationships between living organisms and their physical and biological environment" (McMichael 1993, 40). Hence, an ecological approach can take into account the cyclical, holistic, and interdependent character of relations between bodies, chemicals, and systems of production that, in theory, should ground our understanding of life on a finite planet (McMichael 1993). During the 1970s and 1980s, however, many progressive academics began to challenge certain intellectual trends purporting to employ ecological methodologies to underestimate, obfuscate, and therefore legitimize the increasing devastation associated with human industrialization (Gray and Moseley 2005). As Leslie Gray and William Moseley aptly state in their article "A Geographical Perspective on Poverty-Environment Interactions" (2005): "Cultural ecology and ecological anthropology ... ignore[d] the role of political economy, power and history in shaping human-environmental interactions" (14). By trusting blindly in the ability of ecosystems to "correct" themselves, previous views of ecological modalities failed to capture how new, expansive, and relatively untested production methods – compounded by rising consumption and pollution rates – were irrevocably changing human health in ways that could not simply be corrected through "natural" life processes.

For these reasons, work in the political ecology mode departs from its roots in the discipline of ecology in several critical ways. As demonstrated by Neil Evernden in *The Social Creation of Nature* (1992), the theory is better able to capture "human-environmental interactions" because it recognizes that environmental health is not tangible; it cannot be readily seen or easily quantified (5). Rather, political ecologists challenge the value systems that seek to legitimize the root causes of environmental and health damage (Evernden 1992). For example, they point out the contradiction between the scientific limits of what an ecosystem or a human body might sustainably handle, and what the arena of industrialization touts as manageable levels of chemicals and pollution.

Ronald Wright, in *A Short History of Progress* (2004), notes that a foundational premise of industrialization is the desire for endless growth and profitability. Yet, he also confronts a deeper ideological impediment that spins unrelenting resource depletion, worsening pollution, untested synthetic chemical production, and perpetual consumption as beneficial to all of humankind. Ultimately, it is the all-too-familiar victory of short-term thinking over long-term judgment, as is made beautifully clear in Peter Victor's work, challenging the notion that economic growth is necessary for human progress (Victor 2008).

This is why a feminist political economy of pollution is imperative: it contextualizes the interconnectedness of environmental health harms, chemical production, gender, and consumption within historical and structural findings. For example, when attempting to illustrate the "chains of causality" that threaten health and well-being, Gray and Moseley dissect society at multiple scales of analyses reaching across municipal, state, and global levels of governance (Gray and Moseley 2005). Besides focusing on state institutions, a political economy of pollution simultaneously assesses corporations, academic bodies, interest groups, non-governmental organizations (NGOs), and the web of overlapping interests at work between them. In other words, no actor or organization can be entirely removed from the interconnected system that provides the systemic and day-to-day impetus for health and environmental health harms to take root.

One of the fundamental starting points of this volume is the question of why some people and communities endure higher degrees of risk than others. Within the hierarchical manifestations of capitalism, categories of gender, race, culture, sexuality, religion, physical (dis)-ability, and socio-economic status invariably influence the relative burden of risk (Doyle and Kennelly 2003, 25; Nelson 1990). Yet, it is crucial to begin at a structural level. This way, when coming to terms with the disproportionate health burdens that women and other marginalized groups bear in their day-to-day interactions, we avoid the tendency to focus on any particular single "chemical enemy," and are able to interrogate the systems of production that enable them to continue being produced and consumed.

In many ways, the contributions to this volume reflect the contemporary discourse and literature on gender and environmental health. We highlight the specific paths of women's exposure to chemicals through routes such as air, water, soil, and food, and through occupational exposure and consumer goods, in light of our recognition of the fact that health disparities can be structured, reinforced, and compounded by gendered factors (Buckingham and Kulcur 2009, 659). The category of "women" is not deployed without reflection. Like others working on women's health in Canada, we recognize that the category is complicated by women's many social locations shaped by processes of racialization, ethnicity, age, sex, class, sexuality, citizenship, status/migration experience, and ability (Jackson 2012). In this vein, our analysis probes beyond the boundaries of category to question and explore the ways in which the benefits and burdens of chemical exposures are distributed among women.

The definition of gender has long been a topic of debate among scholars in a wide variety of fields. The conceptual challenge centres on the distinction between gender and sex. In the most general terms, "sex" refers to the biological characteristics that distinguish women's and men's bodies, while "gender" refers to the culturally defined characteristics, differences, and roles that are socially constructed and assigned to women and men (Doyal 2001). The binaries implied by this oversimplification are, of course, false and have been productively challenged by recent social theory and social movements that have broadened our understanding of sex and gender to include intersex, transgender, two-spirited, and other culturally specific expressions of gender (Yee 2011).

The prevalent use of the term "gender" to describe both social and biological determinants of health has meant that the term is often misunderstood, misused, and highly contested. In this volume, our discussions of gender encompass both the physical and socio-cultural definitions and understandings of the term, and accept the fact that while the social and biological aspects of gender are inextricably linked, we can productively try to tease apart the ways in which contemporary pollution acts on and is influenced by both sex and gender. We consider: the social determinants of health model; cultural and policy implications of environmental harms; women's increased exposure to chemicals due to behaviour, lifestyle, and occupation; government legislation and regulation of chemical production, lifespan, and disposal; women's roles in the home and their practices in the everyday management of contaminant exposure; the policy implications of "green consumerism" and other lifestyle modifications designed to reduce chemical consumption; and social and environmental justice approaches to understanding exposure. The treatment by selected authors of the biological aspects of health outcomes can be seen through the analyses of pathways to toxic consumption, fetal development, and women's reproductive health; the treatment of women's life cycles as periods of vulnerability to toxins; and women's disproportionate cancer risks from exposure to synthetic estrogens. We can simultaneously appreciate both the biological and the socio-cultural factors that contribute to women's vulnerability to disproportionate harms. We recognize that gender's visibility has been compromised by factors ranging from institutional gender discrimination, to a blurring of the space between public and private realms, and to a lack of studies that consider the lived environments most regularly inhabited by women (Rahder 2009); we also recognize that in order to fully understand and discern the role that gender plays in determining women's environmental health, we as researchers need to *see* gender (Buckingham and Kulcur 2009, 661 and 669). Doing so requires that we validate the ways in which it affects the distribution of power and resources in society, and consequently influences opportunities to engage, both individually and collectively, with policy reform (Jackson 2012; MacGregor 2006).

What Does a Feminist Method Expose?

In exploring the relationship between exposures to chemicals and women's health, we work to unpack assumptions about gender that permeate environmental health research. At the same time, we highlight the gendered differences in women's exposures to, experiences of, and responses to chemical contamination (Howard and Hollander 1996, 2). The gender lens is useful because it allows us to see how gender – as both a category and a lived reality – influences the ways in which we view the world around us. This includes changing the way we see environmental health issues; the types of questions we ask, the people we study, and the answers we imagine will emerge (Howard and Hollander 1996, 2). Like a flashlight in a dark and cluttered room, a gender lens exposes the *consequences* of risk in the context of the uneven and unequal world in which we live (Howard and Hollander 1996). Within environmental health research – in conjunction with political economy and environmental justice frameworks – a gender lens enables us to investigate where gender has been overstated and where it has been ignored completely. In essence, we work to make gender *visible.*

Leipert and Reutter (2005) identify various "determinants of health" that interact to have an immense impact on the well-being of Canadian women. In this view, gender – in addition to a myriad of other factors, including education, social status, employment and work conditions, support networks, physical environment, biological and genetic factors, and culture – plays a key role in determining women's environmental health (Leipert and Reutter 2005, 241). This perspective is echoed by Buckingham and Kulcur (2009) in their discussion of Kimberle Crenshaw's theory of intersectionality (1991), which posits that gender and various other factors, including race, class, sexuality, and citizenship, all work together to shape women as both individual and social actors. Importantly, this perspective is one that the contributors to this volume both implicitly and explicitly support, as gender is positioned within an interlocking set of oppressions that make up the "conditions of our lives" (Combahee River Collective 1977, 264). This orientation has arguably been evident from the very roots of the women's health movement that are hinted at by the title of this collection. *Our Bodies, Ourselves,* published in 1971 by the Boston Women's Health Collective, examined inequalities and oppressions in all of their intersections and urged a generation of women to move past individual self-help towards collective action.

Thus, feminist political economy understands gender and class as "interrelated systems of power that work through and are continuously (re)constituted by social relations of production and reproduction" (Jackson 2012). The concept of "social reproduction" focuses attention on the critical work that is performed, primarily by women, to support life on a day-to-day basis and to foster, sustain, and encourage a new generation. It demonstrates how both paid employment and unpaid domestic labour are part of the same economic processes "of production and consumption that in combination generate the household's livelihood" (Bezanson and Luxton 2006, 37). Capitalism as a mode of production *depends on* social reproduction, as Vosko (2002) notes, whether it is realized through the gendered division of labour performed in the home or through a transnational, racialized division of labour, as Smith and Stiver make clear in Chapter 11. Thus, throughout this volume, we examine the interconnections between women's exposure to chemicals and their health by "situating (multiply positioned) women in practices of production and reproduction" (Jackson 2012).

At the same time, as we work to expose how gender affects health through various socially prescribed roles, attitudes, values, behaviours, assignments of relative power, and differential levels of authority and control, our approach does not dismiss the biological differences between women and men. These differences often account for the increased burden on women's bodies resulting from toxic chemicals in the environments in which we live. The use of a gender lens to improve conditions of environmental health and to formulate policies that reduce or eliminate chemical production must be sensitive not only to differences in the socio-cultural constructions of gender but also to differences in men's and women's biologies. For example, women's unique biologies may create specific vulnerabilities during critical periods, such as during puberty, lactation, and menopause, completely apart from the burdens women experience related to the possibility that they may pass on harms from chemical exposure to their future children. Further, women experience environments in ways that are rooted in their biologies. For example, as explored in Chapter 8, the ability of environmental chemicals to alter breast tissue and contribute to the development of breast cancer influences how women workers in plastics injection moulding operations experience their workplaces. Researchers, advocates, and policy makers must recognize women's specific embodied needs and experiences and account for biological differences, without allowing this attention to biology to lead us down a familiar path towards essentialist claims of vulnerability that can be used to undermine women's agency and autonomy (Sturgeon 1997).

Our challenge as feminist environmental health activists is to take account of the significance of biological differences between bodies without taking those differences to be natural, or "pre-cultural" (Scott 2009a). In the context of women's health and chemical exposures, we have to delve deeper than a simple "socialization" analysis that explains that pollution affects women differently from men because women's roles in the environment, home, and workplace differ from those of men. This is true but it doesn't tell the whole story. Emerging research shows that serious and important thinking remains to be done about the biological aspects of everyday chemical exposures. Turning our attention to them undoubtedly raises complex questions and tensions for feminists (Sandilands 1999), but it might point out, at a key juncture, that there is a more complete story to be told about why a focus on gender and environmental health matters: contaminants act on bodies, and bodies are sexed (Scott 2009a).

Research in a number of disciplines, including feminist theory of the body (see Alaimo 2010), science and technology studies (see Murphy 2008), and eco-criticism (see Nixon 2011), has begun to explore the phenomena associated with the contemporary production and consumption of chemicals. Nixon's notion of "slow violence," that which "occurs gradually and out of sight, a violence of delayed destruction that is dispersed across time and space" (2), vividly evokes the porosity of borders, "from a somatic, to a bodily, to a transnational scale" (Scott 2012a, 484). As Nixon argues, and Michelle Murphy's notion of "chemical regimes of living" (2008) underscores, the "industrial particulates and effluents [that] live on in the environmental elements we inhabit and in our very bodies ... epidemiologically and ecologically are never our simple contemporaries" (8). In other words, the contemporary pollution harms that are body altering and probably generational in character produce afflictions experienced today – cancer, reproductive problems, developmental difficulties – that could be tied to presently occurring, continuing exposures, or they could be latent manifestations of exposures long past.

Increasing attention to these possible intergenerational impacts of everyday chemical exposures, and the related field of epigenetics, gives rise to the sense that synthetic chemicals in our bodies exhibit "embodied, ongoing percolations" (Nixon 2011, 67) beyond our own lives (Collins 2007). The suspected intergenerational effects of the exposures draws on ideas central to feminist theory of the body and Stacy Alaimo's notion of "trans-corporeality, describing movement and exchange between and across human bodies and nonhuman nature" (2010). According to Michelle Murphy (2008, 696):

> The intensification of production and consumption in recent decades has yielded a chemically recomposed planetary atmosphere to alarming future effect, while it has penetrated the air, waters, and soils to accumulate into the very flesh of organisms, from plankton to humans. Not only are we experiencing new forms of chemical embodiment that molecularly tie us to local and transnational economies, but so too processed food, hormonally altered meat, and pesticide-dependent crops become the material

> sustenance of humanity's molecular recomposition. We are further altered by the pharmaceuticals imbibed at record-profit rates, which are then excreted half metabolized back into the sewer to flow back to local bodies of water, and then again redispersed to the populace en masse through the tap. In the twenty-first century, humans are chemically transformed beings.

Murphy's reference to "alarming future effect" raises the prospect of today's chemical consumption reaching forward, into future generations. These intergenerational equity aspects of our current production and consumption of chemicals have been brought forcefully to the fore by Indigenous activists in Canada. On the Aamjiwnaang First Nation reserve, where the environmental health effects of living beside a major petrochemical cluster have been well documented and include a skewed birth ratio tied to endocrine-disrupting pollution (Mackenzie, Lockridge, and Keith 2005; MacDonald and Rang 2007), the intergenerational aspects of the pollution are accepted, if not understood (Basu et al. 2013; Wiebe 2013).

Chemical Exposure and Indigenous Communities: The Aamjiwnaang First Nation

As mentioned, Indigenous communities in Canada bear more than their fair share of chemical exposures and their associated damaging health effects. The overburdened and overexposed Indigenous communities in this country present a current expression of historical racism, ongoing colonialism, and uneven power relations. Further, within these communities, women often experience both marginalization and feminization of poverty, which impact not just their exposures but also the degree of agency and autonomy they may exercise to mitigate those exposures. In fact, the experience of many Indigenous communities demonstrates the notion central to the environmental justice movement that "disproportionate burdens" are borne by the poor, racialized, and marginalized (Luke 2000).

A community that has fought back against the relentless flow of pollutants into their territory in recent years is the Aamjiwnaang First Nation, which shares an "oversaturated airshed" with Sarnia's "Chemical Valley" – Canada's largest petrochemical complex (Scott 2008; Wiebe 2013). Two members of the Aamjiwnaang community recently launched litigation under the Canadian Charter of Rights and Freedoms to counter the persistent problem presented by the fact that the continuous, low-dose exposures to air pollutants they experience occur within legally sanctioned limits. They argue that the pollution threatens their health and violates their equality rights.[1]

The risks associated with these exposures are well established. In a 1998 review of eleven Canadian cities, Health Canada confirmed that mortality increases as ambient air quality decreases (Burnett, Cakmak, and Brook 1998). The Ontario Medical Association attributed 100 extra deaths, 920 emergency room visits, and 471,000 minor-illness days to air pollution in Sarnia-Lambton alone in 2005 (OMA 2005). Further, much of the pollution in Sarnia's Chemical Valley is known to contain "persistent organic pollutants" that act as endocrine disruptors. Endocrine disruptors have structural similarities to the common sex hormones and are thought to "trick" the body into triggering metabolic, growth, and reproductive changes (Colborne, Dumanoski, and Myers 1996).

In 2001, Canada was the first country to sign and ratify the Stockholm Convention on Persistent Organic Pollutants (POPs) amid worries about the accumulation of toxic chemical residues, such as polychlorinated biphenyls (PCBs), in the breast milk of Inuit women. The convention bans the sale of the twelve most toxic POPs, the "dirty dozen," and aims to reduce their unintentional release to the point of virtual elimination. Biomonitoring data now indicate, however, that the "body burdens" of these chemicals in Canadians are still increasing, and concerns are mounting that their endocrine-disrupting effects are starting to be felt in human populations. The spotlight shines again on Sarnia's Chemical Valley, where members of the Aamjiwnaang First Nation are suffering reproductive and developmental health effects linked to exposures to endocrine disruptors, and the local industry continues to release POPs into the environment as the "unintentional" by-products of intentional economic activity, primarily petrochemical production.

The responses of community residents and advocates in Aamjiwnaang have highlighted the importance of the environmental justice movement in both adopting strategies of resistance and incorporating "precaution" (Scott 2008). The movement seeks to address the social implications of inadequate and discriminatory environmental policies and practices by empowering and educating individuals and communities who bear the greatest burdens of environmental harm. The view of environmental health risk that emerges from the environmental justice movement construes the incidence of harm tied to pollution as not only "significant, intentional, and expected" but also inherent in our current processes of production and consumption (Scott 2008, 296).

Activists from Aamjiwnaang have adopted several strategies common to environmental justice movements to further their cause (Wiebe 2013). In particular, they have employed biomonitoring, or "body burden" testing, which measures the body's total exposure to pollutants over time. They have also engaged in community environmental monitoring, such as through "bucket brigades." Here, groups of residents monitor the air near refineries, chemical factories, and power plants using low-cost grab samplers (O'Rourke and Macey 2003). Their biomonitoring efforts are aimed at demonstrating that even the "safe doses" allowed by current regulations are leading to significant harms to human health, and their deployment of community environmental monitoring is intended to give the lie to the regulators' line that current monitoring systems are adequate, when in fact they perpetuate an environment in which firms pollute beyond safe levels with little threat of punishment.

Connections between Environmental and Reproductive Justice

Activists in Aamjiwnaang and other Indigenous communities across the country are also starting to develop important links between environmental and reproductive justice issues (Hoover et al. 2012; Wiebe and Konsmo, forthcoming). This includes attention to the limits to physical reproductive capacity brought about by environmental contamination, and disproportionate levels of reproductive system cancers – those of the breasts, ovaries, uterus, prostate, and testicles. There are obvious concerns about breast milk contamination and the cultural tensions this creates. There are also broader concerns about social and cultural reproduction as traditional and sacred sites for coming-of-age and rites-of-passage ceremonies are increasingly threatened by pollution and industry. As Hoover and colleagues argue (2012, 1648):

> Concerns about the community's ability to reproduce, whether physically through the birth of healthy children or culturally through the passing on of traditional practices, has sparked interest in the need for environmental health research. We want to expand the definition of reproductive justice to include the capacity to raise children in culturally appropriate ways. For many

Indigenous communities, to reproduce culturally informed citizens requires a clean environment.

Failures in the Current Regulatory Approach

The Aamjiwnaang First Nation example points to a major theme of this volume: that the current regulatory approach fails to capture the essence of contemporary pollution harms. In many ways, this failure derives from the fact that continuous, low-dose exposures to chemicals largely occurs within acceptable legal limits. In other words, the risk assessment approach, based on the idea of thresholds, cannot account for the possibility that chronic low levels of pollution might have real and devastating effects on human health. Further, a meaningful application of precaution in this context must properly consider the effects of pollution from multiple sources and their interactions in the body and over time. Long latency periods between exposure and health effects create scientific and legal uncertainties in linking environmental harms to any particular causal event (Scott 2012b). Most toxic substances have simply not been subjected to systematic epidemiological study, or, where studies have been done, it is often concluded that a substance "might" be hazardous (Brown 2007, 265). At the same time, a precautionary approach is warranted despite the lack of research conclusively proving all harms, given the enormous and significant health interests at stake and our continued dependence on unsustainable production and consumption.

An analysis that pays due attention to the structural and historical bases for pollution inevitably comes to rest at an explanation for the relationship between pollution and environmental health harms that finds those harms to be both chronic and intentional (Scott 2008). It understands pollution to be one of the inherent by-products of ordinary, everyday consumption and production, and it understands that devastating health harms are similarly embedded. On this account, the production of harm in the "ever expanding mosh pit of toxic chemicals" is inextricable from the production of commodities (Steingraber 2010, 103).

Our goal is to expose the political economy of pollution: to question who benefits from and who pays the price for the continued release of carcinogenic, neurotoxic, and endocrine-disrupting chemicals into the environment. Collectively, the chapters in this book begin to piece together a coherent picture. Ultimately, they bring back into focus the role of capital, land use, colonization, race, and the state in our examination of bodies and how they are changing in the context of contemporary pollution. As Sarah Lochlann Jain (2007) states, the aim is to watch "the ways in which gender is constituted and inhabited in relation to industrial capitalism and the distribution of ... its modes of suffering" (506).

Pollution is a "fixed feature" of modern economies (Luke 2000). The production of chemicals, the making of plastics, the refining of oil, and the generation of electricity each have harm and wounding embedded in them. They represent, to a large extent, the *production of pollution.* But just as the production of chemicals, plastics, petroleum, pesticides, and more would be tied to the production of pollution, so the actual consumption of many consumer goods, such as plastics, would be tied to the production of pollution. Using the example of bottled water as explained in Chapter 8, the discarded plastic bottles accumulate in landfills, the leachate eventually ends up escaping into surface waters, which become source waters for drinking, and those endocrine-disrupting chemicals, the "gender-benders," are literally consumed again.

The sheer number of regulatory regimes engaged by the study of the production of pollution and the consumption of chemicals in Canada is almost overwhelming. There are the provincial and territorial air and water pollution regimes, land-use planning laws and policies, waste diversion schemes, and occupational health statutes. Federally, there is a complex regime for assessing and managing the risks of toxic chemicals, and laws governing food and drugs as well as cosmetics and hazardous products. At multiple levels of governance simultaneously, we can find laws regulating the approval, use, and application of pesticides, and laws requiring the reporting and disclosure of toxics use and emissions (see this book's Conclusion). All of this makes the point very clear: in tracking how pollution flows, there are multiple possibilities for leakage, but also for diversion and, ultimately, prevention.

The Chapters in This Volume

The study of gender and environmental health demands a truly interdisciplinary framework, and the chapters in this volume are a testament to the multifaceted nature of these issues. Nevertheless, some common themes run through them. In one way or another, all of the chapters allude to the lack of research and attention to health risks that are a priority for women, particularly marginalized women. Each chapter also ties the unequal distribution of risk to social determinants of health, and all make reference to "precaution" and prescribe regulatory reforms with respect to governing chemicals. They also share an interest in finding ways to pull more women into decision making, to draw more attention to the circumstances in which inequalities occur, and to engage more people, institutions, and policy with change. Beyond these similarities, each chapter presents its own original angle and serves as a clear and explicit reminder that chemical consumption and environmental damage do not impact people uniformly (Buckingham and Kulcur 2009).

Part 1, with the theme of "'Consuming' Chemicals," begins with the contribution of M. Ann Phillips in Chapter 1, "Wonderings on Pollution and Women's Health." Phillips opens with a captivating first-person account of her own daily encounters with chemicals to illuminate the many interconnections between humans and these environments. The narrative reveals the hidden exposures we experience in our everyday lives – without our consent – and the serious limitations on our abilities to predict where such exposures will occur. Phillips concludes with some strong recommendations for action. For her, there is ultimately a need to incorporate women's experiences into further research, to increase awareness of those experiences in the crafting of policy, and to address root causes of inequity and toxicity through collaborative action and responsibility.

In Chapter 2, "Protecting Ourselves from Chemicals: A Study of Gender and Precautionary Consumption," Norah MacKendrick focuses on the emerging practice of precautionary consumption. MacKendrick, a sociologist, conducted research that reveals how women's motivations to practise "green consumerism" often results from a distrust of government health risk assessments for chemicals in common products. She discusses women's often disproportionate responsibilities related to the home and the health of their families. When they engage in precautionary consumption, women choose to buy products that they hope will reduce their family's chemical burdens and associated adverse health effects. This gendered practice offers a second tier of self-protection in response to insufficient regulatory precaution.

Importantly, MacKendrick's study provides a platform for women to explain, in their own words, their selective consumption practices, and leads readers to reflect on their own choices and practices. The data also raise the issue of equity, as women vary in the degree to which they can effectively perform this work, based on financial, geographical, and educational constraints, as well as adequate knowledge and access to alternatives. MacKendrick argues that these inequities ultimately undermine the success of precautionary consumption as an answer to inadequate government regulation, and points to the need for a regulatory approach that would include stricter controls on manufacturing, production, and product labelling.

Picking up on the limitations of current regulatory frameworks, Chapter 3 offers Dayna N. Scott and Sarah Lewis's exploration of "Sex and Gender in Canada's Chemicals Management Plan." The authors discuss the federal government's Chemicals Management Plan (CMP), a regulatory program meant to protect environments and human health from toxic substances. The authors reveal the ways in which the CMP is failing to protect the health of Canadians, and how disproportionate burdens of managing risk often fall on women as a result. Some key reasons for this failure include a focus on chemical risk management rather than prevention, inadequate endpoints for health risk assessment, dated assessment methodologies, significant gaps in data, and a disregard for cumulative and longitudinal effects.

Scott and Lewis provide several recommendations for improving the current policy process. In particular, they argue (1) that the CMP process must become more accessible and transparent, fully engaging the public and stakeholders in decision making; (2) that endpoints for toxicity under the CMP need to be expanded through alternative testing methods to address gendered concerns; and (3) that new data on chemical mixtures need to be generated that take into account up-to-date assessment methodologies, occupational exposures, and long-term monitoring and biomonitoring. Ultimately, the authors assert that the federal government must implement precaution meaningfully by working towards a more stringent, inclusive, and comprehensive regulatory regime for toxic chemicals.

Part 2, "Routes of Women's Exposures," begins with Jyoti Phartiyal's "Trace Chemicals on Tap: The Potential for Gendered Health Effects of Chronic Exposures via Drinking Water." This chapter, drawing on research conducted for the National Network on Environments and Women's Health by Susanne Hamm, explores the health risks associated with chronic low-level exposures to chemicals present in Canadian drinking water, and the ways in which current water regulations may fail to protect human health. The author reveals that it is possible for some chemicals to have no "safe" level of exposure, particularly during key windows of vulnerability in a person's development, and that maximum acceptable concentration (MAC) values provided by government can still result in harm to human health.

Phartiyal discusses the challenges inherent in gathering data to reinforce these understandings, including limitations regarding long latency periods, cumulative/synergistic effects, and implications of gender. She then explores the uses and health effects of three common chemicals found at very low levels in drinking water, and provides specific case studies on contamination in the Canadian context. The chapter concludes with several policy reform recommendations, including the strengthening of federal guidelines for Canadian drinking water quality, increasing water-testing frequency, and reviewing and improving public health education and information. The author argues that health risk assessments need to take into account critical windows of vulnerability related to gender and development, that more research should be done on low-level chemical exposure, and that further biomonitoring studies should be conducted on vulnerable populations, including women.

In Chapter 5, "Consuming DNA as Chemicals and Chemicals as Food," we move from everyday chemical exposures in drinking water to the food we eat. Bita Amani highlights the risks associated with consuming genetically modified (GM) (novel) food. She explores recent developments in biotechnology that require further scrutiny, as they may present unknown and as yet immeasurable health risks for women based on sex and gender roles, their responsibilities in the global food chain, and their status as primary caregivers and nurturers. Amani highlights shifting food production practices resulting from advances in molecular genetics and the practice of patenting that together have spurred the growth in agrobusiness with genetically modified organisms (GMOs) and "Ready" varieties. The proliferation of novel foods and foods with novel traits therefore demands a conceptual shift in focus within critical debates that moves beyond traditional concerns over the presence of various contaminants, additives, toxins, and disease-causing agents in food to the increased risks of chemical consumption with GM foods. Amani concludes that things can change for the better if women's voices are able to penetrate agricultural decision-making mechanisms, if labelling procedures change, and if liabilities rest with producers who profit from GMO exclusivity agreements.

Chapter 6, "Consuming Carcinogens: Women and Alcohol," by Nancy Ross, Jean Morrison, Samantha Cukier, and Tasha Smith, sheds light on the elevated rate of women's alcohol abuse, associated harms, and related cultural and policy implications. Alcohol is described as a toxic chemical. It is the second major risk factor contributing to disease in high-income countries and has been labelled by the World Health Organization as "carcinogenic to humans." Yet, as the authors make clear, the availability and social acceptance of alcohol make these dangers difficult to address.

The chapter looks at how alcohol use – even in small doses – causes distinct health harms for women and subsequent generations at a number of reproductive and developmental stages, including breast and other cancers, Fetal Alcohol Spectrum Disorder, and a weakened immune system. Women experience greater physical harm from alcohol compared with men, and have more rapid progressions of harm. Using twelve social determinants of health, the authors reveal the complex interactions between alcohol consumption, gender, and various environmental factors that might increase an individual's susceptibility. They present several policy and regulatory reforms to prevent morbidity and mortality related to alcohol, including the implementation of firmer regulations and policies governing alcohol advertising, as well as more gender-specific, evidenced-based policy, research and prevention/treatment programming, and universal screening for substance misuse.

Part 3 is called "Hormones as the 'Messengers of Gender'?" Here we allude to the orthodox high school understanding of sex and gender: that sex is determined by genetic factors (XX or XY chromosomes), and that sexual differentiation is driven by hormones. As Nelly Oudshoorn's work (1994) reveals, the "discovery" of hormones early in the twentieth century became celebrated as providing the "missing link" between genetic and physiological models of sex determination. It quickly became accepted that the "intentions of genes must always be carried through by appropriate hormones" (Oudshoorn 1994, 20). Accordingly, hormones assumed the role of the "chemical messengers" of gender.

The research on endocrine disruption is complicated by this orthodoxy. Endocrine disruption is commonly described as follows: "Certain synthetic chemicals share structural features with common sex hormones; these chemicals, or xenoestrogens, mimic hormone action in the body by binding with, and activating, available hormone receptors" (Scott 2009a). As explored in Chapter 9, since the endocrine system is understood as responsible for regulating complex and interconnected physiological processes, synthetic chemicals that interfere with it are thought to have profound and wide-ranging effects on health. Importantly for this volume, the fact that hormones travel in the blood in very small concentrations means that even very low levels of xenoestrogens can disrupt the flow of internal communications. Accordingly, susceptibility to xenoestrogens is thought to depend highly on sex, gender, and the timing of exposures.

In Chapter 7, Maria P. Velez, Patricia Monnier, Warren G. Foster, and William D. Fraser present "The Impact of Phthalates on Women's Reproductive Health: Current State of the Science and Future Directions." The authors introduce the reader to advanced research on gendered exposures to phthalates, a mass-produced group of industrial synthetic chemicals ubiquitous in our surroundings. They discuss how the use of phthalates has most often gained scholarly interest for its repercussions on men's health. Yet, as the authors attest, phthalates may just as readily affect women's endocrine functioning, influencing their psychological, behavioural, reproductive, developmental, and emotional health as well as the health of subsequent generations. Chapter 7 examines how the study of phthalates is complicated by the multiple direct and indirect routes of exposure, leading to an uneven distribution among women depending on their social location. The authors argue that with more research, and through an exploration of progressive reforms being developed in other countries around the world, Canada can take action to prevent exposures to phthalates. They stress that, based on the precautionary principle, efforts to examine the plausible role of phthalates in women's heath need to become a priority for scientists and regulators.

In Chapter 8, Aimée L. Ward and Annie Sasco explore "Plastics Recycling and Women's Reproductive Health." They look at the ways in which plastics recycling – or the lack thereof – has become increasingly relevant to the study of endocrine disruption in women. Noting Canada's dismal regulatory track record, they relate how most plastic waste ends up in landfills, incinerators, and ultimately in groundwater. They also address the diversity of chemical compounds in plastics, which complicates the recycling process and makes the finding of suitable after-markets difficult.

Ward and Sasco outline a range of gendered health risks that accompany corporate negligence around, and state indifference to, plastics recycling. For example, endocrine-disrupting chemicals found in plastic are connected to increased incidences of cancer and reproductive health problems in women. The authors carefully consider the pros and cons of "extended producer responsibility" (EPR) and how EPR might be implemented to properly manage plastics, including changes to product design methods and the creation of after-markets to make use of plastics after their initial life. In accordance with the mantra of "reduce, reuse, and recycle," however, they argue that we must also work on decreasing plastic consumption at the source, and on increasing the efficiency and capabilities of plastics recycling. This involves developing better waste policies and carrying out more research on identifying links to women's reproductive health and the exposure pathways of endocrine disruptors. The authors advocate the establishment of Canadian policies that focus on the systems that produce waste, and that confront the societal and political structures that have led to the currently unsustainable production and consumption of waste.

Chapter 9, Sarah Young and Dugald Seely's "Xenoestrogens and Breast Cancer: Chemical Risk, Exposure, and Corporate Power," presents a formidable body of scientific data connecting escalating levels of breast cancer in Canada to the presence of xenoestrogens in our environment. The authors argue that exposure to estrogen is the most important risk factor in cancer development. This is particularly troubling given that breast cancer is the leading cause of death for middle-aged women in Canada. Young and Seely point to powerful chemical lobbyists, who represent a major obstacle to those fighting for women's health. While operating under the mantra of "endless profitability by any means necessary," major chemical and pharmaceutical corporations distort knowledge production through marketing schemes, sponsorships, and research funding allocations that privilege short-term treatment over long-term preventive programs.

The chapter explores the dangers of xenoestrogens and the policy changes that need to occur to protect population health. The authors guide the reader through four parts: how xenoestrogens contribute to breast cancer risk; their impacts within a social determinants of health model; the political economy of chemicals and cancer; and a refocusing of research and policy on prevention. Recommendations focus on putting the precautionary principle into practice, with the authors emphasizing the need to expand research initiatives in high-risk communities. They discuss how the federal government can play a key role in implementing appropriate regulations, a national risk reduction strategy, and education initiatives for the public. Ultimately, an upstream approach to health is advocated, making prevention the primary goal.

Part 4, "Consumption in the Production Process," explores the unique relationship that workers have with the environment based on occupational exposures. The two chapters in this section reveal the ways in which women workers are exposed to carcinogens and/or endocrine disruptors at rates far greater than the general population. They consider how scientific assessment fails to recognize the unique burden of blue-collar workers, and the health risks associated with their jobs. The study of occupational health is an area fraught with contradiction, and occupational injury and disease often exist as a "hidden problem" (Levenstein and Wooding 2000). Both chapters discuss the intertwined legal, social, and scientific factors affecting women workers.

In Chapter 10, "Plastics Industry Workers and Breast Cancer Risk: Are We Heeding the Warnings?" Margaret M. Keith, James T. Brophy, Robert DeMatteo, Michael Gilbertson, Andrew E. Watterson, and Matthias Beck examine the nature and extent of plastics workers' occupational exposures to carcinogenic and endocrine-disrupting chemicals (EDCs). The chapter presents research on how women working in the plastic injection moulding industry, and particularly in the auto parts sector, are at higher risk of developing breast cancer. By consolidating scientific literature, primary research, and the stories and observations of workers themselves, they elucidate the types of pollution found in automobile factory settings, the historical failures of government to properly regulate workplace exposures, and the adverse health impacts women workers experience as a result. The authors argue that the invisibility of blue-collar workers in policy development reinforces gender and class discrimination.

The chapter emphasizes the inadequacy of existing workplace chemicals testing, particularly given new research on how exposure, even at extremely low levels, can be harmful, and given questions about exposures to complex mixtures. Accordingly, the authors argue that there is a need to re-evaluate the guidelines and regulatory standards for occupational health. There is also a need to acknowledge the connection between workplace exposures and elevated levels of breast cancer. The authors offer increased public inquiries, commissions to examine risk, workers compensation, institutionalized research, prevention campaigns, educational programs, and regulatory changes as ways of addressing current regulatory failures and considering sex and gendered concerns in relation to chemical exposure in the workplace.

Adrian A. Smith and Alexandra Stiver's "Power and Control at the Production-Consumption Nexus: Migrant Women Farmworkers and Pesticides" (Chapter 11) examines production and consumption in the Ontario agricultural sector, and evaluates the significant occupational exposures to pesticides experienced by women migrant farmworkers. Drawing inferences from research on environmental justice and occupational health and safety, the authors argue that the migrant agricultural population faces a greater burden of risk from pesticides than non-migrant workers. The chapter outlines a model of production and consumption in which migrant labourers are often exploited. The authors argue that constraints placed on these workers, such as a lack of power and control over working conditions, occur as a result of processes of racialization, gendering, and the regulation of citizenship status. They point to a critical need for a precautionary approach that restricts pesticide use and improves enforcement mechanisms through collective bargaining.

This volume concludes with a short reflection on the current state of chemical regulation by Dayna Nadine Scott. She argues that we may have crossed a critical threshold to reach a place in which developments in environmental health and science (including the collapse of the notion of a threshold for health effects of certain key chemical exposures); a coalition of interests in women's health, occupational health, and environmental justice; and the willingness of governments to contemplate law reform on the regulation of toxic substances are converging in a way that provides room for greater understanding and social, economic, ecological, and political transformation with regard to issues of toxic exposure. Overall, we strive to provide researchers, policy makers, and advocates with the tools to make use of this moment. We consider it a collaborative effort to expand our collective understanding of the links between social inequity, environmental risks, and the gendered division of health burdens in Canada. We bring together scientific developments, policy options, and legal analysis to develop a critical, engaged theoretical framework for thinking about gender and environmental health. We hope you make good use of it.

Note

1 Notice of Application, Ada Lockridge and Ron Plain, Applicants, Ontario Divisional Court, Court File 528.10 (2010).

References

Adamson, Joni, Mei Mei Evans, and Rachel Stein. 2002. *The Environmental Justice Reader: Politics, Poetics and Pedgogy*. Tucson: University of Arizona Press.

Alaimo, Stacy. 2010. *Bodily Natures: Science, Environment, and the Material Self*. Bloomington: Indiana University Press.

Agyeman, Julian, Peter Cole, Randolph Haluza-DeLay, and Pat O'Riley, eds. 2009. *Speaking for Ourselves: Environmental Justice in Canada*. Vancouver: UBC Press.

Basu, N., D.K. Cryderman, F.K. Miller, S. Johnston, C. Rogers, W. Plain. 2013. *Multiple Chemical Exposure Assessment at Aamjiwnaang*. McGill Environmental Health Sciences Lab Occasional Report 2013-1. Montreal: McGill University.

Batt, Sharon. 2008-09. *Full Circle: Drugs, the Environment, and Our Health*. Toronto: Canadian Women's Health Network. http://www.cwhn.ca/en/node/39366.

Bezanson, Kate, and Meg Luxton, eds. 2006. *Social Reproduction: Feminist Political Economy Challenges Neo-Liberalism*. Montreal and Kingston: McGill-Queen's University Press.

Brown, Phil. 2007. *Toxic Exposures: Contested Illnesses and the Environmental Health Movement*. New York: Columbia University Press.

Buckingham, Susan, and Rakibe Kulcur. 2009. "Gendered Geographies of Environmental Injustice." *Antipode* 41 (4): 659-83, doi:10.1111/j.1467-8330.2009.00693.x.

Burnett, Richard T., Sabit Cakmak, and Jeffrey R. Brook. 1998. "The Effect of Urban Ambient Pollution Mix and Daily Mortality Rates in 11 Canadian Cities." *Canadian Journal of Public Health* 89 (3): 152-56.

Clapp, Jennifer, and Peter Dauvergne. 2005. "Peril or Prosperity? Mapping Worldviews of Global Environmental Change." In *Paths to a Green World: The Political Economy of the Global Environment*, edited by Jennifer Clapp and Peter Dauvergne, 1-17. Cambridge, MA: MIT Press.

Colborn, Theo, Dianne Dumanoski, and John Peterson Myers. 1996. *Our Stolen Future: Are We Threatening Our Fertility, Intelligence, and Survival? A Scientific Detective Story*. New York: Dutton.

Collins, Lynda. 2007. "Environmental Rights for the Future? Intergenerational Equity in the EU." *Review of European Community and International Environmental Law (RECIEL)* 16 (3): 321.

Combahee River Collective. 1977. *The Combahee River Collective Statement*. http://circuitous.org/scraps/combahee.html.

Cooper, Kathleen, and Lorne Vanderlinden. 2009. "Pollution, Chemicals and Children's Health: The Need for Precautionary Policy in Canada." In *Environmental Challenges and Opportunities: Local-Global Perspectives on Canadian Issues*, edited by Christopher D. Gore and Peter J. Stoett, 183-224. Toronto: Emond Montgomery.

Crenshaw, Kimberle. 1991. "Mapping the Margins: Intersectionality, Identity Politics, and Violence against Women of Color." *Stanford Law Review* 43 (6): 1241-99.

Doyal, Lesley. 2001. "Sex, Gender, and Health: The Need for a New Approach." *British Medical Journal* 323 (7320): 1061-63, doi:10.1136/bmj.323.7320.1061.

Doyle, Christy, and Jackie Kennelly. 2003. "An Environmental Framework for Women's Health." In *Head, Heart and Hand: Partnerships for Women's Health in Canadian Environments, Volume 1*, edited by Penny Van Esterik, 17-58. Toronto: National Network on Environments and Women's Health.

Evernden, Neil. 1992. "The Social Use of Nature." In *The Social Creation of Nature*, edited by Neil Evernden, 3-17. Baltimore: The John Hopkins University Press.

Eyles, John, K. Bruce Newbold, Anita Toth, and Tasnova Shah. 2011. *Chemicals of Concern in Ontario and the Great Lakes Basin: Update 2011 – Emerging Issues*. Hamilton: McMaster Institute of Environment and Health.

Gosine, Andil, and Cheryl Teelucksingh. 2008. *Environmental Justice and Racism in Canada: An Introduction*. Toronto: Emond Montgomery.

Gray, Janet, Janet Nudelman, and Connie Engel. 2010. *State of the Evidence: The Connection between Breast Cancer and the Environment/From Science to Action*. 6th ed. San Francisco: Breast Cancer Fund.

Gray, Leslie C., and William G. Moseley. 2005. "A Geographical Perspective on Poverty-Environment Interactions." *Geographical Journal* 171 (1): 9-23, doi:10.1111/j.1475-4959.2005.00146.x.

Hoover, E., et al. 2012. "Indigenous Peoples of North America: Environmental Exposures and Reproductive Justice." *Environmental Health Perspectives* 120: 1645-49.

Howard, Judith A., and Jocelyn Hollander. 1996. *Gendered Situations, Gendered Selves: A Gender Lens on Social Psychology*. Thousand Oaks, CA: Sage Publications.

Jackson, Beth. 2012. "Theory and Methods for Thinking Women." In *Thinking Women: Reforming Health Care,* edited by Pat Armstrong et al. Toronto: Canadian Scholars Press.

Jain, Sarah Lochlann. 2007. "Cancer Butch." *Cultural Anthropology* 22 (4): 501-38, doi: 10.1525/can.2007.22.4.501.

Krupp, Staci Jeanne. 2000. "Environmental Hazards: Assessing the Risk to Women." *Fordham Environmental Law Journal* 12 (1): 111-38.

Leipert, Beverly D., and Linda Reutter. 2005. "Women's Health in Northern British Columbia: The Role of Geography and Gender." *Canadian Journal of Rural Medicine* 10 (4): 241-53.

Levenstein, Charles, and John Wooding. 2000. "Deconstructing Standards, Reconstructing Worker Health." In *Reclaiming the Environmental Debate: The Politics of Health in a Toxic Culture,* edited by Richard Hofrichter, 39-55. Cambridge, MA: MIT Press.

Luke, Timothy W. 2000. "Rethinking Technoscience in Risk Society: Toxicity as Textuality." In *Reclaiming the Environmental Debate: The Politics of Health in a Toxic Culture,* edited by Richard Hofrichter, 239-54. Cambridge, MA: MIT Press.

MacDonald, Elaine, and Sarah Rang. 2007. *Exposing Canada's Chemical Valley: An Investigation of Cumulative Air Pollution Emissions in the Sarnia, Ontario Area*. Toronto: Ecojustice.

Mackenzie, Constance, Ada Lockridge, and Margaret Keith. 2005. "Declining Sex Ratio in a First Nation Community." *Environmental Health Perspectives* 113 (10): 1295-98.

McMichael, Anthony J. 1993. *Planetary Overload: Global Environmental Change and the Health of the Human Species*. Cambridge: Cambridge University Press.

Murphy, Michelle. 2008. "Chemical Regimes of Living." *Environmental History* 13: 695-703.

Nelson, Jennifer J. 2002. "The Space of Africville: Creating, Regulating, and Remembering the Urban 'Slum.'" In *Race, Space, and the Law: Unmapping a White Settler Society,* edited by Sherene H. Razack, 211-32. Toronto: Between the Lines.

Nelson, Lin. 1990. "The Place of Women in Polluted Places." In *Reweaving the World: The Emergence of Ecofeminism,* edited by Irene Diamond and Gloria Feman Orenstein, 173-88. San Francisco: Sierra Club Books.

Nixon, Rob. 2011. *Slow Violence: Environmentalism of the Poor*. Cambridge, MA: Harvard University Press.

O'Rourke, Dara, and Gregg Macey. 2003. "Community Environmental Policing: Assessing New Strategies of Public Participation in Environmental Regulation." *Journal of Policy Analysis and Management* 22 (3): 383-414, doi:0.1002/pam.10138.

OMA (Ontario Medical Association). 2005. *Illness Costs of Air Pollution (ICAP): Regional Data for 2005 (with Projections to 2026)*. https://www.oma.org/Resources/Documents/d2005IllnessCostsOfAirPollution.pdf.

Oudshoorn, Nelly. 1994. *Beyond the Natural Body: An Archaeology of Sex Hormones.* New York: Routledge.

Rahder, Barbara. 2009. "Invisible Sisters: Women and Environmental Justice in Canada." In *Speaking for Ourselves: Environmental Justice in Canada,* edited by J. Agyeman, P. Cole, R. Haluza-DeLay, and P. O'Riley, 81-96. Vancouver: UBC Press.

Sandilands, Catriona. 1999. *The Good-Natured Feminist: Ecofeminism and the Quest for Democracy.* Minneapolis: University of Minnesota Press.

Scott, Dayna Nadine. 2008. "Confronting Chronic Pollution: A Socio-Legal Analysis of Risk and Precaution." *Osgoode Hall Law Journal* 46 (2): 293-343. http://www.ohlj.ca/english/documents/OHLJ46-2_Scott_ConfrontingChronicPollution.pdf.

–. 2009a. "'Gender-Benders': Sex and Law in the Constitution of Polluted Bodies." *Feminist Legal Studies* 17 (3): 241-65, doi:10.1007/s10691-009-9127-4.

–. 2009b."Testing Toxicity: Proof and Precaution in Canada's Chemicals Management Plan." *Review of European Community and International Environmental Law (RECIEL)* 18 (1): 59-76, doi:10.1111/j.1467-9388.2009.00621.x.

–. 2012a. "Book Review: *Slow Violence and the Environmentalism of the Poor,* by Rob Nixon." *Osgoode Hall Law Journal* 50: 479-89.

–. 2012b. "Pollution and the Body Boundary: Exploring Scale, Gender and Remedy." In *Feminist Perspectives on Tort Law,* edited by Janice Richardson and Erica Rackley, 55-79. Abingdon, UK: Routledge.

Steingraber, Sandra. 2010. *Living Downstream: An Ecologist's Personal Investigation of Cancer and the Environment.* Cambridge, MA: Da Capo Press.

Sturgeon, Noel. 1997. *Ecofeminist Natures: Race, Gender, Feminist Theory and Political Action.* New York: Routledge.

–. 1999. "Ecofeminist Appropriations and Transnational Environmentalisms." *Identities: Global Studies in Culture and Power* 6 (2-3): 255-79, doi: 10.1080/1070289X.1999.9962645.

Verchick, Robert R.M. 1996. "In a Greener Voice: Feminist Theory and Environmental Justice." *Harvard Women's Law Journal* 19 (2): 23-88.

Victor, Peter. 2008. *Managing without Growth: Slower by Design, Not Disaster.* Cheltenham, UK: Edward Elgar.

Vosko, Leah F. 2002. "The Pasts (and Futures) of Feminism and Political Economy in Canada: Reviving the Debate." *Studies in Political Economy* 68 (Summer): 55-85.

Wiebe, Sarah Marie. 2013. "Anatomy of Place: Ecological Citizenship in Canada's Chemical Valley." PhD dissertation, University of Ottawa.

Wiebe, Sarah Marie, and Erin Marie Konsmo. Forthcoming. "Indigenous Body as Contaminated Site? Examining Reproductive Justice in Aamjiwnaang." In *Fertile Ground: Reproduction in Canada,* edited by F. Scala and S. Paterson. Montreal and Kingston: McGill-Queen's University Press.

Wright, Ronald. 2004. *A Short History of Progress.* Toronto: House of Anansi Press.

Yee, Jessica, ed. 2011. *Feminism for Real: Deconstructing the Academic Industrial Complex of Feminism.* Ottawa: Canadian Centre for Policy Alternatives.

Part 1: "Consuming" Chemicals

Chapter 1 Wonderings on Pollution and Women's Health

M. Ann Phillips

Women occupy a number of environments on a daily basis. Almost all of these environments contain chemicals of some sort, many of which can cause harm to human health. Gender, race/ethnicity, and class/income level – as well as other social, cultural, political, and economic factors – determine how women are exposed to these chemicals within a given setting. Through air, water, food, soil, and consumer goods, these substances can "trespass" into women's bodies without their consent (Sandra Steingraber, quoted in Schwartz 2012). This chapter explores the prevalence of chemicals in the environment and the many ways in which women are exposed to these chemicals, and questions how environmental exposures may affect women's health. The methodology is based on Dorothy E. Smith's theory of *The Everyday World as Problematic* (1987), which challenges abstract modes of research and action, and prioritizes an individual woman's knowledge of her everyday world as a starting point for inquiry that brings to light a woman's standpoint.

The following narrative of one day in my life in Toronto uses my everyday world as the starting place for this discussion and serves to problematize women's personal experience of everyday chemical exposures. It not only begins to identify some of the chemicals that women may come into contact with in their everyday environments but also suggests how little is commonly known about how these exposures may affect women's health. This narrative provides a starting point for thinking about what can be done to address the issue of everyday chemical exposures and women's health.

Through an exploration of such exposures, I hope to make visible the key processes that connect an increasingly polluted environment with human health, and to highlight women's health as a unique and important consideration within this context. By illustrating the intricate ways in which gender, social location, and the physical environment intersect with and influence chemical exposure, I suggest in this chapter that the current culture of production and consumption has made it quite difficult to dissociate ourselves from the many chemicals around us. I call for specific actions to address potentially toxic exposures, arguing that the most effective way to prevent toxic trespass from occurring involves a shift towards precautionary approaches that involve the reduction, and ultimate removal, of harmful and unnecessary chemicals from our environment.

This chapter prioritizes not only my individual knowledge of my everyday world but also that of twenty-five other women who had varying degrees of awareness about the sources of air and water pollution in their communities and the impacts of chemical exposure on women's health. One hundred twenty-eight women known to me, living primarily in Toronto, were invited to answer a series of questions about their experiences of women's health and chemical exposures. Sixteen women, ranging from twenty-eight to seventy-two years of age, agreed to share information about their everyday experiences of exposures to chemicals by completing a questionnaire, and four of these women agreed to a longer, more in-depth discussion of these issues. As well, nine Chinese immigrant women were willing to share their experiences by answering the same questions in a focus group. These women chose to remain anonymous and, unlike the other women, are identified by number and age instead of first name and age. Together with academic literature, quotes describing the direct experiences of these women's everyday lives provide a framework for discussing issues related to the impacts of chemical exposures on women's health and identifying the actions that women feel can be taken to reduce these impacts.

One Woman's Experience of Everyday Exposures

I roll out of bed and get ready for work. As I jump in the shower, I think about the chlorine in the water and what I might be inhaling through the steam. What other chemicals might I also be exposed to during this time? I brush my teeth with a natural toothpaste. I think about the many products women use on a daily basis for body care. I reflect on the days when I used regular commercial cosmetics – lipsticks, lip-glosses, eye-liners, eye shadows, perfumes, deodorants, make up removers, and face creams. I will never know what trace chemicals I was repeatedly exposing myself to at the time. It was when I started considering the potential effects that low-dose toxic exposures may have on my health that I began to pay attention to the ingredients in the products I was buying. I put on hand lotion labelled as 100 percent natural. I wonder if this is reputable, and if it is possible to significantly reduce chemical exposure through product choice.

Walking to work, I spot a work crew fixing a broken water main. Gallons of water flow down the street and into the sewer drain. The water takes with it the dust, dirt, and grease from the road. I also note the exhaust being emitted from countless cars and buses. Where do these discharges and emissions end up? Are they inhaled by passing pedestrians and cyclists, or consumed by residents? Do they enter the air vents or drinking water sources of buildings? Do some travel further afield to affect other people downwind and downstream? There is a homeless teen sitting on the pavement at nose level with the traffic exhaust. Are constant exposures to dust and exhaust specifically affecting her health and the health of others in her position? Does she have the money, education, or resources to combat, reduce, or avoid these exposures? Or does her every-day basic survival take priority over these concerns?

At work a client has come in to be treated with acupuncture and cupping therapy, a traditional method used in China for back pain. In an effort to make the practice somewhat more environmentally friendly, we replaced isopropyl alcohol with hydrogen peroxide as an antiseptic. However, isopropyl alcohol is still needed for cupping therapy. I have read that in high doses (much higher than what is used in my practice), isopropyl alcohol can be toxic to humans, with the potential of causing developmental and reproductive effects, and carcinogenicity in animals. Many households use this substance for disinfection and most people assume that what they are buying is safe. Should we be worried about low doses of chemicals deemed to be safe?

I go to the grocery store to buy lunch. I look for something fresh, healthy, and if possible organic. Almost everything comes in plastic. As a vegetarian I used to eat a lot of tofu as a protein source until I heard about the high levels of xenoestrogens even in organic tofu, so now I limit my intake. I choose a salad in a take-away container. I wonder when the salad was prepared, how long it has been sitting there, and whether the vegetables have absorbed any chemicals from days surrounded by plastic. I wonder where the ingredients were grown, what types of pesticides, herbicides, and fertilizers were used to grow them, and how much chemical residue remains. How might the regular consumption of everyday foods impact my health over the long term? I wonder how these chemicals may affect the health not only of consumers but also of those working to grow the food. I remember a recent study that found that both acute and cumulative pesticide exposure can lead to depression in those who work in the agricultural setting.

As I walk through the mall back to my office, I take note of the number of workplaces that use or contain toxic chemicals. Many dry cleaners still use volatile organic compounds (VOCs) like perchloroethylene or tetrachloroethylene. Dental offices usually work with fluoride, and some of their fillings contain bisphenol A – isn't that a chemical that Health Canada recently declared toxic due to its endocrine-disrupting potential and associations with early onset of puberty, learning disabilities, and certain cancers? Nail salons use products containing acetone. Hair salons work with bleaches, hair-colouring products, gels, and various VOCs in hair sprays. Photocopy shops generate ozone, chromium, and other particulates. Furniture and clothing stores could have products coated with brominated flame retardants, and off-gassing formaldehyde. I wonder how these low-dose daily cumulative exposures might affect someone's health, especially those who live in the high-rise apartments just above the mall. What amount of exposure, over what period of time, would be needed for health effects to occur?

After work I take the subway downtown. I look around at the dirt and dust and wonder about the level of particulate matter in the subways compared with other indoor or outdoor environments. The panelling has been removed from a section of the roof. Water is dripping from the roof and down to track level. The stucco surface below the panelling has blackened. I wonder whether that is mould and what it is emitting into the air. I enter the train and sit down between two people. The man on my right is wearing a very strong aftershave lotion and the woman on my left smells of cigarette smoke. What is the cumulative effect of inhaling these types of air pollutants every day on the way home from work?

Across from the library is a construction site where a building has been demolished. I wonder what materials were used when the building was originally erected, and whether they were hazardous to human health.

Was there asbestos in the ceilings or leaded paint on the walls? How did the construction team control the dust from the demolition? Who might have been exposed to those particulates, and how might it affect their health?

Later I meet a friend for dinner. He works at raising awareness about the disproportionate impacts of pollution on the health of First Nations peoples. He lives on a highly polluted reserve. I think back to when I last visited him in his community. There was a benzene release by local industry. The industry labelled the release "accidental," and issued a community advisory three hours later. I reflect on the many individuals living near or downwind from the plant who were unaware of that event and were unknowingly exposed to the chemical. What of the women, men, and children who live on reserve who experience these exposures on a regular basis?

As I drive home from dinner, I pass a working-class neighbourhood where many new immigrants live. How might the intersecting factors of gender, class, race/ethnicity, and safety affect a woman's exposure to chemicals in this area, and how might availability and access to appropriate health services affect her capacity to address or avoid the effects of these exposures?

When I get home from work, I call my sister-in-law to schedule a visit. She works as a professional housecleaner and uses conventional cleaners and bleaches on a regular basis. I wonder what kind of chemical exposures she experiences in her regular workday. I also think of my mother, who has always worked full-time in the home and has been primarily responsible for domestic duties, including washing dishes, cleaning, cooking, and laundry, and has been regularly exposed to all those everyday household cleaning products. Even though I have talked to my housecleaner sister-in-law and my mother about using non-toxic products, they still choose to use chlorine bleach, chemical scouring powders, and other chemical cleaners. Both of these paid and unpaid work environments expose these women, like many others, to a number of hazardous chemicals. I wonder what studies have been done on the long-term, cumulative effects of these types of "workplace" exposures on women's health?

Later in the evening, I sit in a cafe with my partner, our two laptops back-to-back at the table as he tells me about his day. There is a sea of computers in the room as people sit, talk on their cell phones, and nurse their drinks. I think about the effects of Wi-Fi and electromagnetic radiation on women's health, and where all the used computers and cell phones will go when we no longer need them. What will happen to all the chemicals they contain? Who will be exposed to them?

Industrialization and the Changing Global Environment

Researchers suggest that the contamination of our environment is a manifestation of the toxic elements of our culture. Chemical production and use are seen by many as an inevitable part of progress, and have become an accepted part of life. Industrial and technological development over the past fifty years has led to dramatic changes in the global environment, with worldwide growth in the production and use of synthetic chemicals (Ellwood 2008). Increasingly, the air we breathe, the water we drink, the food we eat, and the products we buy to sustain us contain chemicals that can affect human health. For example, a number of chemical pollutants have been detected in our water supplies, including solvents, plasticizers, pesticides, herbicides, nitrates, chlorine, fluoride, heavy metals, and radionucleotides (see Chapter 4). These substances derive from many sources, including agricultural runoff, industrial discharges, the proper and improper use and disposal of household products, and the water purification process, where even chemicals such as chlorine and fluoride have been linked to human health concerns (Ayoob and Gupta 2006; Peterson 2008). Pollution is also prevalent in our atmosphere, as a result of burning firewood, household mould, and industrial and vehicular emissions (Environment Canada 2012). It is becoming increasingly apparent that current processes of industrialization, production, and consumption locally and globally have resulted in pollution's being seen as inevitable rather than the exception. Several of the survey respondents expressed their own experiences of feeling surrounded by pollution and toxic chemicals:

Every community has air and water pollution in varying degrees.
– LINDA, 49

It is very depressing to think that we are exposed to so many chemicals and that it is just 'normal.' – KATHRYN, 28

The rivers [in my hometown in China] are so polluted that people are afraid to eat the fish. Also, the housing is so close to industry that people are using whatever water is coming out of the industries to water the plants and the crops. – FOCUS GROUP PARTICIPANT #8, 69

The prevalence of chemicals and of air and water pollution means that people are being exposed to hazardous substances on a regular basis. Many studies have now demonstrated the persistence of chemicals in people, with tests detecting measurable levels of various toxins within the body (Environmental Defence 2006; Health Canada 2010).

Pollution and Human Health

> *As an individual, and an individual who recognizes she is a fellow drop in humanity's ocean, we all impact each other and our environment; a lack of respect in any direction reveals a lack of respect for ourselves. What are we saying when we poison ourselves?* – LEE, 32

As Lee points out, we as individuals and as a collective are responsible for the level of chemicals in the environment, and can therefore be seen as poisoning ourselves. There is growing evidence demonstrating the connection between environmental contamination and adverse effects on human health. In its 2006 report *Preventing Disease through Healthy Environments: Towards an Estimate of the Environmental Burden of Disease,* the World Health Organization (WHO) revealed that 85 of the 102 major diseases and injury categories identified worldwide are the result of environmental risk factors (Prüss-Üstün and Corvalán 2006). The authors also estimated that nearly one-quarter of the total disease burden and death in the human population can be attributed to the environment. In a systematic review of the literature on estimates of the global burden of disease from chemicals, Prüss-Üstün and colleagues (2011) found that 8.3 percent of total death worldwide and 5.3 percent of disability-adjusted life years (DALYs) were attributable to environmental exposure and management of selected chemicals in 2004. They concluded that these figures were probably an underestimate of the actual disease burden, since only a few chemicals for which data were available were included in the review, and chemicals with known health effects, such as dioxins, cadmium, and mercury, or chronic exposure to pesticides, could not be included in the study because of incomplete data and information. WHO estimates that in Canada approximately 13 percent of all preventable diseases are due to environmental causes, and that two major factors – outdoor air pollution on the one hand, and water, sanitation, and hygiene on the other hand – account for 7 percent of total reported preventable deaths and 4 percent of total reported preventable DALYs per 1,000 persons (WHO 2009).

In their 2011 review, Prüss-Üstün and colleagues estimated the health impact of urban air pollution to have been roughly 1,152,000 deaths worldwide in 2004. Based on an analysis of census tract data, the Ontario Medical Association's Illness Cost of Air Pollution model estimated that there were 9,500 premature deaths due to smog in 2008 (OMA 2008). In 2006, the corresponding figure had been only 5,940, but it was also estimated that there were 17,070 hospital admissions and 60,640 emergency department visits due to smog (OMA 2006). Traffic-related air pollution has been associated with the development and exacerbation of asthma, atophy, bronchitis, and chronic coughs (Meng et al. 2007; Schwartz 2004). Exposure to traffic-related air pollution has recently been shown to be a risk factor for cardiovascular disease (Chen et al. 2013).

Immigrant women from China now living in Toronto emphasized their experiences and concerns about links between environmental contaminants and health effects in the following ways:

> *In my hometown, as a high school student we used to drink the water from the rivers and the wells. Since 1995 we can't drink the well water anymore.* – FOCUS GROUP PARTICIPANT #3, 33

> *I lived in the rural areas and when they spray with pesticides they ended up in the water and people cannot drink it.* – FOCUS GROUP PARTICIPANT #7, 40

> *My family has lots of allergies and itchiness. There are many people suffering from this and they don't know why. I am concerned with what is in the water. I have read reports of people throwing things like birth control pills down the toilet and I wonder if these chemicals are affecting my health.* – FOCUS GROUP PARTICIPANT #7, 40

Despite the growing academic evidence, and the awareness of women's everyday experiences that support them, connections between chemical exposure and health outcomes continue to be disputed by some scientists, governments, and institutions. As Perrotta and Macfarlane (2002, 3) note, "it is very difficult to prove that environmental exposures to chemical and physical agents are causing cancer because of the difficulties involved in estimating personal exposures and because of confounding exposures."

Connecting Human Health and the Environment

Conventional biomedicine has lagged in making connections between human health concerns and the environment. Environmental health is a relatively new field as biomedicine is based on the conceptualization of the body as a machine, with health often characterized as the presence or absence of specific diseases or physical conditions. Biomedical practices tend to study humans in isolation, removing them from the natural world and other social, economic, cultural, and political contexts in which human bodies are located. Although this perspective is still widespread in many modern scientific settings, there have been some significant shifts in worldview concerning the connections between human health and the environment.

Following the Second World War, WHO (1946, Preamble) defined health as "a state of complete physical, mental, and social well-being, and not merely the absence of disease or infirmity." From this perspective, health is recognized as essential in order for people to realize their potential, satisfy their needs, and cope with changes in the environment. In order for an individual to attain a "healthy" status, she requires a number of socio-economic and political resources and physical capacities. This "ecosystem" or "holistic" health model, now adopted by many health authorities and organizations, recognizes that humans are connected to complex physical, social, political, and economic environments, and that our health is intertwined with the health of these environments (McCally 2000). Such an approach helps us to better understand the many interactions humans have with their surroundings, and the ways in which environmental contamination can affect health and well-being.

As human beings, we are interfaces. Our bodies are places of interactivity and engagement, intermingling and becoming a part of our surroundings. For example, when babies are born, their bodies are over 70 percent water, a percentage that remains relatively constant as they age (US Geological Survey 2011). Water represents a significant part of our bodies, and we need it to survive. The water we drink becomes our blood, our cerebrospinal fluid; it lubricates our cells and joints, transports nutrients throughout our bodies, and flushes toxins from our system. As Steingraber (2001, 66-67) relates:

> *I drink water, and it becomes blood plasma, which suffuses through the amniotic sac and surrounds the baby – who also drinks it. And what is it*

before that? Before drinking water, amniotic fluid is the creeks and rivers that fill reservoirs. It is the underground water that fills wells. And before it is creeks and rivers and ground water, amniotic fluid is rain. What I hold in my hands is a tube full of raindrops. Amniotic fluid is also the juice of oranges that I had for breakfast, and the milk that I poured over my cereal, and the honey I stirred into my tea ... When I look at amniotic fluid, I am looking at rain falling on orange groves ... The blood of cows and chickens is in this tube. Whatever is inside hummingbird eggs is also inside my womb. Whatever is in the world's water is here in my hands.

This description exemplifies the many ways in which human bodies are intricately connected to the external environment at all times. As a result, environmental changes and deteriorating air and water quality can often act as predictors of human health. By breathing, eating, and drinking, we are continuously introducing substances into our bodies that become part of our internal physiology and cellular structure, and that can affect our health and development (Hackney et al. 1978; May 2003; Welshons et al. 2003).

Effects on Women's Health: Gendered Risk of Exposure

While pollution and chemical exposures pose risks to the health of all people, it is likely that the ways in which women are exposed to environmental contaminants, and the effects of those exposures, differ from those of men. Before scientists began researching connections between environmental contaminants and human health, women's health advocates were suggesting that increases in breast cancer, infertility, and birth defects were symptomatic of the deterioration of the physical environment (Diamond and Orenstein 1990).

Unfortunately, decades later, factors contributing to a woman's susceptibility to chemical exposure continue to be poorly understood, are often given limited attention, or are neglected altogether in health risk assessments, exposure research, and the research and policy decisions of political, economic, and social institutions involved in environmental health (Hatch 2000; Weiss 2011). As Thompson (2003, 110) states:

> Studies to determine "safe" levels of exposure to toxic substances are performed on healthy young men, without requiring evidence of their applicability to women, and women's health issues, like chemical sensitivity, are given a low research priority: "The types

> of health problems women have are not recognized or compensated, creating a vicious circle where women's occupational health problems are not taken seriously, therefore not recognized, therefore do not cost enough to matter." [Messing 1998, 13]

The neglect of gendered concerns in research and policy development in and of itself makes women vulnerable to current and future toxic exposure, and leads to women's health and lives being affected in significant ways (Hatch 2000). One of the survey respondents remarked:

> *Issues like inequality, discrimination, unequal distribution of power and poverty, and lack of power affect a woman's employment, education, where she lives, how she feels about herself, and how she interacts with society. Since health encompasses a whole person, her health will be affected by all of these. Does she have access to alternative health care? Is she even aware that there are alternatives? Does she have a doctor? Is she covered by OHIP? This is just the tip of the iceberg.* – LEE, 32

Lee's perspective acknowledges the ways in which social factors influence women's health, and suggests that a woman's body can act as an important indicator of the state of the environment.

The physical, social, political, and economic factors that influence women's health and well-being are known as "social determinants of human health," a term that distinguishes between the more familiar biological and genetic factors and the uniquely "social" determinants of health, such as income, division of labour, education, social support networks, geographic location, or culture (Wilkinson and Marmot 2003). Although there are many slightly different definitions of the social determinants of health, it is widely recognized that income and the social gradient are key determinants of health worldwide.

These social determinants, also discussed in the other chapters in this volume, influence a woman's vulnerability or susceptibility to chemical exposures, their ability to avoid or address these exposures, and their access to the resources, education, and support systems needed to do so (Olden and Newbold 2000; Rahder and Peterson 2000). Only by considering the physical, social, economic, cultural, and political contexts in which pathogens and toxins flourish will institutions be able to fully understand why particular groups, such as women (and specific groups of women among them also), are more likely than others to experience adverse health effects (Rahder and Peterson 2000). Some of the factors contributing to a woman's unique vulnerability to pollution in the environment are discussed in more detail below.

Biological Differences

Women may experience unique physiological impacts from chemical exposures, as they tend to be smaller, with thinner skin, a higher percentage of body fat, and a different hormonal makeup than men (Women's Foundation of California 2003). These biological traits can result in women's being more vulnerable to particular health issues and patterns of chronic disease (Olden and Newbold 2000). For example, Cantarero and Aguirre (2010) relate epidemiological findings suggesting that women's exposures to pesticides are associated with biological disruptions, including menstrual disturbances, reduced fertility, and pregnancy-related problems (226). Moreover, they note (224):

> A gender finding of interest in connection with environmental estrogens is that women are more susceptible than men to auto-immune conditions. Women's heightened immune responses to both foreign and self-antigens appear to account for their greater preponderance of auto-immune disease. An important difference is that usually women have a higher percentage of body fat than men and this has been associated with a larger storage of lipophilic chemicals.

Some of the women who responded to the survey were also aware of the impact of these physiological differences:

> *As a woman, I would say I am more exposed to chemicals because I have more fatty tissue for them to accumulate in – so there is more effective exposure that way. Similarly, my biology may render me differently susceptible to pollution – i.e., hormone disruptors will have a different sort of impact on a female body, even though men can get breast cancer. I don't think what I do as a woman, however, exposes me any more.*
> – KATHRYN, 28

Toxic exposure can be particularly significant during specific periods of development, known as "critical windows of vulnerability," which are highly sex-dependent (see also Chapters 2, 3, 4, 7, and 11;

Rice and Barone 2000). Contact with chemicals during these windows, including pregnancy, puberty, and menopause, can cause permanent and irreversible harm to the developing body and brain, and can lead to a number of health issues further down the road (Grandjean and Landrigan 2006; Selevan, Kimmel, and Mendoza 2000).

More and more studies are revealing the unique health effects on women of toxic exposure during these windows. Exposure during key windows of vulnerability not only affects the health of adult women but can also pose particular risks to the health of the fetus (Selevan, Kimmel, and Mendoza 2000; Women's Foundation of California 2003). As a result, any discussion of women's health must also consider the health of their children, as women's bodies are the first environments for the growing fetus. As Cantarero and Aguirre stated (2010, 225):

> During pregnancy certain synthetic chemicals stored in a woman's body have the ability to cross the placenta, where they have the potential to cause birth defects or other more subtle damage to the development of the fetus. It is estimated that background concentrations of dioxins and PCBs in industrialized areas in Western Europe account for subtle congenital disorders such as hyperactivity and lowered IQ in 10 percent of newborns.

Many studies have demonstrated the significant effect of maternal chemical exposure on the health of infants and children. There are also many birth cohort studies of prenatal and early childhood exposures to lead, mercury, pesticides, and PCBs that show that they may affect the neurological development of children (Jurewicz, Polańska, and Hanke 2013). Maternal exposures to air pollutants also have an effect on the fetus. A 2007 study of births in Massachusetts and Connecticut found that four types of air pollution (carbon monoxide, nitrogen dioxide, and two classes of airborne particles) correlate with low birth weight. The study indicated that pregnant women exposed to even moderate amounts of these pollutants have babies with lower birth weights than women in areas with cleaner air (Barry 2007).

As children continue to develop and grow older, their health can be influenced by the exposures they experienced inside and outside of the womb. Babies develop faster in their first year than at any other point in their lives; eat more food, drink more fluids, and breathe more air per unit body weight than an adult; and have a greater ratio of surface area to body mass (CPCHE 2005). As a result, not only are young children more biologically sensitive to chemical exposure but they also face the combined health effects of maternal and personal exposures at a young age. These cumulative exposures can increase a child's risk of long-term health conditions and diseases, and have long-term effects on children's intellectual and motor abilities and behaviour (Braun et al. 2006; Grandjean and Landrigan 2006). A very high correlation has been found between umbilical cord blood lead levels and mental development of children at age two (Bellinger et al. 1987), and elevated levels of lead in children's teeth lead to deficits in psychological and classroom performance in children (Needleman et al. 1979). These biological effects of chemicals on children's health and intellectual capacities also affect their mothers, as women are usually the primary caregivers.

Mergler (2012) proposes that there is a life cycle effect of exposures to neurotoxic substances on girls and women. At the far end of that spectrum, the hormonal changes associated with menopause may increase women's risk for environmental health-related problems. Most of the total lifetime body burden of substances like lead are stored in the bone. During menopause, when calcium and other minerals are released from the bone, covalent substances like lead are also released and mobilized into the blood, where they can be transferred into soft tissue such as the central nervous system, and affect cognitive and motor functions (Berkowitz et al. 2004).

Besides the specific effects associated with the windows of vulnerability, research has shown that particulate matter of 10 microns or less has an effect on women's health after menopause, even at levels below regulated air quality standards. A 2008 study on air pollution suggests suppressed immune function in post-menopausal women living within 150 metres of major arterial roadways (Williams et al. 2009). Another study carried out in thirty-six cities in the United States in 2007 indicated a higher risk of cardiovascular disease in women aged fifty to seventy-five living closer to sources of outdoor air pollution. The higher the level of particulate matter measured in the air, the higher the overall risk of cardiovascular events, including strokes and heart attacks (Miller 2007).

Social Differences

Beyond biological differences and susceptibilities, it is important to consider how gendered relations of power and systemic gender inequality can impact women's and men's exposures to environmental chemicals. These inequalities are often exemplified by gender-specific roles, responsibilities, expectations, opportunities, and constraints. Many of the contaminants women encounter are found in their daily environments, which tend to be distinct as a result of ongoing gendered divisions of labour. Paid work, and its accompanying occupational exposures, is significantly gendered in nature (Armstrong and Armstrong 2010; Messing 1998). Certain professions, such as teaching, nursing, hairdressing, and aesthetics, are primarily female professions, whereas other occupations, such as construction work, mining, and corporate management, continue to be primarily male. Women often take on lower-paying jobs and have less control over their working environment, as well as limited resources to protect themselves from unwanted chemical exposures (see Chapters 10 and 11).

Survey respondents suggested that there are many women working in precarious, low-wage jobs in their neighbourhoods, and that the women experiencing the greatest exposures include:

> *apple farmers, and those working for cleaning companies that use chemical based cleansers.* – MARA, 42

> *Women who use cleaning fluids, especially those who clean for a living; women who work in factories or drive buses or are on the road a lot.*
> – KATHERINE, 60

Another woman spoke of her own work-related exposures and their effects:

> *I have worked in Canada as a paint chemist in research and development for industrial paints for 17 years, and was exposed to many toxic substances including solvents, lead pigments and isocyanates, among other toxic substances. Though I had some sensitivities and allergies prior to that, it would appear that my work exposure contributed to my present state of health. My sensitivities and allergies increased dramatically and my responses were very pronounced. As a result, I was forced to retire and address my medical situation.* – SANDRA, 58

Even within the same workplace, women and men exposed to the same substances may experience differential health impacts. Women's exposure patterns are less frequently studied than men's, and it is notable that there are very few studies on women's exposures during pregnancy.

Chemical Exposure and Domestic Labour

Women are also exposed to harmful chemicals through domestic labour. They are still primarily responsible for the majority of unpaid household labour, including cleaning, preparing meals, doing laundry, and caring for children and the elderly (Diaz 2006; Women's Foundation of California 2003). As a result, they tend to bear the burden of exposure to chemicals associated with domestic chores:

> The women ... are very much affected by air pollution. It is they who spend a great deal of time in the house, cleaning it, because of the dirtying influence of vehicle exhaust ... Caring for those who are ill from pollution and cleaning furniture and maintaining households is labour that is left up to women, and the time they need to spend on these activities increases because of pollution. (Diaz 2006, 40)

Indoor chemical exposures are connected to VOCs, carbon monoxide, pesticides, and particulate matter. Women who juggle both paid and unpaid employment are at even greater risk, as they experience the combined exposures resulting from toxins present in their workplaces and in their homes (Luxton and Corman 2001).

Consumerism and Women's Health

Because women are primarily responsible for domestic labour and household duties, they are often in charge of household consumption and purchases (Barletta 2006). A number of products are marketed specifically towards women, playing on gendered roles and expectations (see Chapters 2 and 3). This can be seen most clearly in the sale of body care products and cosmetics, many of which contain toxic chemicals that women are unaware of. As a result, women are often disproportionately exposed to toxic chemicals from product use. They must navigate the pressures of gender norms to use certain products, and they may experience toxic chemical exposures as a result.

The survey respondents remarked on the social pressure to use harmful products, and the difficulty of finding products that are free of toxic ingredients:

> *Standards of beauty are different for women and men. As women, we are expected to make ourselves beautiful, to use cosmetics, to 'make up' for*

> *whatever we may be lacking in terms of not fitting into a standard of beauty that is designed to please the eye of the beholder, the male beholder, and is widely publicized by advertising and other media. Trying to uphold or conform to standards of beauty designed for the male eye continues to be a stressor for many women, particularly for young women.* – ROSE, 60

> *Doing laundry, the dishes, and even bathing with the wrong products, is exposing women to chemicals.* – DIANNE, 37

Intersecting Identities and Vulnerability to Exposure

Women who face gender discrimination in conjunction with other factors, such as lack of income or racial discrimination, can be more vulnerable to chemical exposure.

> *This is a very important issue, the relationship between health and wealth. It isn't only the exposures from the big plants or from agriculture.* – ROSE, 60

A significant social determinant affecting toxic exposure is a woman's socio-economic status (see Chapters 3, 10, and 11). Class and income are key predictors of health around the world. Those with money, resources, and power tend to be in better health than those who are marginalized or poor (Wilkinson and Marmot 2003). Individuals in poverty often lack the power or capacity to avoid or address contexts that lead to poor health, such as toxic exposure, and they experience higher rates of illness and mortality, with limited access to adequate health care and other health services (Wright 2006). As Thompson states (2003, 111):

> Privileged people insulate themselves from environmental problems in many ways including consumerism and moving away from them. The result is that privileged people, and those with decision-making power, are often spatially distant from environmental hazards and do not share the health risks to the same degree as marginalized members of society. Toxic development is more likely in poor neighborhoods, as they, relatively speaking, lack resources, knowledge of risks, and political representation to organize effective resistance, as the poor expend more of their energy and resources on mere survival.

A study called *An Examination of Pollution and Poverty in the Great Lakes Basin* found a correlation between areas experiencing elevated air pollution and high rates of poverty (Pollution Watch 2008). Indeed, evidence is building that toxic exposure is often place-based and unevenly distributed, with impoverished neighbourhoods bearing the brunt and suffering the consequences of environmental pollution, without the power or resources to escape it. For example, one survey respondent remarked:

> *If you're broke, it's hard to move somewhere cleaner. I've noticed that the public garbage bins in Forest Hill are always clean, as if they are scrubbed down regularly. The ones in my blue-collar neighborhood are grimy and frequently overflowing.* – CARRIE, 30

At present there continues to be a gendered disparity in wages, with women earning substantially less than men. As a result, women constitute a large portion of the world's poor, and are therefore more susceptible to chemical exposure and less able to prevent exposure from occurring (Rahder 2009). Women who are single parents and do not have the assistance or support of a partner experience even greater challenges in managing the stress and health impacts associated with poverty. As one survey respondent noted:

> *Women who are single parents on low income often live in areas of high pollution and in rundown public housing. They are the ones most affected, as they often do not have the means to be heard. Racial discrimination can be a problem. Poor dietary habits, lack of ongoing medical care, lack of social services, low income, insufficient education, all combine to be very negative factors in a woman's life under these circumstances. Not only is the woman affected – if there are children, they are negatively impacted as well. Women in this social class can often be ignored by society, doctors and politicians.* – SANDRA, 58

Interconnections between Gender, Class, and Race

Discriminatory practices of racism, classism, and sexism, as well as processes of colonization, have worked to benefit some segments of society while disadvantaging others. This system has led to the exploitation of land, resources, and people, and the degradation of the natural environment. Women who experience oppression on multiple fronts because of intersecting aspects of their identities – such as women of colour, Aboriginal women, or women with disabilities – may experience the added effects of discriminatory practices and processes, which can contribute to greater marginalization as well as greater frequency of and vulnerability to chemical exposure (Thompson 2004). When asked which women were more likely to be exposed to chemical pollution, the survey respondents remarked:

> *I'm sure that less educated and immigrant women work amidst uglier substances than I do.* – KATHERINE, 60

> *We all are [exposed to chemicals]. But poorer women are more so, due to the fact that cheaper housing is in dirtier places.* – CARRIE, 30

Indigenous peoples represent a segment of the population that faces marginalization on multiple scales. Around the world, Indigenous communities experience a disproportionate share of pollutants, toxins, and hazardous waste in their territories. These environmental concerns, combined with processes of racism, colonization, and sexual discrimination, have put the health of Indigenous peoples, particularly the health of the women, at risk. Many Aboriginal communities in Canada suffer ongoing contamination of their lands and often lack access to clean air and water, resulting in elevated chemical exposures among community members (Hanrahan 2003). For example, the continuous, low-dose exposures to contaminants experienced by the Aamjiwnaang First Nation in Sarnia's "Chemical Valley" have led to a declining ratio in the number of boys born for every girl (Ellwood 2008; Scott 2008; Introduction to this volume).

Aboriginal women may be at particular risk of chemical exposure as a result of their traditional roles in their communities. Fish and game, which represent a large part of the diet of women and their families, are often contaminated by a number of toxins, such as mercury, PCBs, and persistent organic pollutants (POPs), that persist in the local lands and water. Studies have revealed elevated levels of PCBs in the breast milk of nursing women from the Akwesasne First Nation as a result of eating contaminated fish (Hwang et al. 2001). In many Inuit communities, organochlorine compounds are concentrated in the aquatic food chain and Inuit mothers exhibit the highest body burden (Dewailly et al. 1993). Environmental contamination continues to affect the health and livelihoods of these women in multiple ways. According to Suzanne, a

31-year-old survey respondent and Mohawk woman from the Tyendinaga First Nation:

> *Women are responsible for the water. If the waters in the rivers and lakes are getting contaminated, so are the waters in our body and we don't know what we are transferring to the babies. The value of water, our moon ceremonies that are not done as often, they get lost too. We forget that is our responsibility. We have forgotten. I don't know if that is directly due to contaminants. Most women don't want to see contaminants in the air. Most of the time they are not aware. That is why it is such a huge issue in our communities. Lot of our wells are contaminated. The bay, we found benzene, toluene, PCBs, mercury and arsenic, in the bay – either here or in the surrounding areas.*

Steps Forward: Chemical Exposures, Awareness, and Action

> *A lot of the systems would be set up differently if women had more say. Education, health and the way in which disputes are resolved would probably be more community based, future thinking with less red tape. We might see more creativity on the environment front as well as the way society supports those raising a family – the next generation.*
> – YVETTE, 30

Women Taking Action

Many women have been stepping up to the plate to address chemical exposures on an individual, family, and community level for quite a while. More women are becoming aware of the health impact of the chemicals they and their family consume, and are taking precautions to reduce exposures to these contaminants. Survey respondents revealed this concern and motivation for action and described some of the steps they are taking to limit chemical exposure, including:

> *Eating fresh local produce, driving rarely, drinking water, walking the dog, hardly ever taking pills (prescription drugs).* – KATHERINE, 60

> *Trying to look at labels more carefully; cooking more of my own food, as opposed to buying precooked; using natural products; and reducing the amount of plastics.* – EVELYN, 51

I make or buy my own cleaning, hair, and skin care products. I stay away from places of strong pollution if I can help it. I boost my immunity when I know I will be exposed or have been exposed to such environments by adding supplements, and sending love and healing energy to my being. I take breaks when needed. – YVETTE, 30

When I am able, and where there is a relevantly connected issue in the media online, I leave comments. People are starting to blog about it. – LINDA, 49

Many women have also begun to work in groups at the community level to protect environmental and human health. The environmental justice movement, primarily headed by women at the grassroots, seeks to address the social implications of discriminatory environmental policies and practices through the empowerment and education of communities that bear the greatest consequences of policies and practices. Through organized advocacy and action, women are demanding a change in patriarchal structures so that women's health needs and concerns can be appropriately addressed, and voicing their concerns and ideas about what needs to be done by industry, labour, and government to reduce the impact of chemical exposures on women's health.

The Need for Action on a Broader Scale

As my initial narrative that problematized my everyday world reveals, I have attempted to be conscious of potential exposures in my environment. I have a strong constitution. I take as many precautions as I can. But I recognize that regardless of these actions, I will never be able to reduce my exposures to zero, or predict the places where those exposures might occur. And what about women who are not as well educated or who are more vulnerable or susceptible to exposure? What about pregnant women, fetuses in utero, babies and young girls, and elderly women? What about the canaries in the coal mine, the women who have already experienced the consequences of toxic exposure, women with multiple chemical sensitivities and other conditions that affect multiple systems in the body? What about women with low socio-economic status who live in polluted neighbourhoods and cannot afford to move away? What about immigrant women working in factories, as farm workers, or as cleaning ladies? What about Aboriginal women living on reserve and dealing with disproportionate exposures to industrial air pollution and water contamination? How well are they able to mediate their toxic exposures in conjunction with other stresses in their lives, and how do these factors interact to affect their immediate and long-term health?

While individual women can make certain changes to their behaviour and lifestyle to reduce the risk of exposure, the prevalence of toxic chemicals in the environment makes it clear that individual actions can go only so far. As one of our respondents said:

> *My life now revolves around avoiding the consumption of chemicals, while corporations force more and more into our lives, our environment, our bodies. The sense of the inability to protect ourselves, children, and other life forms, knowing how harmful these chemicals are is all-consuming, as well as the knowledge that 'fighting' the infusion of more and more petro-chemicals for profit seems hopeless. But it must be continued in spite of this ... 'consuming chemicals' is the most important issue in our lives, since those chemicals are killing us, slowly but surely.* – BEV, 60

Women are clearly becoming more aware of what they can do, but placing full responsibility for avoiding chemicals on women's shoulders is not acceptable. In order to truly protect women's health and the environment, an alternative approach to chemicals policy and management at the government level is vital. Industry and government must begin to take responsibility for the toxins that are being absorbed into our environments and bodies. Policy makers have a part to play in confronting the toxic elements of our current economic, social, and political systems, and need to work with communities, particularly with women, to create effective pollution reduction and prevention strategies.

Where Do We Go from Here?

What steps can industry, government, and the public take to address chemical exposure, particularly in relation to women's health? The following are some important measures that would protect women's health and contribute to a safer and healthier environment and society as a whole:

- *We need gender-sensitive and sex-specific research* that asks questions that apply to both females and males and takes gender into account

(Messing et al. 2003). We also need research that continues to provide qualitative and quantitative data on the links between contaminants, the environment, and women's health, as well as the cumulative effects of chemical exposures. We also need community-based participatory research that involves women at the grassroots in determining questions of relevance to women, and we need research that works to inform policy decisions and promote action.

- *We need to raise awareness* at the individual, community, organization, and government level about the risks of chemical exposure, how policy reforms can work to reduce exposures and ensure greater environmental and human health, and how people can get involved in making change. Part of this awareness involves listening to and considering women's everyday stories and experiences, and their unique knowledge about exposure, health, and prevention. Through awareness comes advocacy and action, where individuals from all walks of life work together to pressure industry and government to reduce the risk of chemical exposure.
- *We need public policies* that work to protect not only women's health but the health of everyone, through stricter pollution reduction and prevention, monitoring, and enforcement practices. Part of this work involves government's *embrace of the precautionary principle* in policy decisions and assessments. Policy makers do not serve the public or protect women's health when their focus is on managing rather than preventing pollution. In many cases, government has chosen to undertake limited and costly cleanup programs instead of mitigating emissions and discharges at the source. It is time to embrace a more precautionary approach in policy decisions. The precautionary principle states that when an activity threatens public health or the environment, measures to avoid adverse effects are warranted, even if there is scientific uncertainty about the nature and extent of harm that may occur (UNESCO 2005, 14). Such an approach gives priority to the protection of ecological systems and human health: instead of waiting for proof of harm before taking action on a chemical, decision makers err on the side of caution and work to anticipate and avoid harm.
- *We need to address all forms of toxicity* at their root by confronting inequity, injustice, sexism, racism, classism, and other forms of discrimination and oppression that influence the health of women and other members of society.

- *We need to recognize the interconnected nature of our environments,* and realize that by taking actions to improve and promote women's health, we are also taking actions to improve and promote the health of the earth and of all living things. Humans are not in control of the environment; we are one part of it, no more or less than any other living being. What we do to our environments we are doing to ourselves.
- *We need to live more simply, respect ourselves, and respect each other.*

Acknowledgments

Thank you to the twenty-five women who took the time to talk with me and complete the survey.

Bibliography

Armstrong, Pat, and Hugh Armstrong. 2010. *The Double Ghetto: Canadian Women and Their Segregated Work.* 3rd ed. Toronto: Oxford University Press.

Ayoob, Sulaiman, and Ashok Kumar Gupta. 2006. "Fluoride in Drinking Water: A Review on the Status and Stress Effects." *Critical Reviews in Environmental Science and Technology* 36 (6): 433-87, doi:10.1080/10643380600678112.

Barletta, Marti. 2006. "Women Control about 80 Percent of Household Spending: A Look at the Numbers." http://www.trendsight.com/content/view/40/204/.

Barry, Patrick. 2007. "Pregnancy and Pollution: Women Living in Areas with Poor Air Quality Have Babies with Lower Birthweights." *Science News* 171 (17): 261, doi: 10.1002/scin.2007.5591711706.

Bellinger, D., et al. 1987. "Longitudinal Analyses of Prenatal and Postnatal Lead Exposure and Early Cognitive Development." *New England Journal of Medicine* 316: 1037.

Berkowitz, Gertrud S., Mary S. Wolff, Robert H. Lapinski, et al. 2004. "Prospective Study of Blood and Tibia Lead in Women Undergoing Surgical Menopause." *Environmental Health Perspectives* 112 (17): 1673-78. http://www.ncbi.nlm.nih.gov/pmc/articles/PMC1253658/.

Braun, J., et al. "Exposures to Environmental Toxicants and Attention Deficit Hyperactivity Disorder in US Children." *Environmental Health Perspectives* 114 (12): 1904-9.

Cantarero, Lourdes, and Isabel Yordi Aguirre. 2010. "Gender Inequities in Environment and Health." In *Environment and Health Risks: A Review of the Influence and Effects of Social Inequities,* edited by the World Health Organization, 217-37. Copenhagen: World Health Organization. http://www.euro.who.int/__data/assets/pdf_file/0003/78069/E93670.pdf.

Chen, H., M.S. Goldberg, R.T. Brunett, et al. 2013. "Long-Term Exposure to Traffic-Related Air Pollution and Cardiovascular Mortality." *Epidemiology* 24 (1): 35-43, doi:10.1097/EDE.0b013e318276c005.

CPCHE (Canadian Partnership for Children's Health and the Environment). 2005. *Child Health and the Environment: A Primer.* Ottawa: CPCHE. http://www.healthyenvironmentforkids.ca/sites/healthyenvironmentforkids.ca/files/cpche-resources/Primer.pdf.

Dewailly, Eric, Pierre Ayotte, Suzanne Bruneau, et al. 1993. "Inuit Exposure to Organochlorines through the Aquatic Food Chain in Arctic Quebec." *Environmental Health Perspectives* 101: 618-20.

Diamond, Irene, and Gloria Orenstein. 1990. *Reweaving the World: The Emergence of Ecofeminism*. San Francisco: Sierra Club Books.

Diaz, Jackeline Contreras. 2006. "Women at Disadvantage: Pollution in Quito, Ecuador." *Women and Environments International Magazine* 70/71 (Spring/Summer): 40-41.

Ellwood, Wayne. 2008. "This Toxic Life." *New Internationalist* (415). http://www.newint.org/features/2008/09/01/keynote-plastic/.

Environment Canada. 2012. *Pollution Sources*. http://www.ec.gc.ca/.

Environmental Defence. 2006. *Polluted Children, Toxic Nation: A Report on Pollution in Canadian Families*. Toronto: Environmental Defence. http://environmentaldefence.ca/sites/default/files/report_files/PCTN_English%20Web.pdf.

Grandjean, P., and P.J. Landrigan. 2006. "Developmental Neurotoxicity of Industrial Chemicals." *Lancet* 368 (9553): 2167- 78.

Hackney, Jack D., Fredrick C. Thiede, William S. Linn, et al. 1978. "Experimental Studies on Human Health Effects of Air Pollutants. IV. Short-Term Physiological and Clinical Effects of Nitrogen Dioxide Exposure." *Archives of Environmental Health* 33 (4): 176-81.

Hanrahan, Maura. 2003. "Water Rights and Wrongs." *Alternatives* 29 (1): 31-34.

Hatch, Maureen C. 2000. "Centers Needed to Study Women's Environmental Health." *Environmental Health Perspectives* 108 (7): A10-11.

Health Canada. 2010. "Results of the Canadian Health Measures Survey Cycle 1 (2007-9)." http://www.hc-sc.gc.ca/ewh-semt/pubs/contaminants/chms-ecms/index-eng.php.

Hwang, Syni-An, Bao-Zhu Yang, Edward F. Fitzgerald, et al. 2001. "Fingerprinting PCB Patterns among Mohawk Women." *Journal of Exposure Analysis and Environmental Epidemiology* 11 (3): 184-92, doi:10.1038/sj.jea.7500159.

Jurewicz J, Polańska K., and W. Hanke. 2013. "Chemical Exposure Early in Life and the Neurodevelopment of Children: An Overview of Current Epidemiological Evidence." *Annals of Agricutural and Environmental Medicine* 20 (3): 465-86.

Luxton, M., and J. Corman. 2001. *Getting by in Hard Times: Gendered Labour at Home and on the Job*. Toronto: University of Toronto Press.

May, Elizabeth. 2003. "The Environment: A Place to Visit." *Women and Environments International Magazine* 60/61: 8-9.

McCally, Michael. 2000. "Environment and Health: An Overview." *Canadian Medical Association Journal* 163 (5): 533-35.

Meng, Ying-Ying, Michelle Wilhelm, Rudolph P. Rull, et al. 2007. "Traffic and Outdoor Air Pollution Levels Near Residences and Poorly Controlled Asthma in Adults." *Annals of Allergy, Asthma and Immunology* 98 (5): 455-63, doi:10.1016/S1081-1206(10)60760-0.

Mergler, Donna. 2012. "Neurotoxic Exposures and Effects: Sex and Gender Matter! Hänninen Lecture 2011." *Neurotoxicology* 33 (4): 644-51. http://dx.doi.org/10.1016/j.neuro.2012.05.009; http://www.cinbiose.uqam.ca/upload/files/Mergler_2012_Neurotox_gender__sex.pdf.

Messing, Karen. 1998. *One-Eyed Science: Occupational Health and Women Workers*. Philadelphia: Temple University Press.

Messing, K., L. Punnett, M. Bond, K. Alexanderson, J. Pyle, S. Zahm, D. Wegman, S.R. Stock, and S. de Grosbois. 2003. "Be the Fairest of Them All: Challenges and Recommendations for the Treatment of Gender in Occupational Health Research." *American Journal of Industrial Medicine* 43: 618-29.

Miller, Kristen. 2007. "Women in Polluted Areas at Higher Risk of Cardiovascular Disease." *Journal of Environmental Health* 69 (10): 70.

Needleman, H.L., Charles Gunnoe, Alan Leviton, Robert Reed, Henry Peresie, Cornelius Maher, and Peter Barrett. 1979. "Deficits in Psychological and Classroom Performance of Children with Elevated Dentine Lead Levels." *New England Journal of Medicine* 300: 689-95.

Olden, Kenneth, and Retha R. Newbold. 2000. "Women's Health and the Environment in the 21st Century." *Environmental Health Perspectives* 108 (S5): 767-68.

OMA (Ontario Medical Association). 2006. "Smog's Excess Burden on Baby Boomers: Aging Population Most Vulnerable to Smog." https://www.oma.org/Resources/Documents/SmogBoomersReport.pdf.

–. 2008. "Local Premature Smog Deaths in Ontario." https://www.oma.org/.

Perrotta, Kim R., and Ronald R. Macfarlane. 2002. *Ten Key Carcinogens in Toronto Workplaces and Environment: Assessing the Potential for Exposure.* Toronto: Toronto Public Health. http://www.toronto.ca/health/pdf/cr_summaryreport.pdf.

Peterson, Gregory W. 2008. "Water: Essential for Life." *Original Internist* 15 (1): 21-25.

Pollution Watch. 2008. *An Examination of Pollution and Poverty in the Great Lakes Basin.* Toronto: Canadian Environmental Law Association and Environmental Defence. http://www.cela.ca/publications/examination-pollution-and-poverty-great-lakes-basin.

Prüss-Üstün, Annette, and Carlos Corvalán. 2006. *Preventing Disease through Healthy Environments: Towards an Estimate of the Environmental Burden of Disease.* Geneva: World Health Organization. http://www.who.int/quantifying_ehimpacts/publications/preventingdisease.pdf.

Prüss-Üstün, Annette, Carolyn Vickers, Pascal Haefliger, et al. 2011. "Knowns and Unknowns on Burden of Disease Due to Chemicals: A Systematic Review." *Environmental Health* 10 (9). http://www.ehjournal.net/content/10/1/9.

Rahder, Barbara. 2009. "Invisible Sisters: Women and Environmental Justice in Canada." In *Speaking for Ourselves: Environmental Justice in Canada,* edited by Julian Agyeman, Peter Cole, Randolph Haluza-DeLay, and Pat O'Riley, 81-96. Vancouver: UBC Press.

Rahder, Barbara, and Rebecca Peterson. 2000. *An Environmental Framework for Women's Health.* A discussion paper for the National Network on Environments and Women's Health (NNEWH). http://www.cewh-cesf.ca/PDF/nnewh/environmental-framework.pdf.

Rice, D., and S. Barone Jr. 2000. "Critical Periods of Vulnerability for the Developing Nervous System: Evidence from Human and Animal Models." *Environmental Health Perspectives* 108 (S3): 511-33.

Schwartz, J.D. 2012. "Powerful Interests: A Conversation with Sandra Steingraber." *Sage Magazine,* http://www.sagemagazine.org/powerful-interests-a-conversation-with-sandra-steingraber/.

Schwartz, Joel. 2004. "Air Pollution and Children's Health." *Pediatrics* 113 (S-4): 1037-43.

Scott, Dayna Nadine. 2008. "Confronting Chronic Pollution: A Socio-Legal Analysis of Risk and Precaution." *Osgoode Hall Law Journal* 46 (2): 293-344. http://www.ohlj.ca/english/documents/OHLJ46-2_Scott_ConfrontingChronicPollution.pdf.

Selevan, S.G., C.A. Kimmel, and P. Mendoza. 2000. "Identifying Critical Windows of Exposure for Children's Health." *Environmental Health Perspectives* 108: 451-55.

Smith, Dorothy E. 1987. *The Everyday World as Problematic: A Feminist Sociology.* Toronto: University of Toronto Press.

Steingraber, Sandra. 2001. *Having Faith: An Ecologist's Journey to Motherhood.* Cambridge: Perseus Publishing.

Thompson, Shirley. 2003. "From a Toxic Economy to Sustainability: Women Activists Taking Care of Environmental Health in Nova Scotia." *Canadian Woman Studies* 23 (1): 108-14.

–. 2004. "From Environmental Ill to Environmental Health." *Women and Environments International Magazine* 62/63: 40-42.

UNESCO (United Nations Educational, Scientific and Cultural Organization). 2005. *The Precautionary Principle: World Commission on the Ethics of Scientific Knowledge and Technology.* Paris: UNESCO. http://unesdoc.unesco.org/images/0013/001395/139578e.pdf.

US Geological Survey. 2011. *Water Science for Schools: The Water in You.* http://ga.water.usgs.gov/edu/propertyyou.html.

Weiss, B. 2011. "Same Sex, No Sex, and Unaware Sex in Neurotoxicology." *Neurotoxicology* 32: 509-17.

Welshons, Wade V., Kristina A. Thayer, Barbara M. Judy, et al. 2003. "Large Effects from Small Exposures. I. Mechanisms for Endocrine-Disrupting Chemicals with Estrogenic Activity." *Environmental Health Perspective* 111 (8): 994-1006.

WHO (World Health Organization). 1946. *WHO Definition of Health.* Preamble to the Constitution of the World Health Organization as adopted by the International Health Conference, New York (19-22 June). http://www.who.int/about/definition/en/print.html.

–. 2009. "Country Profile of Environmental Burden of Disease: Canada." http://www.who.int/quantifying_ehimpacts/national/countryprofile/canada.pdf.

Wilkinson, Richard, and Michael Marmot, eds. 2003. *Social Determinants of Health: The Solid Facts.* 2nd ed. Copenhagen: World Health Organization.

Williams, Lori, Cornelia M. Ulrick, Timothy Larson, et al. 2009. "Proximity to Traffic, Inflammation, and Immune Function among Women in the Seattle, Washington Area." *Environmental Health Perspectives* 117 (3): 373-78. http://www.ncbi.nlm.nih.gov/pmc/articles/PMC2661906/.

Women's Foundation of California. 2003. *Confronting Toxic Contamination in Our Communities: Women's Health and California's Future.* San Francisco: Women's Foundation of California. http://sandiegohealth.org/women/wfc/confronting_toxic_contamination.pdf.

Wright, Rosalind J. 2006. "Health Effects of Socially Toxic Neighborhoods: The Violence and Urban Asthma Paradigm." *Clinics in Chest Medicine* 27 (3): 413-21.

Chapter 2 Protecting Ourselves from Chemicals: A Study of Gender and Precautionary Consumption

Norah MacKendrick

Exposure to chemical substances occurs through typical everyday activities, such as eating, breathing, drinking water, and using cosmetics or common household goods. Many of these consist of environmental contaminants, including brominated flame retardants, heavy metals, and pesticides, which are ubiquitous in the environment of industrialized countries. Exposure to these contaminants begins at conception and continues through the life course, such that individuals have little awareness and limited control over their exposure (Sexton, Needham, and Pirkle 2004). In response to health concerns associated with exposure to environmental contaminants, community groups, advocacy organizations, and consumers themselves have been calling for the implementation of a "precautionary" policy framework for the management of toxic substances. National policy frameworks in Canada have not followed a precautionary approach (CELA 2006; Denison 2007), although some elements of a more precautionary framework may now be emerging (see Chapter 3). Nevertheless, the shift towards more precautionary regulation is a recent one, meaning that chemicals policy in Canada is best described as a patchwork of precautionary and non-precautionary approaches (see also CSM and CELA 2008).[1]

The precautionary principle, referred to in Chapter 1, allows for restrictions on human activities suspected to pose a threat to the environment or human health by stating that proof of safety must be established before new technologies are introduced (O'Riordan, Cameron, and

Jordan 2001). This principle stands in contrast to the conventional regulatory regime, which, until fairly recently, places restrictions only on proof of *harm*. The focus of this chapter is on precaution as it is enacted at the individual level, by consumers purchasing environmentally friendly products. I call this phenomenon "precautionary consumption" (MacKendrick 2010).

Because women are disproportionately involved in household tasks, particularly decisions regarding household consumption and family health and nutrition (Johnson and Johnson 2008; Sayer 2005), precautionary consumption (PC) represents a *gendered* practice. We know very little about how PC is experienced as a "normal" aspect of everyday life by women seeking to protect their own health and that of their families. Moreover, we are only beginning to consider the extent to which PC is seen as an effective method of controlling one's exposure to chemicals, and whether this attempt at reclaiming control is associated with a lack of trust in the regulatory system responsible for managing chemical substances. Finally, there is a lot to learn about the factors that might influence PC, such as socio-economic status, access to markets for "chemical-free" or non-toxic alternatives,[2] and general awareness of chemical risks – including access to information on chemical safety and information regarding appropriate choices of alternative products.

In this chapter, I draw both from the literature on gender and caring work and from qualitative empirical data gathered from focus groups consisting of "precautionary consumers." I examine the role of women in the everyday management of exposure to contaminants, and discuss the broader policy implications of selective consumerism and lifestyle modifications as mechanisms for managing this exposure. In what follows, I provide an overview of PC as an approach for dealing with chemical exposures and situate it within a gender and caregiving perspective. I then summarize the major findings from focus groups and conclude with a discussion of the policy implications of these findings.

What Is Precautionary Consumption?

In the context of chemical exposures, PC consists of purchasing products labelled as "non-toxic," "natural," or "certified organic," and adopting lifestyle changes, such as adhering to particular diets that purport to rid the body of contaminants. This has also been described as an "inverted quarantine" (Szasz 2007) or consumption fallacy (Altman et al. 2008), terms that refer to self-protection from perceived external threats by using specialized goods and services such as bottled water, sunscreen, and home security systems. "Precautionary consumption" is the preferred term in this case, as we are interested in how *precaution* informs self-protection.

PC appears to be a reaction against institutionalized risk assessments that have deemed a product "safe enough" because conclusive proof of harm could not be established. A shopper may select natural dish detergent over the conventional brand because of concerns about chemicals (such as, for example, triclosan or sodium laureth sulfate) that are not restricted under standard government safety assessments. In doing so, this shopper is acting in a precautionary way because her choices are directed towards the product that provides greater certainty that chemical ingredients are *safe* for human health and the environment. She is, in effect, attempting to circumvent the conventional consumer landscape, which is characterized by a lack of precautionary decision making by chemical producers, product manufacturers, and governments charged with overseeing chemical safety.

In the last decade, PC has come to represent a common strategy for negotiating exposure to chemicals. In the Canadian news media, consumer products are increasingly identified as a source of harmful toxins (MacKendrick 2010). The growing sale of non-toxic products reflects this trend (e.g., Vora 2012), and individuals value being able to take a precautionary approach to everyday chemical exposure, as evidenced by the push for legislation requiring that all categories of consumer products (especially cosmetics and cleaning products) have their ingredients fully disclosed on product labels (e.g., Toxic Free Canada 2011).

There are several possible explanations for the growth of PC as a strategy for contaminant avoidance. First, it is likely a response to growing concern about "body burdens," or the internal chemical load carried by organisms that can be detected through biomonitoring analyses – a method for analyzing animal tissue and fluids for biological markers of contaminant exposure.[3] Body burdens have been the subject of several high-profile campaigns by prominent environmental organizations.[4] Biomonitoring is a powerful new informational tool that reveals the extent to which our regulatory regimes have not been able to prevent a whole host of known and suspected toxins from accumulating in the tissues of our bodies (Health Canada 2012).

Second, chronic regulatory backlogs and decades of non-precautionary chemicals management in Canada provide a rationale for a more precautionary approach to consumption. Regulatory agencies overseeing the manufacture and use of chemicals in Canada have not consistently adopted a precautionary approach.[5] Rather, when scientific evidence suggested that certain chemicals could pose a risk to human health or the environment, government agencies did not restrict the production and distribution of these chemicals until they had gathered sufficient evidence providing *conclusive* proof of harm (CELA 2006; Denison 2007). Furthermore, because of the decision to grandfather chemicals that were in use when the Canadian Environmental Protection Act[6] was first passed, as well as the chronic backlogs in chemical testing, as many as 23,000 substances currently in use in Canada have only recently undergone preliminary safety assessments (CELA 2007, 7).[7] Nevertheless, these regulatory backlogs are now being partially addressed, and restrictions are being placed on "high-priority" chemicals, such as bisphenol A in baby products.[8]

Importantly, PC depends on access to information about potentially harmful chemicals. In the past several years, numerous guides have been published providing advice to consumers on contaminant avoidance. These include, for example, *The Safe Shopper's Bible: A Consumer's Guide to Nontoxic Household Products, Cosmetics, and Food* (Steinman and Epstein 1995); *The Toxic Sandbox: The Truth about Environmental Toxins and Our Children's Health* (McDonald 2007); *What's in This Stuff?: The Hidden Toxins in Everyday Products and What You Can Do about Them* (Thomas 2008); and *Super Natural Home: Improve Your Health Home and Planet One Room at a Time* (Greer 2009).

Hundreds of websites also provide information on chemical safety and provide tips on PC. One of the most extensive is a cosmetics database published by the Environmental Working Group (EWG), a non-profit organization based in Washington, DC. One can search through the EWG database to find ingredient lists for thousands of cosmetic products, including descriptions of harmful health effects associated with each ingredient (EWG 2014a). The EWG website also publishes summaries of recent studies in environmental health (EWG 2014b). In Canada, Environmental Defence has published several high-profile reports discussing chemical body burdens in adults and children, along with various toxic-free consumer guides.[9] Toxic Free Canada produces one of the more comprehensive guides to non-toxic living, called

CancerSmart 3.1: The Consumer Guide (Griffin 2011). This guide addresses toxins throughout the home, including in cleaning products, children's toys, household insecticides, and cosmetics. Lifestyle magazines, such as *Green Living* and *Organic Living Magazine,* also regularly publish features on non-toxic consumption and lifestyles.

Finally, agencies and organizations involved in public health issues are taking measures to advise families on environmental health best practices. Toronto Public Health, for example, has published a guide informing parents and pregnant women on how to avoid exposure to harmful chemicals in food and the home (TPH 2003). The Canadian Partnership for Children's Health and Environment, a non-profit organization consisting of nearly a dozen partner organizations involved in children's health, has produced a pamphlet called *Playing It Safe: Childproofing for Environmental Health* (CPCHE 2008). This pamphlet provides tips on protecting children from exposure to chemicals in food and everyday consumer products.

Women, Caregiving, and Environmental Health

Concern for environmental health as part of a woman's caregiving role has not yet been studied in the public health or sociological literature. In what follows, I argue that a woman's role in managing individual- and family-level exposure to environmental contaminants relates to her role as a primary caregiver and her responsibility for making most household consumption decisions.

The sociological literature recognizes that in most families women are the primary caregivers, even when they are part of the paid labour force (Allen and Sachs 2007; DeVault 1991; Hochschild 2003). As a result, despite the growing share of caregiving shouldered by men, women continue to contribute to most of the domestic labour, including child care, food preparation, shopping, and cleaning (Arendell 2000; Craig 2006; Lindsay 2008; Sayer 2005). A significant aspect of this work involves managing family health through monitoring children's well-being and development, and planning healthy meals for the family (Beagan et al. 2008; DeVault 1991; Hays 1996). Women therefore frequently act as "gatekeepers" for family health (Calabretta 2002). This role develops as early as cohabitation and marriage (Lupton 2000), but especially during pregnancy, when women begin to identify themselves as mothers (Markens, Browner, and Press 1997). As one recent Canadian study suggests, women's disproportionate involvement in food preparation and shopping – or "foodwork" – is explained by their desire to protect family health by maintaining proper nutrition (Beagan et al. 2008).

Women's magazines have also captured the role of women as managers of individual and family health. In these magazines, health columns often prominently feature and reinforce good health as both a moral imperative and individual responsibility (Roy 2008). Finally, public health organizations (e.g., CPCHE 2008; TPH 2003) increasingly present precautionary mechanisms for reducing personal and family-level exposure to potentially harmful chemicals as a basic part of managing family and maternal health. Many of the suggested actions mirror those already accepted as part of regular household cleaning and shopping for family meals. These include, for example, removing shoes before entering the home, keeping cleaning products away from children, washing children's toys carefully, and selecting foods low in chemical residues.

Women's disproportionate involvement in PC also relates to their primary role in household consumption decisions (Zukin and Maguire 2004). Domestic consumer products and food are thought to constitute part of the private sphere (Andrews and Talbot 2000) and are considered integral to motherhood (Cook 1995; Taylor, Layne, and Wosniak 2004). US marketing data estimate that women make approximately 85 percent of household consumption decisions (Holland 2010).

While much of the gender and mothering literature establishes that women are primarily involved in managing family health, there is little empirical work showing how women are responding to concerns about environmental health, and whether PC is viewed as a reasonable strategy for protecting against harmful chemical exposures. For example, practising PC depends on the social location of an individual, including factors such as socio-economic status, race, education, ability, geographic location, and access to services (Clow et al. 2009; Public Health Agency of Canada 2010). PC can therefore raise equity concerns, as women vary in their capacities to engage in the selection of, and to gain access to, "safe" products based on time availability, level of education, and household income. Consequently, three main questions guide the analysis and discussion in this chapter:

1 Do women interpret precautionary consumption as a gendered practice? If so, how do women explain why this practice is gendered?

2 To what extent are modifications to consumption patterns and lifestyle perceived as viable strategies for controlling exposure to chemicals?
3 What are the perceived social and economic constraints on PC?

Drawing from focus group discussions involving individuals who regularly use non-toxic products, I use the remainder of this chapter to describe how PC is viewed as a gendered practice and as part of the management of day-to-day exposure to chemical risks.

Focus Groups

Focus groups were selected as the main form of empirical data collection in order to benefit from interactive discussions. The discussion format of focus groups lends itself to exploratory work, as moderators can gauge the discussion and shift questions to probe the more fruitful ideas that emerge (Bloor et al. 2001). By having participants discuss ideas as a group, we gain insight into the experience of exposure avoidance, including the uncertainties and anxieties involved in making product choices and adapting lifestyles. The focus groups were not meant to be a representative sample of precautionary consumers, nor was the goal of data collection to achieve saturation but to generate preliminary and exploratory data.

Participants were selected using a combination of convenience and purposive sampling strategies that targeted individuals *already attuned to contaminant risks and engaging in various forms of PC*. They were recruited through the posting of notices in major health food stores in downtown Toronto and on environmental health websites and listservs specific to the Toronto area. Notices described the general objectives of the study and the purpose of the focus groups, and offered modest financial compensation for participation (a coupon towards purchases at a food store of their choice). Participants were screened to ensure that they met basic criteria (live in central Toronto, shop for non-toxic products, and have a general awareness of contaminant risks without specializing in related fields either professionally or academically).

Fifty-three individuals responded to recruitment notices, 91 percent of them female. Because of screening criteria, and in some cases scheduling conflicts, only twenty individuals were accommodated as focus group participants. Most participants were female (85 percent) and married (60 percent). Participants reflected a range of age groups, with seven participants in the 20-29 range, seven in the 30-39 range, and four above the age of 40. Only four participants indicated that they had children, meaning that the groups consisted mainly of women in their reproductive years but with no children.

Participants were asked to discuss whether they saw modifications to consumption patterns and lifestyles in response to environmental health threats as a gendered practice, and whether they believed that such modifications represented an effective method for controlling individual exposure to chemicals. They were also asked to reflect on the factors that both enable and restrict access to alternative products. Finally, participants were asked about their level of trust in the regulatory regime responsible for protecting Canadians from toxic substances. Prior to the start of the discussion, all participants completed a short questionnaire to provide additional information on PC behaviour and lifestyle, knowledge of contaminant risks, and socio-economic status.

All focus groups were digitally recorded, and all recordings were transcribed verbatim after the discussions were completed. Transcriptions were then coded to identify key themes (see, for example, Macnaghten and Myers 2004). As much as possible, the words of participants were used to help describe themes observed in the focus group discussions. Below, participants are not identified by name but through a code (e.g., A-3).

Precautionary Consumption as a Gendered Practice

Participants were asked to reflect on whether interest and involvement in environmental health was gendered. At first this was posed as a general question regarding "Who shops the way you do? What kinds of people are concerned about shopping for non-toxic products?" As a follow-up, participants were asked whether they thought women were more interested in environmental health matters. Most participants believed that women were more concerned than men about chemical exposures, owing to their involvement in domestic, reproductive, and caregiving responsibilities, with some participants noting that more fathers of young children were becoming involved in managing children's exposure. Participants generally agreed that women were more involved in the "the realm of food and personal care products and cleaning" (Participant D-3). In the words of Participant G-2: "I think in an average household the woman is more likely to buy the groceries and make those purchases. It's more in her realm, and then if she's caring for a child she's more likely to care about the child's health and be more interested in it."

Some participants also noted that women were generally more concerned about maintaining good health and proper nutrition. Participant A-3 explained: "I just noticed in my own little circle that women seem to be more engaged with health issues and more engaged with being proactive around their own health than the men." Participant A-1 referred to women's awareness of their susceptibility to environmental contaminants: "I think women are more concerned about their health, as well ... We have to deal with, from a very early age, our physical selves and how our environment impacts us."

Importantly, for several participants, women's awareness of environmental health was linked to reproduction and future motherhood:

> *When you're of child-bearing age – I mean, I don't have kids yet but I hope to in the next couple of years – I'm definitely more aware of my vitamin intake or my folic acid intake ... my body is also a vessel for somebody else's life and you know, in the back of your mind you think that you have a responsibility to be healthy and to make the [right] choices that will eventually affect a child.* – PARTICIPANT D-1

> *I think that must affect all women whether we end up having children or not, just that underlying ability to create another life, it just must create a fundamentally different perspective on our responsibility, our fundamental responsibility.* – PARTICIPANT B-1

Participant A-3 reflected on how becoming a mother affected her awareness of chemical exposures: "I think it deepened it ... Yeah, I was quite a worrier. [laughs] I would say that it definitely deepened. I'd be a mess now as a new mother with the toys that are out there and all the ones that keep getting recalled." She later attributed her attention to issues surrounding health and contaminants to the gendered division of labour within her household. While acknowledging that her husband had similar concerns about chemical exposures, she explained that household tasks were divided in a deliberate way to enable her to manage family-level exposure to chemicals: "You know, we do have our rules, right? And they've evolved over time by who's better at what because things had to get done. [laughs] And I guess this is something I made myself be better at."

In some cases, participants attributed an interest in environmental health to having children. Several noted that in families with young children, parents often buy organic for their children but forgo organic products for themselves:

> *I think for some people it hits home when they have kids. They suddenly start caring about the future because their kid is going to be in it and their kid is the most important thing, and some people don't wake up until that moment. So it can sometimes change people ... My friend, when she was very poor, she bought organic for her baby but not for herself. As soon as she started earning more she bought organic for everyone but there was a time when she couldn't.* – PARTICIPANT A-2

A male participant (F-2) also noted this as a reason he shopped for non-toxic products, and explained that "it seems more important now to buy organic and go out of our way to pay a higher price to buy better food for [our child] now than for us." Participant C-2 agreed, noting that "I've seen a lot of friends that don't buy organic for themselves ... and they just say, 'you know what, we'll buy organic for the baby.'"

The Experience of Precautionary Consumption

When asked about the effectiveness of modifying consumer habits and lifestyles in response to concerns about environmental health, some participants described a more precautionary approach to consumption as part of a personal responsibility to protect themselves. Building on a discussion about distrust of federal regulators, Participant E-1 explained it this way: "There's a lot of personal responsibility, as well, involved, even though we should be able to trust the people who are making these decisions for us, but I think you just have to kind of go with your own instincts." Participant B-1 stated: "I believe the onus is on me to make an educated choice and do research to find out what I'm getting myself into."

The emphasis on personal responsibility was often informed by a sense of precaution. In reference to federal regulatory bodies, Participant A-3 explained: "I think their tolerance level for what is safe is a lot looser and higher than mine would be if I knew enough about it." Participant E-1 noted that "erring on the side of caution, I think, is what I tend to do" when trying to make a choice between a conventional and an alternative product.

Finally, participants felt that having some choice among a range of non-toxic commodities was necessary to enable them to mediate their exposure to chemicals. As Participant G-2 put it, "you can't fulfill your personal responsibility without having the personal power of choice." Participant D-1 explained: "It's good to have choices, and there's definitely more choice now than there used to be. So, even within the supermarket chains they have – Shoppers Drug Mart sells fragrance-free dish detergent, whereas a few years ago they probably wouldn't have."

To further explore views of PC as a viable strategy for self-protection from chemical exposures, participants were asked to describe the task of shopping for alternative products. For most, shopping at a health food store – or similar "alternative" retailer – was described as enjoyable, revealing how the act of shopping in a more precautionary way can provide a sense of fun but also offer reassurance and a sense of control over chemical exposures. Participant B-1 felt that shopping at specialty retailers "takes the stress away and it becomes fun. I love shopping. I go to Kensington [Market] ... I get to talk to the people in the shops, it's a really great experience." Participant F-1 observed that specialty retailers offer "a non-marketing environment, the lighting is friendly, it's a marketplace experience, [and] you're always running into people that you know."

All participants expressed considerable trust in alternative retailers, which contrasted significantly with their assessment of conventional grocery stores and government regulators. Part of this trust came from the relative transparency in production practices provided by the labelling and certification systems that are used in many organic, natural, and non-toxic commodities. Certification systems were described as offering a shortcut to finding safer products. As Participant A-2 put it: "I don't know the names of any of the chemicals that I need to avoid, so if it's got a long fancy name [then] I wouldn't want to pick it up ... I know organic. If something's organic then I'm going to buy that."

Participants explained that they rely on organic and health food retail venues to filter out the undesirable products, revealing the considerable trust they have in these retailers to sell the "right" products. Participant B-1 explained that she shops at "certain stores that I know carry good stuff and have certain concerns. It just simplifies the process so much." Similarly, Participant A-1 noted that "once you get a routine and you have places that you trust, it makes [shopping] like second nature." Participants belonging to a local food co-op felt especially confident in the choices made by the organization because "things are sort of pre-sorted, pre-filtered for my concerns" (Participant F-1), and "there's a committee of unpaid people who do a lot of research about the products that are sold there … [and] probably care about it as much as I do" (Participant A-3). As these participants reveal, alternative retail environments provide a feeling that shopping can be both enjoyable and safe, as they can "switch off" their own consumer vigilance and trust that products have met a set of standards that closely approximates their own precautionary preferences.

In contrast, shopping at conventional stores was described as "stressful, because there's a lot to consider" (Participant E-1), and like being in a "battle zone … and it's really, really hard for me" (Participant G-2). When discussing how she feels when shopping at conventional grocery stores, Participant B-1 replied that she hates grocery stores "with a passion. I just feel like I'm walking into a place that hates food. [group laughter] I really do! It's just so inhospitable, it feels inhumane on so many levels."

In reference to shopping at a conventional grocery store, Participant D-1 explained that she felt frustrated when alternatives were not readily available, "and I find it weird that you can go to a grocery store where there [are] forty types of yogurt and I can't find one that doesn't have gelatin and non-food substances in it." Participant D-3 explained that shopping at conventional stores was frustrating: "I find it frustrating a lot … I get frustrated by the lack of choice." The contrast between descriptions of alternative and conventional retail environments points to how the market for non-toxic products has adapted itself to the concerns of the precautionary consumer by providing greater transparency and certainty regarding what a product contains and how it has been produced. Focus group participants craved greater control over what they were putting on and into their bodies, and conventional retailers and products were perceived as a threat to this control.

Although shopping for non-toxic products seemed enjoyable and second-nature for many participants, a few described the process of informing themselves about chemical risks as anxiety producing: "If you know too much, or you learn too much, then you just scare yourself silly … I don't really feel like I have the power to change the whole system" (Participant D-1). Participant C-2 revealed that she did not share her knowledge about chemical risks with people around her: "Yeah, and I actually choose not to tell my friends about a lot of this stuff because I don't want to ruin their happy life, honestly. I just feel that at a certain point you don't want to know almost, but once you know you can't un-know."

These responses illustrate how PC is motivated by fears about health consequences from consuming conventional products, and how an individual's precautionary shortcuts and heuristics offer comfort by giving them a greater sense of control over personal exposure to perceived threats.

Constraints on Securing Alternatives

In all three focus groups, participants were asked to discuss some of the factors that limit one's ability to avoid exposure to chemicals in consumer products. In all focus groups, participants highlighted the importance of access to non-toxic products, and saw access as being determined by socio-economic status, proximity to retailers that carry these products, and availability of time to invest in shopping in a more careful and deliberate way. Although most participants did not face material constraints in accessing these products, some described their ability to pick and choose among alternative products as a luxury, reflecting on the available time and money for researching the "right" choice and the affordability of these products. Participant D-1 said: "In some ways it feels like a luxury to be able to be educated about your choices and to make choices … [and] when you want to buy organic, it's usually way more expensive." Another participant, who was living on a modest student income, described the stress of trying to shop in a more precautionary way because her monthly food bill exceeded her limited means:

> *I would say income and price is a huge limit for me. Just thinking about the last several times I've been grocery shopping with my partner and every time we get to the register and we say 'how did it get that expensive again, we were trying so hard!' and every time we spend more than we were intending [to], even though we're buying mostly vegetables. So it's constantly an issue. Constantly. I almost never buy organic or animal cruelty-free meat because it's so expensive, it's ridiculous. So I would never buy one chicken, a whole chicken. It's very expensive.*
> – PARTICIPANT E-3

Some respondents cited region and neighbourhood as a barrier to PC. Downtown Toronto (where most of the respondents lived) was described as providing easy access to non-toxic products, but neighbourhoods and communities outside of the city were viewed as providing far more limited access. As Participant F-1 explained, in reference to people living outside of Toronto, "if their local grocery store doesn't have organic local options, then they don't have as much control as we do in a big city."

Time was another limiting factor noted by many participants: "If you have a busy life then you may not pass by the place where it's convenient to get natural products," as one participant (D-3) put it. She reflected on her own privileged position, not only in terms of income and access to products but in having the time to prepare her own food:

> *People are so busy – if they have two jobs or if they have children or they have other things in their life – people don't necessarily have the time to set aside to cook from scratch. I cook a lot of my food from scratch but it takes time and it takes planning and it's a bit of a luxury. I know there are some people who are on a low income and they have to fit it into their life and schedule. And when they live in an urban setting and they work a job, like a 9-5 job, they don't have as much control over their time. So they resort to buying packaged foods because they think it's the best option.* – PARTICIPANT D-3

Since none of the participants reported facing serious material constraints, responses to this question reflected what participants believed the barriers for others might be. Consequently, these responses reveal only a partial picture of the actual socio-economic limitations on PC.

Discussion

Focus group discussions reinforced an understanding of PC as a gendered activity. Participants described non-toxic consumerism as a health-related activity mainly undertaken by mothers and women in their reproductive years who are worried about reproductive health and children's health. Many of the female participants considered chemical avoidance to be part of family health and domestic responsibilities, duties that disproportionately fall on women as a result of gendered divisions of labour that place caretaking and health protection in the private sphere. The groups spoke about PC as a response to a sense of future responsibility to children and families. Many of the unmarried and childless women in the focus groups were thinking ahead to their future role as primary caregivers, and considered PC as a necessary measure to safeguard reproductive health and the well-being of future generations.

PC also represented a form of self-protection in response to insufficient precaution at the regulatory level. Limited access to alternative products caused anxiety and frustration, and lack of access to these products was viewed as a form of inequality. The emphasis on self-protection highlights how personal consumption has become a key mechanism for managing health risks, such as chemical body burdens. Chemical exposure was seen as inadequately managed through collective and systemic means (such as regulatory change and institutionalized risk assessments). Instead, participants perceived management of risk as an individual's responsibility, achieved through more careful and responsible shopping. PC requires the deployment of personal resources, including the time to inform oneself about the most precautionary consumer choices and lifestyle practices, financial resources to afford alternative products, and access to retailers that carry these products.

The precautionary consumers who participated in this study had relatively high socio-economic status, where even those with lower household incomes had fairly high levels of educational attainment. As many of these participants noted, a person's social location, including factors such as socio-economic status, time commitment, geographical location, and level of education, influences one's awareness of the potential health risks associated with chemical exposures, most especially the ability to actualize a more precautionary shopping routine. Low-income women and those without the time or knowledge to interpret product labels and make precautionary choices are at a great disadvantage in a regulatory system that favours consumer self-protection (see also MacKendrick 2011). Moreover, in rural areas, access to retailers for alternative goods is limited. Even for middle-class women living in well-serviced urban neighbourhoods, carrying out PC adds to the domestic workload and may require a woman to choose between the needs of the family, household expenses, time, and health considerations (Buckingham and Kulcur 2009; Luxton, Rosenberg, and Arat-Koç 1990; MacGregor 2006).

Policy Directions

Despite efforts to improve the management of chemical substances in Canada, the current consumer landscape continues to reflect decades of non-precautionary management of these substances. Individuals concerned about chemical risks are alarmed by what is currently available on store shelves, and are not reassured by the recent implementation of more precautionary measures – namely, the categorization and risk assessment procedures through Canada's Chemicals Management Plan. In the absence of precautionary policies that prevent potentially harmful chemicals from entering the environment, the ability to engage in PC represents a privatized privilege. Because PC depends on sufficient education and consumer literacy to interpret a complex product label, the time to compare products and seek out alternatives, and the income to pay the price premium for non-toxic goods and organic foods, low-income women are at a significant disadvantage in regulatory systems in which the onus is on consumer vigilance. When time and money are limited, the most affordable and easily accessible products are the only available option. Furthermore, in regions in which non-toxic options are not available, conventional products are the only "choice," regardless of income or time constraints.

Consequently, PC represents an uneven and ad hoc approach to protecting environmental health compared with precautionary policy mechanisms. Universal protection from potentially harmful chemical exposures thus requires the enactment of precaution at the regulatory level, rather than at the level of the individual consumer. Public institutions must enact more stringent precautionary standards for chemicals management in order to maintain environmental and human health. Such action requires collective and coordinated efforts to *prevent pollution at the outset,* not only through a more precautionary approach to chemicals management but also through stricter controls on chemical manufacturing and more updated risk assessments that effectively and efficiently halt the production of dangerous chemicals.

Notes

1 As described in Chapter 3, the Government of Canada completed a preliminary categorization of the 23,000 chemicals on the Domestic Substances List in 2006. Of these, 4,300 were identified as requiring further assessment and 200 were considered to be of

high-priority concern. These 200 chemicals were determined to be "(a) persistent, bioaccumulative, and inherently toxic to the environment and that are known to be in commerce in Canada; and/or (b) A high hazard to humans and as having a high likelihood of exposure to individuals in Canada" (Government of Canada 2006, 27). The government then implemented a "Challenge" program as part of a Chemicals Management Plan for these 200 chemicals (see http://www.chemicalsubstanceschimiques.gc.ca/challenge-defi/index-eng.php). The Challenge process includes some elements that could be described as precautionary. More specifically, even though conclusive scientific evidence of harm has not yet been established for many of these chemicals, government regulators have in some cases restricted the use of these substances (e.g., bisphenol A). The Chemicals Management Plan also allows for public comment and stakeholder feedback. Finally, in 2006, the Pest Control Products Act incorporated more precautionary measures for regulating pesticide use in Canada, due in large part to concerns about children's health. Overall, however, with the above exceptions, a precautionary approach has not consistently been followed in the regulation of certain chemicals. For example, advocacy groups have noted serious inconsistencies and gaps in the regulation of chemicals falling under the Hazardous Products Act (R.S.C., 1985, c. H-3) and the Food and Drugs Act (R.S.C., 1985, c. F-27) (e.g., CELA 2008). More specifically, the use of certain phthalates identified as toxic under the Canadian Environmental Protection Act, 1999 (S.C. 1999, c. 33) has not been restricted in cosmetics under the Food and Drugs Act, and several chemicals that have been restricted in other jurisdictions (e.g., certain phthalates banned in Europe) have not been restricted by Canadian regulators (see CELA 2008, 6-7).

2 In this chapter, I use "non-toxic" and "alternative" to refer to certified organic food products and consumer products that are marketed as having only "natural" ingredients or have labels indicating that they are free of potentially harmful compounds (e.g., phthalate-free, BPA-free, paraben-free, etc.).

3 Bioaccumulation of certain contaminants – such as DDT and PCBs – in humans and animals has been studied for the past fifty years. Since the late 1990s, however, biomonitoring technology has improved significantly, enabling the detection of chemical compounds not previously known to bioaccumulate (Sexton, Needham, and Pirkle 2004). Biomonitoring is also able to detect chemicals that are absorbed temporarily by the body but do not bioaccumulate, such as bisphenol A, which is excreted from the body (Dekant and Völkel 2008).

4 Major global organizations that have incorporated a discussion of body burdens as part of anti-toxic campaigns include Greenpeace, Environmental Defense (Canada and USA), and World Wide Fund for Nature (WWF). Canadian organizations include Toxic Free Canada, Canadian Environmental Law Association, Breast Cancer Action Montreal, and Canadian Association of Physicians for the Environment, among others.

5 A precautionary approach is mandated by the Canadian Environmental Protection Act but, as several policy critics have noted, it has rarely been applied (CELA 2006; Denison 2007).

6 *Canadian Environmental Protection Act, 1999,* S.C. 1999, c. 33 (http://laws-lois.justice.gc.ca/eng/acts/C-15.31/index.html).

7 Preliminary safety testing refers to the *categorization* of chemicals – or more accurately "substances" – as persistent, bioaccumulative, or inherently toxic (PBiT). All substances currently in use in Canada are listed in the Domestic Substances List (DSL) and are regulated under the Canadian Environmental Protection Act. Prior to 2006, as many as 23,000 of substances on the DSL had not been categorized as PBiT. This represented a major regulatory backlog, although one that was cleared in 2006 when the Government of Canada announced that it had completed its review and that it had identified 4,300

chemicals as meeting one or more of the categorizations of PBiT. Of these, just under 400 met all three categories and will undergo further screening (CELA 2007, 20).

8 See CELA 2007, 19-20, and Government of Canada 2012.

9 For a list of guides, visit http://environmentaldefence.ca/campaigns/toxic-nation/tips-and-guides/toxicnation-guides.

References

Allen, Patricia, and Carolyn Sachs. 2007. "Women and Food Chains: The Gendered Politics of Food." *International Journal of Agriculture and Food* 15 (1): 1-23.

Altman, Rebecca Gasior, Rachel Morello-Frosch, Julia Green Brody, et al. 2008. "Pollution Comes Home and Gets Personal: Women's Experience of Household Chemical Exposure." *Journal of Health and Social Behavior* 49 (4): 417-35, doi:10.1177/002214650804900404.

Andrews, Margaret R., and Mary M. Talbot, eds. 2000. *All the World and Her Husband: Women in Twentieth-Century Consumer Culture.* New York: Cassell.

Arendell, Terry. 2000. "Conceiving and Investigating Motherhood: The Decade's Scholarship." *Journal of Marriage and Family* 62 (4): 1192-1207, doi:10.1111/j.1741-3737.2000.01192.x.

Beagan, Brenda, Gwen E. Chapman, Andrea D'Sylva, et al. 2008. "'It's Just Easier for Me to Do It': Rationalizing the Family Division of Foodwork." *Sociology* 42 (4): 653, doi:10.1177/0038038508091621.

Bloor, Michael, Jane Frankland, Michelle Thomas, and Kate Robson. 2001. *Focus Groups in Social Research.* Thousand Oaks, CA: Sage Publications.

Buckingham, Susan, and Rakibe Kulcur. 2009. "Gendered Geographies of Environmental Justice." *Antipode* 41 (4): 659-83, doi:10.1111/j.1467-8330.2009.00693.x.

Calabretta, N. 2002. "Consumer-Driven, Patient-Centered Health Care in the Age of Electronic Information." *Journal of the Medical Library Association* 90 (1): 32-37.

CELA (Canadian Environmental Law Association). 2006. *Confidentiality and Burden of Proof under the Canadian Environmental Protection Act (CEPA). Submission to the House of Commons Standing Committee on Environment and Sustainable Development.* http://s.cela.ca/files/uploads/551_CEPA_BofP.pdf.

–. 2007. *European and Canadian Environmental Law: Best Practices and Opportunities for Co-operation.* Toronto: CELA. http://www.cela.ca/sites/cela.ca/files/555_EU.pdf.

–. 2008. *Regulating Toxic Substances in Consumer Products: Response to the Discussion Paper on Canada's Food and Consumer Safety Action Plan.* Toronto: CELA. http://s.cela.ca/files/uploads/599_Toxic_Products.pdf.

Clow, Barbara, Ann Pederson, Margaret Haworth-Brockman, and Jennifer Bernier, eds. 2009. *Rising to The Challenge: Sex- and Gender-Based Analysis for Health Planning, Policy and Research in Canada.* Halifax: Atlantic Centre of Excellence for Women's Health.

Cook, Daniel Thomas. 1995. "The Mother as Consumer: Insights from the Children's Wear Industry, 1917-1929." *Sociological Quarterly* 36 (3): 505-22.

CPCHE (Canadian Partnership for Children's Health and Environment). 2008. *Playing It Safe: Childproofing for Environmental Health.* http://www.healthyenvironmentforkids.ca/img_upload/13297cd6a147585a24c1c6233d8d96d8/Brochure_English_8_pages.pdf.

Craig, Lyn. 2006. "Does Father Care Mean Fathers Share? A Comparison of How Mothers and Fathers in Intact Families Spend Time with Children." *Gender and Society* 20 (2): 259-81, doi:10.1177/0891243205285212.

CSM and CELA (Chemical Sensitivities Manitoba and Canadian Environmental Law Association). 2008. "Response to Draft Risk Assessment Results for Chemicals

Management Plan Industry Challenge Batch 3 Substances." *Canada Gazette Part 1* 142 (34). http://s.cela.ca/files/uploads/629_CMP_Bat3_CSM.pdf.

Dekant, Wolfgang, and Wolfgang Völkel. 2008. "Human Exposure to Bisphenol A by Biomonitoring: Methods, Results and Assessment of Environmental Exposures." *Toxicology and Applied Pharmacology* 228 (1): 114-34, doi:10.1016/j.taap.2007.12.008.

Denison, Richard A. 2007. "Not that Innocent: A Comparative Analysis of Canadian, European Union and United States Policies on Industrial Chemicals." Washington, DC: Environmental Defense and Pollution Probe. http://www.edf.org/documents/6149_NotThatInnocent_Fullreport.pdf.

DeVault, Marjorie L. 1991. *Feeding the Family: The Social Organization of Caring as Gendered Work*. Chicago: University of Chicago Press.

EWG (Environmental Working Group). 2014a. "EWG Skin Deep Cosmetic Safety Database." http://www.ewg.org/skindeep/.

–. 2014b. "Research." http://www.ewg.org/research/.

Government of Canada. 2006. "Government Notices." *Canada Gazette Part 1* 140 (49).

–. 2012. *Government of Canada Protects Families with Bisphenol A Regulations*. Ottawa: Government of Canada. http://www.chemicalsubstanceschimiques.gc.ca/challenge-defi/batch-lot-2/bisphenol-a/index-eng.php

Greer, Beth. 2009. *Super Natural Home: Improve Your Health Home and Planet One Room at a Time*. New York: Rodale Books.

Griffin, S. 2011. *CancerSmart 3.1: The Consumer Guide*. Vancouver: Toxic Free Canada. http://www.toxicfreecanada.ca/articlefull.asp?uid=136.

Hays, Sharon. 1996. *The Cultural Contradictions of Motherhood*. New Haven, CT: Yale University Press.

Health Canada. 2012. *Second Report on Human Biomonitoring of Environmental Chemicals in Canada: Results of the Canadian Health Measures Survey*. Ottawa: Government of Canada.

Hochschild, Arlie Russell. 2003. *The Second Shift*. New York: Penguin.

Holland, Stephanie. 2010. "Marketing to Women: Quick Facts." http://she-conomy.com/report/facts-on-women/.

Johnson, Jennifer A., and Megan S. Johnson. 2008. "New City Domesticity and the Tenacious Second Shift." *Journal of Family Issues* 29 (4): 487-515, doi:10.1177/0192513X07310313.

Lindsay, Colin. 2008. "Are Women Spending More Time on Unpaid Domestic Work than Men in Canada?" Statistics Canada, Catalogue 89-630-X.

Lupton, Deborah. 2000. "'Where's Me Dinner?': Food Preparation Arrangements in Rural Australian Families." *Journal of Sociology* 36 (2): 172-86, doi:10.1177/144078330003600203.

Luxton, Meg, Harriet Rosenberg, and Sedef Arat-Koç. 1990. *Through the Kitchen Window: The Politics of Home and Family*. 2nd ed. Toronto: Garamond Press.

MacGregor, Sherilyn. 2006. *Beyond Mothering Earth: Ecological Citizenship and Politics of Care*. Vancouver: UBC Press.

MacKendrick, Norah. A. 2010. "Media Framing of Body Burdens: Precautionary Consumption and the Individualization of Risk." *Sociological Inquiry* 80 (1): 126-49, doi:10.1111/j.1475-682X.2009.00319.x.

–. 2011. "The Individualization of Risk as Responsibility and Citizenship: A Case Study of Chemical Body Burdens." PhD dissertation, University of Toronto.

Macnaghten, Phil, and Greg Myers. 2004. "Focus Groups." In *Qualitative Research Practice*, edited by Clive Seale, Giampietro Gobo, Jabur F. Gubrium, and David Silverman, 65-79. Thousand Oaks, CA: Sage Publications.

Markens, Susan, Carole H. Browner, and Nancy Press. 1997. "Feeding the Fetus: On Interrogating the Notion of Maternal-Fetal Conflict." *Feminist Studies* 2: 351-72.

McDonald, Libby. 2007. *The Toxic Sandbox: The Truth about Environmental Toxins and Our Children's Health*. New York: Penguin.

O'Riordan, Tim, James Cameron, and Andrew Jordan. 2001. "The Evolution of the Precautionary Principle." In *Reinterpreting the Precautionary Principle,* edited by Tim O'Riordan, James Cameron, and Andrew Jordan, 1-34. London: Cameron May.

Public Health Agency of Canada. 2010. "What Determines Health?" http://www.phac-aspc.gc.ca/ph-sp/determinants/index-eng.php#health_stat.

Roy, S.C. 2008. "'Taking Charge of Your Health': Discourses of Responsibility in English Canadian Women's Magazines." *Sociology of Health and Illness* 30 (3): 463-77, doi:10.1111/j.1467-9566.2007.01066.x.

Sayer, L.C. 2005. "Gender, Time and Inequality: Trends in Women's and Men's Paid Work, Unpaid Work and Free Time." *Social Forces* 84 (1): 285-303.

Sexton, Ken, Larry Needham, and James Pirkle. 2004. "Human Biomonitoring of Environmental Chemicals." *American Scientist* 92 (1): 38-45, doi:10.1511/2004.1.38.

Steinman, D., and S.S. Epstein. 1995. *The Safe Shopper's Bible: A Consumer's Guide to Nontoxic Household Products, Cosmetics and Food*. New York: Wiley Publishing.

Szasz, Andrew. 2007. *Shopping Our Way to Safety: How We Changed from Protecting the Environment to Protecting Ourselves*. Minneapolis: University of Minnesota Press.

Taylor, Janelle S., Linda L. Layne, and Danielle F. Wozniak. 2004. *Consuming Motherhood*. New Brunswick, NJ: Rutgers University Press.

Thomas, Pat. 2008. *What's in This Stuff? The Hidden Toxins in Everyday Products and What You Can Do about Them*. New York: Penguin.

Toxic Free Canada. 2011. "Toxic Free Canada List of Campaigns." http://www.toxicfreecanada.ca/campaign.asp.

TPH (Toronto Public Health and Small World Communications). 2003. *Hidden Exposures: Informing Pregnant Women and Families about Harmful Environmental Exposures*. Toronto: South Riverdale Community Health Centre and Toronto Public Health. http://www.healthyenvironmentforkids.ca/resources/hidden-exposures-booklet.

Vora, Shivani. 2012. "Starting Early, and Young." *New York Times,* 28 March.

Zukin, Sharon, and Jennifer Smith Maguire. 2004. "Consumers and Consumption." *Annual Review of Sociology* 30 (1): 173-97.

Chapter 3 Sex and Gender in Canada's Chemicals Management Plan

Dayna Nadine Scott and Sarah Lewis

Chemical substances are found everywhere in our environment. As Chapter 1 makes clear, whether it be at home, outdoors, or in the workplace, we are continuously coming into contact with various chemicals through our air, water, food, cosmetics, clothes, personal care products, and everyday household items (Cooper, Vanderlinden, and Ursitti 2011; Program on Reproductive Health and the Environment 2008). As our detection methods improve, we are increasingly forced to confront the evidence of these exposures: biomonitoring studies now show that nearly everyone has measurable amounts of almost all known toxic chemicals stored somewhere in their bodies (CDC 2013; Environmental Defence 2009; Statistics Canada 2012).

At the same time, we are witnessing a rise in the incidence of a number of diseases and disorders in men and women. These include cancers, irreversible developmental and neurodevelopmental syndromes, reproductive disorders, and a number of autoimmune diseases. Many scientists, environmental groups, and health practitioners suggest that the rising incidence of many of these disorders and diseases can be tied to chemical exposures in our environment (Cooper, Vanderlinden, and Ursitti 2011).

There is also growing evidence that these exposures affect different bodies in different ways, because people's lives and health are influenced by both biological (sex-related) and social (gender-related) factors. Not only do all people possess different vulnerabilities to exposure based on biology, but they also face different health risks based on gendered practices, socio-economic and cultural circumstances, structural disparities in access to basic resources, varied health-seeking behaviours, and different responses from health systems, leading to diverse health outcomes (Clow et al. 2009). An analysis of chemical exposures, their biological and social impacts, and the implications for the evaluation and management of toxic substances should therefore be undertaken with sex and gender considerations in mind. The Canadian federal government has a clear primary responsibility and role in ensuring a safe environment and healthy population.

Chemicals Management in Canada

As mentioned in the Introduction, although there are several federal, provincial, and municipal regulatory frameworks that influence the degree to which Canadians come into contact with chemicals, the Canadian Environmental Protection Act, 1999 (CEPA) is the primary legislative tool for managing toxic chemical risks. It includes specific requirements for assessing and managing approximately 23,000 substances currently used in commerce or being released in Canada in significant quantities. These substances are listed in an inventory called the *Domestic Substances List (DSL),* and are jointly assessed by the Minister of the Environment and the Minister of Health to determine whether they meet the definition of "toxic" under CEPA ("CEPA-toxic"). The process of categorization of the DSL, which took place between 1999 and 2006, involved the identification of low-, medium-, and high-priority substances, based on whether they are persistent, bioaccumulative, and inherently toxic (PBiT), or present significant potential for human exposure.

On 8 December 2006, following the completion of the categorization process, the federal government introduced the Chemicals Management Plan (CMP) (Environment Canada 2012; Government of Canada 2013a). The CMP, a joint initiative of Health Canada and Environment Canada, is aimed at improving the degree of protection against hazardous chemicals in Canada and ensuring their proper management through a number of proactive measures (Government of Canada 2013b). The CMP focuses on chemicals flagged in the categorization process as potentially harmful. These substances include those that were put into commercial use before 1987 without ever being subjected to a health and environmental assessment. The stated intention of the CMP is to provide a basis for sound and effective public and environmental health policies, interventions, and control measures through substance identification, tracking of exposures, monitoring, and surveillance.

The CMP mandates the evaluation of substances to determine whether they meet the criteria for CEPA-toxicity and should be added to the List of Toxic Substances in Schedule 1 of the Act (Environment Canada 2012; Government of Canada 2013b). The decision to add a substance to Schedule 1 is based on several factors, including whether a substance meets the ecological criteria of persistence, potential for bioaccumulation, and inherent toxicity (PBiT), and presents a significant risk of exposure; and whether the substance is identified as posing considerable hazard to human health based on available evidence on carcinogenicity, mutagenicity, developmental toxicity, or reproductive toxicity (Environment Canada 2012; Government of Canada 2013b).

Adding a chemical to Schedule 1 gives the federal government the authority under CEPA to place restrictions – known as "risk management measures" – on the substance. Measures can include, among others: *regulations* limiting substance-related activities or substance concentrations in the environment; *pollution prevention plans* that outline actions to prevent or minimize the creation or release of pollutants; *environmental emergency plans; guidelines* to recommend a concentration for toxic substances; and *voluntary codes of practice.* It is important to note that adding a chemical to Schedule 1 does not necessarily require further action on the part of the government to restrict or manage the substance (Scott 2009).

One of the most contentious elements of the CMP has been the Ministerial Challenge program ("the Challenge"). The program called on chemical manufacturers, importers, and industrial users to provide new information about the properties, uses, releases, and management of 200 high-priority chemical substances that are PBiT and present a high likelihood of exposure (Government of Canada 2006, 2013b). Regulators indicated that they would find substances CEPA-toxic under the Challenge, based on a "weight-of-evidence" approach and the precautionary principle, unless industry submitted evidence convincing them otherwise.

Environment Canada and Health Canada began by drafting a risk assessment (also called a screening assessment) for each substance to evaluate exposure and harm to human health and the environment. This was done using a voluntary questionnaire, technical substance profiles, and mandatory surveys, as well as data from original literature, assessment documents, stakeholder research reports, literature reviews, and computer modelling records. In the assessment, the government developed conclusions about the toxicity of the substance and decided whether or not to list it on Schedule 1 (Government of Canada 2013b). Decisions to add a substance to Schedule 1 based on harm to human health were made based on a comparison of known and estimated chemical exposure and effect, as well as an assessment of the government's confidence that the data set was complete (Government of Canada 2013c; Health Canada 2008). This first phase of the CMP was completed in 2011.

A second phase, which began in 2011 and spans five years, is meant to build on the government's stated commitment to protect the health of Canadians and the environment by further improving product safety in Canada; completing assessments of 500 substances across nine categories, including phthalates; and investing in additional research for substances such as bisphenol A and flame retardants, substances that affect hormone function, and substances that affect the environment (Government of Canada 2013b; Health Canada 2011). In addition, approximately 1,000 substances will be addressed through other initiatives, including rapid screening assessments of substances that pose little or no risk (Government of Canada 2013b; Health Canada 2011).

Despite more optimistic beginnings (see Scott 2009), our analysis after one completed phase shows that the CMP process has resulted in a relatively modest number of substances being designated as "CEPA-toxic" and added to Schedule 1 (Environment Canada 2013). Further, and more disappointing, is that even for the substances added to Schedule 1, corresponding risk management measures have been inadequate and slow in coming. The result is that the ultimate goal of the CMP – reductions in exposures to harmful substances for everyone – is not likely to be met in the near future. As we argue in the next section, the failure to effectively regulate toxic substances has disproportionate impacts on women.

How the CMP Is Failing Canadian Women

Sex and gender are important considerations in the assessment and regulation of toxic substances, as male and female bodies respond to harmful chemicals in different ways, and men and women tend to have distinct patterns of use and exposures to chemicals based on their particular social location. Women have a unique susceptibility to chemicals as a result of subtle sex-specific differences in biochemical pathways, hormones, metabolism, body fat composition, blood chemistry, and the size of body tissues (Arbuckle 2006; Buckingham and Kulcur 2009; Clow et al. 2009; Public Health Agency of Canada 2009). As discussed in Chapters 2 and 3, the timing of exposures can further influence the effect of chemicals on a woman's body, with emerging epidemiological evidence showing that women can be more biologically vulnerable to certain exposures during *critical windows of vulnerability* (Batt 2009; Cooper and Vanderlinden 2009; Eyles et al. 2011; Gray, Nudelman, and Engel 2010). These windows, which include the prenatal period, early life, puberty, pregnancy, lactation, menstruation, and menopause, represent times of development or hormonal activity in which women are more sensitive or susceptible to chemical exposures, and their ability to adapt to these exposures may be compromised. Additionally, while traditional toxicology has been based on the understanding that the greater the dose of a toxic substance, the greater the harm, new research points to low doses of some chemicals having more severe effects than high doses, especially during critical windows of vulnerability (Chapter 10; Kortenkamp 2008; Kortenkamp et al. 2011; Vandenberg et al. 2012). Exposures during these critical windows can interrupt hormonal processes and can lead to health problems such as chronic disease, disorders, and developmental or reproductive problems in a woman, her fetus, children, and subsequent generations (Butter 2006; Gray, Nudelman, and Engel 2010; CHE 2011; Reuben 2009).

Exposure to contaminants at various developmental stages is also strongly influenced by social, economic, and cultural factors. As noted in Chapters 1 and 2, a woman's risk of exposure depends on her social location, which is characterized according to what the experts call the "social determinants of health." These include: socio-economic or occupational status, race or ethnicity, sexual orientation, education, age, language, living conditions or geography, nutrition, and access to safe drinking water (Chakravartty 2010; Cooper, Vanderlinden, and Ursitti 2011; Gupta and Ross 2007; Hamm 2009; MacGregor 2010; Public Health Agency of Canada 2011; Scott and Stiver 2009). Sometimes a woman will be disadvantaged in multiple areas of her life, compounding her risk of chemical exposure. For example, women's work in the domestic sphere, a space that is largely unsupervised and unregulated, often brings them into direct contact with chemicals. In this case, avoiding or minimizing exposures requires navigation among the needs and health of the family, economics, time, and environmental considerations (Buckingham and Kulcur 2009). Additionally, women constitute a large percentage of the country's poor (Rahder 2009). Poverty and low social status put women in these circumstances at greater risk of exposure to environmental contaminants and less likely to be involved in decision making about environmental health issues.

All of these factors are shaped by gender norms framed by social institutions such as the media, academia, and health care systems that define, reproduce, and often justify different expectations and opportunities for women, men, girls, and boys (Clow et al. 2009; MacGregor 2010). The current CMP process does not explicitly acknowledge the unique set of considerations raised by these sex and gender differences with respect to chemical exposures and effects, or consider how women's social or economic situations might influence their capacity to manage chemical exposure and risk. As a result, the differential impacts women experience from contact with chemical substances are overlooked in assessments. Final decisions about chemical use in Canada often fail to take into account possible long-term health implications for women. It is important that the CMP undertake an analysis that recognizes women as a vulnerable group, and work to confront the reasons that chemical evaluation cannot be a one-size-fits-all practice.

Burdens of Managing Risk Fall on Women

To date, the CMP process has fallen short: it has not delivered on its mandate to reduce and ultimately eliminate toxic chemical exposures, with the aim of safeguarding human health. Instead, the work of reducing exposures seems to have fallen on individual Canadians. This can be seen in the frequent calls for effective "labelling" of consumer products containing toxic substances, a staple demand of groups seeking policy change around toxics (Boyd 2010; David Suzuki Foundation 2012; Deacon 2011; Smith and Lourie 2009). These demands are usually couched in the language of a "consumer's right to know" about the contents of the products they use, so that they may make informed decisions about their purchases. In the end, however, labelling serves only short-term needs and ultimately encourages practices of *precautionary consumption,* discussed by Norah MacKendrick (2010; Chapter 2). Instead of chemicals being eliminated from various consumer products based on the risk of exposure to them and their potential for harming human health, the responsibility is placed on the consumer to make decisions on products based on what they believe is healthy and safe (Altman et al. 2008; Chapter 2).

Because women are often the primary caregivers within the home and family and usually control household consumption, the burden of this individualized regulatory regime and the duty to make informed choices often fall on them. This practice reinforces women's socially prescribed roles as providers for the household, adding to their "care burden" from both a physical and emotional perspective, and contributing to the gendered divisions of labour and exploitation of women's unpaid work in the home (Buckingham and Kulcur 2009; MacGregor 2010; Picchio 1992; The Source for Women's Health 2013).

Further, practices of precautionary consumption cannot guarantee reduced exposures or fewer adverse health outcomes. Exposures to a certain chemical could theoretically be avoided by staying away from certain labelled products, but exposures to the same chemical might occur when the chemical is found as an additive or residue in other consumer products that are unlabelled (Chapter 2). The stakes for women are high, as personal care and cleaning products are heavily marketed towards – and used by – women, thereby increasing their exposure to harmful chemicals.

As noted by MacKendrick (2011; Chapter 2), precautionary consumption raises equity concerns: we know that women will vary in their capacities to engage in informed decision making about product purchases based on levels of education, income, language proficiency, scientific literacy, time, and geography. As a result, precautionary consumption is more likely to occur within groups with higher socioeconomic and educational status who are able to obtain "green" alternatives that are not affordable to all Canadians – a particular concern for women who make up the majority of Canada's poor (Chakravartty 2009; MacKendrick 2011). Further, labelling is a policy response that fails to address the production of chemicals and their use in manufacturing processes that may have impacts on workers or communities in which those facilities are located, as explored in Chapters 10 and 11.

Canadian chemicals regulation should be more sensitive to how burdens of exposure might be placed on women as a result of their societal roles and responsibilities. Regrettably, this type of social analysis is missing from assessments of chemicals within the CMP, ultimately encouraging risk management to fall disproportionately onto women.

Regulatory frameworks, as well as campaign priorities of environmental non-governmental organizations (ENGOs), need to shift focus from individual action to ideas of collectivized care that emphasize public decision making and government policy, and work to support and protect all women (MacGregor 2010).

How the CMP Is Failing Canadian Women

Various elements of the CMP neglect sex and gender considerations. The current assessment process under the CMP employs inadequate endpoints and dated assessment methodologies, suffers from several data gaps, and is hindered by a lack of legislative requirements for examining cumulative and longitudinal chemical effects (Lewis 2011). Further, fragmentation in the regulatory regime means that occupational exposures are not included in overall exposure estimates. Finally, there are limitations in public participation, inadequate risk management measures, and a lack of commitment to implement a genuinely precautionary approach. As a result of these weaknesses in the process, health issues unique to women are not appropriately recognized or researched, women's voices have been left out of important decisions about their health, and Canada is failing to advance towards a safer and more inclusive chemicals management regime. Only by acknowledging and properly addressing these gaps in the assessment and management of chemicals will the federal government be able to provide a comprehensive tool that adequately integrates sex and gender concerns.

Inadequate Endpoints

A toxicity hazard endpoint is a biological event used to determine when a change in the normal function of the human body occurs as a result of toxic exposure. Such an event can include the growth of cancerous tumours or the development of reproductive irregularities (e.g., infertility, miscarriage). Under CEPA, the government is required to assess substances for carcinogenicity, mutagenicity, developmental toxicity, and reproductive toxicity (the endpoints for assessing a chemical's effect on human health) (Government of Canada 2013b). Critics allege that the government's assessment and management decisions have been based almost exclusively on carcinogenicity and have neglected endpoints that may be more important for women's health (Tilman 2010a). These include neurodevelopmental impacts and hormonal and endocrine-disrupting effects.

Endocrine-disrupting chemicals (EDCs) are structurally similar to hormones and are capable of triggering changes in how cells and organs function, with an impact on a diverse array of metabolic, growth, and reproductive processes in the body (European Commission 2013; Schwartz and Korach 2007; TEDX 2013; US EPA 2013; US FDA 2013). In recent years, there has been a proliferation of evidence related to the significant impact of EDCs on the health of Canadians (Hanahan and Weinberg 2000; Thornton 2000; vom Saal and Sheehan 1998). Evidence has demonstrated that besides causing physical changes to the body, EDCs can have an impact on the imprinting of genes (epigenetic changes) (Crews and McLauchlan 2006; TEDX 2013). Changes in gene expression through chemical exposure can lead to detrimental impacts on a person's health and the health of their children, contributing to the development of cancer and other diseases later in life (Crews and McLauchlan 2006; Eyles et al. 2011; Gray, Nudelman, and Engel 2010; Kloc 2011; TEDX 2013). A growing body of knowledge suggests that epigenetic effects extend to gender differences in brain function and behaviour. Many psychiatric disorders, such as depression, that are controlled by hormones and often manifest after critical windows of vulnerability, tend to affect women disproportionately (Crews and McLauchlan 2006).

A number of studies have also demonstrated a relationship between early life exposures to EDCs and the dramatic increase in the incidence of contested illnesses and disorders over the last two decades, especially in women (Brown 2007; Crews and McLachlan 2006; Gray, Nudelman, and Engel 2010; Moss and Teghtsoonian 2008; Program on Reproductive Health and the Environment 2008; TEDX 2013; see Michael Orsini, 2014 on 'health social movements'). Breast and thyroid cancer, multiple chemical sensitivity or "environmental illness," fibromyalgia, and autoimmune diseases produce symptoms that are often ignored or poorly understood by traditional medical practitioners, have delayed diagnoses, and result in women having unequal access to health care services and qualifying for or acquiring insurance and disability allowances (Butter 2006; Genius 2010; Moss and Teghtsoonian 2008).

Despite these understandings, we have detailed knowledge about exposure to endocrine disruptors and their potential health effects for only a handful of substances under the CMP (e.g., bisphenol A, or BPA), even though endocrine-disruptive potential is suspected for many more substances (Cooper and Vanderlinden 2009; Environment Canada and Health Canada 2008b; TEDX 2013). Additionally, endocrine disruption potential was not explicitly requested in mandatory surveys sent to industry to collect information on the extent and nature of manufacture, import, export, and use of a particular Challenge substance being considered "CEPA-toxic"; nor was it required under the voluntary questionnaire, an invitation to interested stakeholders to submit additional data relating to the use of these substances (CELA/CSM 2010b; Environment Canada 2010; Tilman et al. 2010). As a result, risk assessments continue to overlook understudied diseases that disproportionately affect women. It is important that additional data be gathered for alternative toxicity hazard endpoints, even if it means conducting new biomonitoring studies and laboratory tests, in order to account for sex- and gender-specific effects of chemical exposure (Crews and McLachlan 2006; de Leon, Richardson, and Madray 2010).

Use of Dated Assessment Methodologies

New theoretical considerations regarding toxicology and modes of action of various chemicals challenge pre-existing assessment methodologies currently used under the CMP. Current risk assessments are based on two key assumptions: (1) the greater the dose of chemical exposure, the greater the harm to human health, and (2) human bodies can safely accommodate some degree of chemical exposure based on the idea of thresholds. New research has shown, however, that a number of chemicals, including EDCs, can cause adverse health impacts at low doses, can increase risk at any level of exposure, and can have different modes of action (e.g., epigenetic effects) that lead to diverse health outcomes (Hanahan and Weinberg 2000; Thornton 2000; Vandenberg et al. 2012; vom Saal and Sheehan 1998). As a result, the accepted assessment approach is inadequate for ensuring the safety of Canadians, and the health of women in particular.

Gaps in Research Data

Inconsistent and insufficient data coupled with analytical shortcomings in risk assessment documents contribute to difficulty in ascertaining health outcomes that are sex- and gender-specific. There are numerous data gaps in information collected for high-priority chemicals with respect to hazard, exposure scenarios, and use applications. Many questions still go unanswered in assessments, including what constitutes a high and low dose, the timing of exposures, the delayed effects of exposure, and confounding variables. Adverse effects of a substance are often acknowledged but rarely explored. Some studies have been found to be deficient or of low reliability based on highly uncertain modelling data, or because they did not follow scientific protocol, yet they have still been judged to be of satisfactory confidence (Tilman 2010b; Tilman et al. 2010). In a number of cases, risk assessments have been critiqued for the practice of filling information gaps with informed guesswork, and using discretion where information was limited, leading to no-risk conclusions and justifying a refusal to regulate (Cooper and Vanderlinden 2009; Gray, Nudelman, and Engel 2010). Obtaining comprehensive scientific research on environmental exposure is made more difficult by the absence of a well-recognized Canadian institute of environmental health research or equivalent body that could provide sufficient funds for research and the training of researchers in environmental exposures.

Additionally, studies used by Health Canada to assess chemicals under the CMP rarely focus on gender-specific responses to exposure, and physical effects on women are often measured only in relation to the health of the fetus and newborns. A large percentage of research in the lab is still done using male rats and mice, even in the study of diseases that disproportionately affect women (Beery and Zucker 2011; Hughes 2007; Mergler 2012; Pigg 2011; Wald and Wu 2010). This dependency, and the lack of critical reflection on research practices in clinical studies, may hamper efforts to understand the unique biological effects of chemical exposure on women, and to tackle diseases that affect women more than men through more inclusive science and health policy.

Finally, although some assessments separate data on chemical use or exposure based on sex (also known as sex- and gender-disaggregated data), the government has yet to employ these data in a meaningful way, and continues to apply a mean all-person daily intake approach for many substances (Environment Canada and Health Canada 2010). There is a critical need for more disaggregated data to be made available, to be incorporated into a greater number of chemical assessments, and to be sufficiently considered in final decisions (Clow et al. 2009; Mergler 2012). These weaknesses in data collection and methodology highlight the need for comprehensive monitoring and new research that is current and that addresses sex and gender concerns (Chakravartty 2010). The absence of such information, however, should not prevent the government from taking action to protect Canadians from these chemicals more fully through precautionary measures.

No Legislative Requirements to Consider Possible Cumulative, Synergistic, Longitudinal, or Delayed Effects

Research has demonstrated that exposure to a mixture of chemicals can be much more toxic than exposure to chemicals on an individual basis (Eyles et al. 2011; Program on Reproductive Health and the Environment 2008). With the exception of a limited number of chemical ingredients in pesticides under the Pest Control Products Act,[1] chemicals management policy in Canada remains committed to the unsatisfactory and narrow practice of examining the effects of chemicals one at a time, failing to consider the real-world circumstances of exposure to multiple chemicals. Assessments rarely acknowledge that certain chemicals might interact with other chemicals in the environment to produce effects that none could produce on its own, and that cumulative or aggregate impacts are possible in relation to other environmental stressors. All of this is relevant and important as we characterize actual exposures and assess toxicity (Boyd 2003; Eyles et al. 2011; Ginsburg and de Leon 2007; Tilman et al. 2010). In particular, the potential for multiple exposures to chemicals with a common mechanism of toxicity (or mode of action) calls for attention to the effects of mixtures (Scott 2008). Some substances that belong to the same chemical class or family may have similar toxicity impacts and use patterns, and additive or cumulative effects for these chemicals need to be included in federal chemical assessments (CELA/CSM 2010a). This is of particular concern when considering gender and the cumulative chemical exposures women may experience on a daily basis as a result of the particular environments they occupy and the products they use in and outside the home.

Additionally, little research has been done on the longitudinal effects of chemical exposure on health and the environment. Other than the Maternal-Infant Research on Environmental Chemicals (MIREC) study, a longitudinal study following exposures in pregnant women, the government has not demonstrated a commitment to long-term biomonitoring initiatives that could deliver reliable evidence about the effects of prenatal exposures to chemicals on the health of individuals later in life (Arbuckle et al. 2013; MIREC 2013). This contrasts with the number of more in-depth studies in other jurisdictions, such as the National Children's Study being developed in the United States, that will follow the effects of exposure on the children of 100,000 families across the country from conception to the age of 21 (Eunice Kennedy Shriver National Institute of Child Health and Human Development of the National Institutes of Health 2012).

New data need to be generated on the cumulative, synergistic, and long-term effects of chemicals, and mixtures that contain potentially toxic chemicals, in order to understand how chemical interactions affect sex and gender. There is also a need for more long-term exposure monitoring/biomonitoring studies that show the effects of exposures to specific chemicals at various windows of vulnerability, how those exposures affect gendered development and health later in life, and how a phase-out of a chemical can lead to a weakened association between exposure and negative health impacts (Cooper and Vanderlinden 2009).

Lack of a Regime for Occupational Exposure

Occupational exposures to chemical substances can play a considerable role in exacerbating health disparities. Despite evidence that workers are getting sick from exposure, and more studies that show a connection between occupational exposure and disease, there continues to be widespread exclusion of women from many occupational health studies and a lack of longitudinal data on women's exposures in the workplace (Messing and Stellman 2006). This is of concern when gender discrimination in the workplace and gendered divisions in labour often lead to inequalities in risk of occupational exposure for women workers (Chapters 10 and 11; Messing et al. 2003; Quinn 2011). Because women are disproportionately in the low-income bracket, they are more likely to take on precarious employment and are found in greater numbers in hazardous work environments (Noack and Vosko 2011). Occupations like automotive plastics manufacturing, agriculture, aesthetics, cleaning and housekeeping services, and health care bring women workers into direct contact with a number of harmful carcinogenic substances and EDCs, such as solvents, flame retardants, pesticides, and detergents, sometimes at levels above what is considered safe (Chapter 10). Workers are also often exposed to a mixture of chemicals that may have harmful additive or synergistic effects on health. A shortage in research on women's health and exposure in the workplace makes it difficult to evaluate relationships between gender-specific disease and occupational exposures, especially during windows of vulnerability (Thompson et al. 2005).

These types of industrial workplaces also tend to be less regulated and have poor health and safety protections. The precarious nature of this work can have a chilling effect on efforts to improve occupational health because of the fear of job loss (Lewchuk, Clarke, and de Wolff 2008). Many of these jobs are held by new immigrants or racialized groups who enjoy fewer legal protections, lower rates of unionization, and less access to health care than the general population. These factors limit the ability of many to protect themselves from exposure or to seek medical care in response to chemically induced health problems (BOLS 2005; Gray, Nudelman, and Engel 2010; Jackson 2004).

Finally, aside from general terms set out in national occupational health and safety legislation and provisions in Quebec's Occupational Health and Safety Act (2013)[2] providing for preventive leave during pregnancy, very little attention has been paid to pregnant women's exposures in the workplace and the possible dangers to the health of the fetus (Julvez and Grandjean 2009). More research needs to be done in this area to fully understand the effects of in utero exposures on the health of individuals later in life. (See, for example, the growing body of literature relating to the concept of fetal origins of adult disease [FOAD].)

Workplace health and safety regulation is currently divided between federal and provincial jurisdiction. A small percentage of the Canadian workforce is covered by federal legislation (e.g., the Canada Labour Code; Canada Occupational Health and Safety Regulations) for work that is outside the authority of provincial legislatures (Canadian Centre for Occupational Health and Safety 2013). Federal jurisdiction over occupational health and safety has also been established in certain cases where federal and provincial regulations overlap (*Bell v. CSST,* SCC 1988).[3] All other workplaces fall under provincial health and safety legislation, such as the Ontario Occupational Health and Safety Act.[4] This division in regulatory powers has contributed to a federal chemicals management regime that fails to capture occupational exposure in the creation of policy concerning potentially toxic chemicals, and excludes workplace exposures in risk assessments of chemicals under the CMP. As a result, negligible attention is paid to how the combination of exposures in the workplace, the home, and the external environment might increase hazards to workers and harm to human health (Chapter 10; Gray, Nudelman, and Engel 2010). In order to adequately address the risks of toxic exposures and to allow for better correlations between exposures and health outcomes, it is important that occupational exposures be fully and meaningfully included in assessments of chemicals under the CMP, and their impact on health incorporated into understandings of risk.

Consequences of Information Gaps

The abovementioned gaps in assessment information lead to weaknesses in evaluations by the federal government to ascertain whether a chemical is safe. The margin-of-exposure evaluation within an assessment attempts to determine whether a chemical is safe by calculating the difference between its estimated exposure level and the estimated threshold at which the chemical is considered harmful to human health. This evaluation rarely takes into account the ways in which sex and gender considerations might influence margin values (see Figure 3.1). For example, the estimated threshold of a chemical can be much lower for women depending on the timing of exposure (if the exposure occurs during critical windows of vulnerability). Moreover, the estimated exposure level can be elevated as a result of a woman's domestic

FIGURE 3.1 The ways in which various factors can lower the margin of exposure

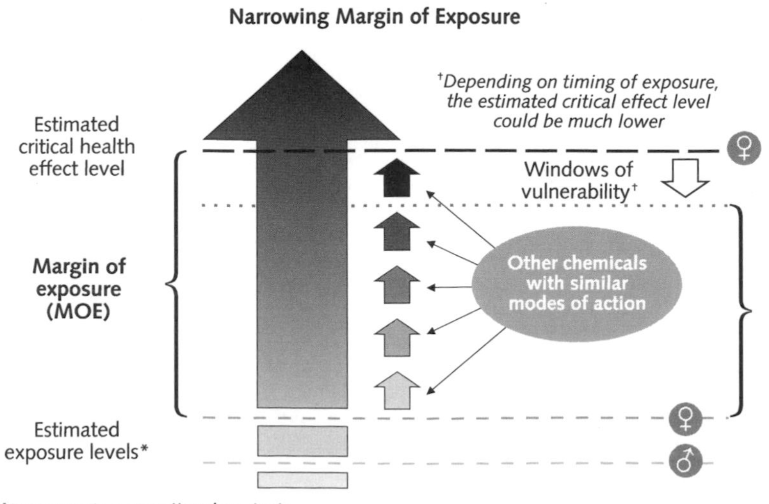

responsibilities and her disproportionate contact with gender-specific products. Exposure levels would be even higher if additional toxic loads, such as occupational exposure, were included in chemical assessments. The margin of exposure evaluation also fails to take into account chemicals that have similar modes of action, including chemicals with similar structures or substances with common mechanisms of toxicity that, when combined in the body, could exacerbate toxic effects. Considering all of these variables, it is likely that in many cases, the margin of exposure is vastly overestimated, especially in relation to women's exposures.

Restricted Public Participation

An emphasis by government and ENGOs on individual consumption strategies to mediate chemical exposure encourages women to assume that their contribution to regulating chemicals is by "doing good shopping" for safe product choices (Boyd 2010; David Suzuki Foundation 2012; Deacon 2011; Government of Canada 2011; Health Canada 2013a; Smith and Lourie 2009). This practice does not ensure the protection of Canadians from exposures to many harmful substances or from the additive effects of chemical mixtures, and assumes that women have the education and funds required to make safe choices (MacKendrick 2011).

For women to be truly engaged in the CMP process, transparency in the decision-making practices of government officials concerning chemicals regulation is paramount. The public requires a better understanding of what is found in products, the potential risks of chemical exposure in everyday products, and how regulators have analyzed and compared the costs and benefits of potential risk management actions for each toxic chemical, in order to allow for meaningful participation in government decision making on chemicals management and regulation (de Leon, Richardson, and Madray 2010). For example, the structure of the CMP allows for important decisions about the toxicity of chemicals to be based on confidential information or studies that are not peer-assessed (Scott 2009; Tilman et al. 2010). Restricting public access to information hinders others from obtaining adequate information on a substance and assessing the quality of the data provided. This lack of transparency affects the ability to make decisions that adequately address the health and well-being of Canadians and the integration of understandings of sex and gender into the risk assessment process.

An Emphasis on Risk Management over Pollution Prevention

The federal government made a clear commitment to pollution prevention in chemicals management with its 1995 federal Pollution Prevention Strategy and its positioning of pollution prevention as a cornerstone of CEPA 1999. Unfortunately, to date, few pollution prevention plans have been proposed under the CMP for high-priority substances found to be toxic, with rare exceptions such as BPA (Environment Canada and Health Canada 2008a). The government's approach to risk management so far reveals a preference for non-regulatory measures that have weak legal effect, focusing action on end-of-the-pipe solutions, and generally aiming to maintain continuous chemical use with only slight reductions in releases (Chakravartty 2010; de Leon, Richardson, and Madray 2010). An observable trend in risk management over recent years is the increasing use of Significant New Activity (SNAc) Notices. This mechanism allows for the preservation of the regulatory status quo in relation to a substance that could pose harm but for which at present there is not a significant level of exposure in Canada. It involves imposing a requirement on industry to submit additional information for existing chemicals under the CMP whenever a new activity is proposed that could increase exposure potential and contribute to a substance becoming CEPA-toxic (Government of Canada 2012). Another common tool has been the addition of chemicals to the Cosmetic Ingredient Hotlist, a record of prohibited and restricted cosmetic ingredients published by Health Canada (Health Canada 2013b). Once a chemical is placed on the hotlist, the government can require industry to remove the ingredient from a formulation, reduce the concentration of the ingredient, provide evidence that the product is safe for its intended use, or confirm that the product is labelled as required.

Such actions are not adequate for addressing hazards and risks posed by chemicals. These approaches do not require industry to submit data on vulnerable populations (such as women), chronic toxicity, endocrine disruption potential, neurotoxicity, or cumulative/synergistic effects that might differentially affect women's health. In addition, these mechanisms provide little information on what they involve, have only limited opportunities for the public to engage in subsequent assessments, and can permit the continued use of a range of toxic chemicals (CELA/CSM 2010c; de Leon, Richardson, and Madray 2010). Non-regulatory, end-of-pipe risk management is inadequate in achieving the overall goal of the CMP to eliminate or reduce toxic chemicals at the source (production, sale, and use), identify safe alternatives, or remove inefficiencies in industrial processes (Government of Canada 2013b).

Failing to Apply Precaution

The need to use precaution has been recognized since CEPA was enacted in 1999. As mentioned in Chapters 1 and 2, the precautionary principle asserts that a lack of full scientific certainty should not be used as a reason to postpone cost-effective measures to prevent threats of harm to human health or the environment (United Nations 1992). While new policies concerning chemical use and exposure, such as Europe's REACH program, have been lauded globally for embodying a truly precautionary orientation, critics argue that, to date, Canada's regulation of toxic chemicals has failed to meaningfully apply the precautionary principle (Boyd 2003; Cooper and Vanderlinden 2009; Health and Safety Executive 2013; Scott 2009). For the most part, assessments and responses to risk continue to be reactive, based on the assumption that risk is unavoidable and that human bodies can accommodate some degree of chemical exposure. As a result, the focus has been on risk management over precaution. It is seen as acceptable for there to be delays in responding, or refusals to act, based on gaps in the research data. Despite the government's transparency about the many uncertainties regarding chemical exposure and harm in its assessments, it has rarely taken preventive measures in the face of these uncertainties, thereby allowing existing exposures to continue (Cooper and Vanderlinden 2009).

Additionally, designations of toxicity under CEPA require both a potential for exposure and a potential for harm, so that even a substance demonstrating a high probability of harm at any exposure level will not be listed as toxic if estimates of exposure are currently considered low (Environment Canada 2012; de Leon, Richardson, and Madray 2010). Further, even if a substance is listed as toxic, no mandatory risk management measures will flow from this designation. These fundamental weaknesses of CEPA undermine the ability of advocates to demand precautionary action in the face of risks to human health and the environment from toxic exposures, especially when it comes to considerations of sex and gender.

Moving towards Chemicals Management that Works for Women

The government must begin to work towards a chemicals assessment and management regime that is more responsive to issues of sex and gender and more inclusive and comprehensive in addressing women's disproportionate risks and burdens. In light of evidence tying environmental chemical exposures to the rising incidence of diseases and disorders, the federal government has an obligation to take precautionary action to prevent illnesses in all Canadians, including those that have a disproportionate impact on women.

The CMP process needs to encourage increased public engagement by presenting information on chemical substances in a more understandable and accessible format (de Leon, Richardson, and Madray 2010). The government should establish a process to enhance public transparency in any notifications regarding new substances or future use, and develop reporting that is targeted to specific communities and subgroups (such as women). It should support women both technically and financially in mobilizing around chemical prevention and management before final decisions are made regarding the use of substances. The government should also include organizations focused on women's health in advisory and technical groups related to the assessment and management of chemicals (Altman et al. 2008; de Leon, Richardson, and Madray 2010).

Further, it is paramount that the endpoints for toxicity under the CMP be expanded through alternative testing methods to address gendered concerns. A study in the United States by the National Academy of Sciences (NAS) recommends screening chemicals based on toxicity pathways linked to the development of disease, rather than relying on traditional toxicology or epidemiological studies that focus exclusively on overt disease endpoints. These early biological indicators of harm, such as interference with cellular signalling, hormone disruption, or alterations in gene expression, occur "upstream" of disease endpoints and can potentially be evaluated using in vivo and in vitro cell-based tests (NAS/NRC 2007; Program on Reproductive Health and the Environment 2008; Schwarzman and Janssen 2010). To accurately evaluate the potential of a chemical to raise the risk of a woman-specific illness, toxicity tests need to: (1) assess the impact of chemical exposure during a variety of life stages, including gestation, puberty, pregnancy, and post-menopause; (2) account for increased susceptibility due to genetic variation, underlying disease, or exposure to other chemicals and environmental stressors; and (3) account for other disparities in the incidence of the disease, such as those that might derive from ethnicity or processes of racialization (Program on Reproductive Health and the Environment 2008; Schwarzman and Janssen 2010).

Other important actions the government should take to fully address sex and gender in chemicals assessment, management, and regulation include:

- generating new data and creating legislative requirements regarding cumulative and synergistic chemical effects, in order to understand exposures to a mixture of chemicals (e.g., some provincial workplace legislation uses a formula to calculate multiple exposures to a chemical – see Schedule 1 of O. Reg. 833 of the Ontario Occupational Health and Safety Act)
- including occupational exposures in overall exposure estimates
- updating chemical assessments using new understandings of toxicology to articulate how human harm relates to high and low dose, the timing of exposure, the delayed effects of exposure, and confounding variables
- supporting more long-term monitoring and biomonitoring studies to track the effects of exposures to specific chemicals and how they affect gendered development and health.

Most importantly, as argued in other chapters, the federal government must find ways to meaningfully implement precaution in its regulation of chemicals: this is the only way to fully protect the environment and human health from the effects of toxic substances (Cooper and Vanderlinden 2009; de Leon, Richardson, and Madray 2010). This requires designating as CEPA-toxic chemicals that are not necessarily in use, manufactured, or imported into Canada but that have potentially harmful ecological and health impacts or may be hazardous to human health (de Leon, Richardson, and Madray 2010). Such action might ultimately involve the redesigning of CEPA legislation to place greater emphasis on the hazard of a substance rather than its potential exposure. A hazards identification approach would detect a chemical's effect on key events in biological processes known or suspected to raise the risk for developing a specific disease or disorder, thereby guiding policy makers in making more informed decisions about what chemicals merit regulation (Schwarzman and Janssen 2010). Priority would be given to those chemicals that have these preliminary indicators of hazard to the development or progression of a specific disorder or disease. In conjunction with these amendments, the federal government should shift its current approach from chemicals management to pollution prevention measures that eliminate or significantly reduce exposures to toxic chemicals over time. This could include developing federal toxic chemical substitution and toxic use reduction programs, as well as green chemistry strategies linked to the CMP (de Leon, Richardson, and Madray 2010).

With the appropriate legislative reform, CEPA could more effectively incorporate sex- and gender-based concerns into the CMP process. Such reforms could ultimately provide more universal forms of protection, eliminate some of the burden of responsibility that currently falls on women, and prompt more effective engagement of the public in challenging the production of harmful chemicals through a stringent, inclusive, and comprehensive regulatory regime.

Notes

1 *Pest Control Products Act,* S.C. 2002, c. 28 (http://laws-lois.justice.gc.ca/eng/acts/P-9.01/index.html).
2 *An Act Respecting Occupational Health and Safety,* R.R.Q., c. S-2.1 (http://www2.publicationsduquebec.gouv.qc.ca/dynamicSearch/telecharge.php?type=2&file=/S_2_1/S2_1_A.html).
3 *Bell Canada v. Quebec* (Commission de la santé et de la sécurité du travail). 1988, 1 SCR 749, 1988 CanLII 81 (SCC) (http://canlii.ca/t/1ftgp).
4 *Occupational Health and Safety Act,* R.S.O. 1990, c. O.1 (http://www.e-laws.gov.on.ca/html/statutes/english/elaws_statutes_90o01_e.htm).

References

Altman, Rebecca Gasior, Rachel Morello-Frosch, Julia Green Brody, et al. 2008. "Pollution Comes Home and Gets Personal: Women's Experience of Household Chemical Exposure." *Journal of Health and Sociological Behaviour* 49 (4): 417-35, doi:10.1177/002214650804900404.

Arbuckle, Tye E. 2006. "Are There Sex and Gender Differences in Acute Exposure to Chemicals in the Same Setting?" *Environmental Research* 101 (2): 195-204. doi:10.1016/j.envres.2005.08.015.

Arbuckle, Tye E., William D. Fraser, Mandy Fisher, et al. 2013. "Cohort Profile: The Maternal Infant Research on Environmental Chemicals Research Platform." *Paediatric and Perinatal Epidemiology* 27 (4): 415-25, doi:10.1111/ppe.12061.

Batt, Sharon. 2009. "Full Circle: Drugs, the Environment, and Our Health." In *Push to Prescribe: Women and Canadian Drug Policy,* edited by Anne Rochon Ford and Diane Saibil, 185-205. Toronto: Canadian Scholars' Press.

Beery, Annaliese K., and Irving Zucker. 2011. "Sex Bias in Neuroscience and Biomedical Research." *Neuroscience and Biobehavioural Reviews* 35 (3): 565-72, doi:10.1016/j.neubiorev.2010.07.002.

BOLS (Bureau of Labor Statistics). 2005. "11. Employed Persons by Detailed Occupation, Sex, Race, and Hispanic or Latino Ethnicity." http://www.bls.gov/cps/cpsaat11.pdf.

Boyd, David. 2003. *Unnatural Law: Rethinking Canadian Environmental Law and Policy.* Vancouver: UBC Press.

–. 2010. *Dodging the Toxic Bullet: How to Protect Yourself from Everyday Environmental Health Hazards.* Vancouver: Greystone Books.

Brown, Phil. 2007. *Toxic Exposures: Contested Illnesses and the Environmental Health Movement.* New York: Columbia University Press.

Buckingham, Susan, and Rakibe Kulcur. 2009. "Gendered Geographies of Environmental Injustice." *Antipode* 41 (4): 659-83, doi:10.1111/j.1467-8330.2009.00693.x.

Butter, Maureen E. 2006. "Are Women More Vulnerable to Environmental Pollution?" *Journal of Human Ecology* 20 (3): 221-26.

Canadian Centre for Occupational Health and Safety. 2013. "Canadian Government Departments Responsible for OH&S." http://www.ccohs.ca/oshanswers/information/govt.html.

CDC (Centers for Disease Control and Prevention). 2013. *Fourth National Report on Human Exposure to Environmental Chemicals: Updated Tables, September 2013.* Atlanta: Centers for Disease Control and Prevention. http://www.cdc.gov/exposurereport/pdf/FourthReport_UpdatedTables_Sep2013.pdf.

CELA/CSM (Canadian Environmental Law Association and Chemical Sensitivities Manitoba). 2010a. *NGO Comments on Final Assessment for 1,4- Dioxane: A Response to Canada Gazette Part 1, 144, 7 – March 31, 2010 – Batch 7 of the Industry Challenge of the Chemicals Management Plan.* http://www.cela.ca/sites/cela.ca/files/723.RA1%2C4-dioxane%28Batch7%29.pdf.

–. 2010b. *NGO Comments on Final Screening Assessment and Proposed Risk Management Approach Documents for Selected Batch 9 Chemicals: A Response to Canada Gazette Part 1, 144, 38 – September 18, 2010 on Industry Challenge Chemicals of the Chemicals Management Plan.* http://www.cela.ca/sites/cela.ca/files/754_-_CELA_and_CSM_submission_final_RA_and_draft_RM.pdf.

–. 2010c. *Response to Canada Gazette Part 1, 144, 31 (July 31, 2010) – NGO Comments on Publication of Final Decision after Screening Assessment of Substances – Batch 8.* http://www.cela.ca/sites/cela.ca/files/742%20-%20CELA%20and%20CSM%20subm%20Bat%208%20RM%20%28Sept%2029%202010%29.pdf.

Chakravartty, Dolon. 2010. *Toward a Sex- and Gender-Based Analysis of the Chemicals Management Plan: A Report of Project Activities.* Final draft. Prepared for the National Network on Environments and Women's Health, Toronto.

CHE (Collaborative on Health and the Environment). 2011. *Environmental Exposures, Infertility, and Related Reproductive Disorders: An Update.* http://www.healthandenvironment.org/infertility/peer_reviewed.

Clow, Barbara, Ann Pederson, Margaret Haworth-Brockman, and Jennifer Bernier, eds. 2009. *Rising to The Challenge: Sex- and Gender-Based Analysis for Health Planning, Policy and Research in Canada.* Halifax: Atlantic Centre of Excellence for Women's Health.

Cooper, Kathleen, and Loren Vanderlinden. 2009. "Pollution, Chemicals and Children's Health: The Need for Precautionary Policy in Canada." In *Environmental Challenges and Opportunities: Local-Global Perspectives on Canadian Issues,* edited by C.D. Gore and P.J. Stoett, 183-224. Toronto: Emond Montgomery.

Cooper, Kathleen, Loren Vanderlinden, and F. Ursitti. 2011. *Early Exposures to Hazardous Chemicals/Pollution and Associations with Chronic Disease: A Scoping Review.* A report from the Canadian Environmental Law Association, the Ontario College of Family Physicians, and the Environmental Health Institute of Canada. http://www.healthy

environmentforkids.ca/sites/healthyenvironmentforkids.ca/files/EarlyExpandCD ScopingReview-lowres.pdf.

Crews, David, and John A. McLachlan. 2006. "Epigenetics, Evolution, Endocrine Disruption, Health and Disease." *Endocrinology* 147 (6): S4-10, doi:10.1210/en. 2005-1122.

David Suzuki Foundation. 2012. *Cleaners Survey Executive Summary*. Vancouver: David Suzuki Foundation. http://www.davidsuzuki.org/publications/downloads/2012/CS_executive_summary.pdf.

De Leon, Fe, Mary Richardson, and Sandra Madray. 2010. "Re: Risk Management Under the CMP." Letter prepared for the Canadian Environmental Network Toxics Caucus. http://s.cela.ca/files/716.CMP_Risk_Management_letter.pdf.

Deacon, Gillian. 2011. *There's Lead in Your Lipstick: Toxins in Our Everyday Body Care and How to Avoid Them*. Toronto: Penguin.

Environment Canada. 2010. "Challenge Questionnaire." http://www.ec.gc.ca/ese-ees/default.asp?lang=En&n=30C5D26E-1.

–. 2012. *Canadian Environmental Protection Act, 1999 (CEPA 1999)*. http://www.ec.gc.ca/lcpe-cepa/default.asp?lang=En&n=26A03BFA-1.

–. 2013. *Canadian Environmental Protection Act, 1999 Annual Report for April 2011 to March 2012: Part 5 Controlling Toxic Substances*. http://www.ec.gc.ca/lcpe-cepa/default.asp?lang=En&n=0EB06C79-1&offset=6&toc=show.

Environment Canada and Health Canada. 2008a. *Proposed Risk Management Approach for Phenol, 4,4' -(1-methylethylidene) bis (Bisphenol A)*. http://www.ec.gc.ca/ese-ees/default.asp?lang=En&n=6FA54372-1.

–. 2008b. *Screening Assessment for The Challenge: Phenol, 4,4' -(1-methylethylidene) bis-(Bisphenol A)*. http://www.ec.gc.ca/ese-ees/default.asp?lang=En&n=3C756383-1.

–. 2010. *Screening Assessment for the Challenge: Phenol, (1,1-dimethylethyl)-4-methoxy-(Butylated hydroxyanisole)*. http://www.ec.gc.ca/ese-ees/default.asp?lang=En&n=6E4A53B5-1.

Environmental Defence. 2009. *Pollution in People: Toxic Chemical Profiles of 11 Adults and 5 Families across Canada*. Toronto: Toxic Nation. http://environmentaldefence.ca/sites/default/files/report_files/BodyBurdenTestingreport.pdf.

Eunice Kennedy Shriver National Institute of Child Health and Human Development of the National Institutes of Health (NIH). 2012. *The National Children's Study: Study Overview*. http://www.nationalchildrensstudy.gov/about/overview/Pages/default.aspx.

European Commission. 2013. *What Are Endocrine Disruptors?* http://ec.europa.eu/environment/chemicals/endocrine/definitions/endodis_en.htm.

Eyles, John, K. Bruce Newbold, Anita Toth, and Tasnova Shah. 2011. *Chemicals of Concern in Ontario and the Great Lakes Basin: Update 2011 – Emerging Issues*. Hamilton, ON: McMaster Institute of Environment and Health.

Genius, Stephen J. 2010. "Sensitivity-Related Illness: The Escalating Pandemic of Allergy, Food Intolerance and Chemical Sensitivity." *Science of the Total Environment* 408 (24): 6047-61, doi:10.1016/j.scitotenv.2010.08.047.

Ginsburg, Jessica, and Fe de Leon. 2007. "ENGO Letter on Significant New Activity Dated February 14, 2007 – Re: Canada Gazette, Part 1, 140, 49, December 9, 2006 Notice of Intent to Amend the Domestic Substances List to Apply the Significant New Activity Provisions under Subsection 81(3) of the Canadian Environmental Protection Act, 1999 to 148 Substances." Canadian Environmental Network Toxics Caucus. http://s.cela.ca/files/724.Final%20CELA%20and%20CSM%20CMP%20draft%20RA%20batch%209%28May%2019%202010%29.pdf.

Government of Canada. 2006. "Notice of Intent to Develop and Implement Measures to Assess and Manage the Risks Posed by Certain Substances to the Health of Canadians and their Environment." *Canada Gazette, Part 1* 140 (49). http://publications.gc.ca/gazette/archives/p1/2006/2006-12-09/pdf/g1-14049.pdf.

–. 2011. *Chemicals and Your Health*. http://www.chemicalsubstanceschimiques.gc.ca/fact-fait/chem_health-chim_sante-eng.php.

–. 2012. "The Significant New Activity (SNAc) Approach." http://www.chemicalsubstanceschimiques.gc.ca/plan/approach-approche/snac-nac-eng.php.

–. 2013a. "Categorization." http://www.chemicalsubstanceschimiques.gc.ca/approach-approche/categor-eng.php.

–. 2013b. "Chemicals Management Plan." http://www.chemicalsubstanceschimiques.gc.ca/plan/index-eng.php.

–. 2013c. "Risk Assessment." http://www.chemicalsubstanceschimiques.gc.ca/approach-approche/assess-eval-eng.php.

Gray, Janet, Janet Nudelman, and Connie Engel. 2010. *State of the Evidence: The Connection between Breast Cancer and the Environment/From Science to Action*. 6th ed. San Francisco: Breast Cancer Fund.

Gupta, Shamali, and Nancy A. Ross. 2007. "Under the Microscope: Health Disparities within Canadian Cities." In *Health Policy Research Bulletin: People, Place and Health*, edited by Health Canada, 23-28. Issue 14. http://www.hc-sc.gc.ca/sr-sr/alt_formats/hpb-dgps/pdf/pubs/hpr-rps/bull/2007-people-place-gens-lieux/2007-people-place-gens-lieux-eng.pdf.

Hamm, Susanne. 2009. *The Gendered Health Effects of Chronic Low-Dose Exposures to Chemicals in Drinking Water*. Toronto: National Network on Environments and Women's Health. http://www.womenandwater.ca/pdf/NNEWH%20water%20contaminants.pdf.

Hanahan, Douglas, and Robert A. Weinberg. 2000. "The Hallmarks of Cancer." *Cell* 100 (1): 57-70, doi:10.1016/S0092-8674(00)81683-9.

Health Canada. 2008. "Screening Health Assessment of Existing Substances." http://www.hc-sc.gc.ca/ewh-semt/contaminants/existsub/screen-eval-prealable/index-eng.php.

–. 2011. "Harper Government Takes Action for Consumer Product Safety." Health Canada news release. http://hc-sc.gc.ca/ahc-asc/media/nr-cp/_2011/2011-128-eng.php.

–. 2013a. *Healthy Living: It's Your Health*. http://www.hc-sc.gc.ca/hl-vs/iyh-vsv/index-eng.php.

–. 2013b. "List of Prohibited and Restricted Cosmetic Ingredients ('Hotlist')." http://www.hc-sc.gc.ca/cps-spc/cosmet-person/indust/hot-list-critique/index-eng.php.

Health and Safety Executive. 2013. *Registration, Evaluation, Authorisation and Restriction of Chemicals (REACH)*. http://www.hse.gov.uk/reach/index.htm.

Hughes, Robert N. 2007. "Sex Does Matter: Comments on the Prevalence of Male-Only Investigations of Drug Effects on Rodent Behaviour." *Behavioural Pharmacology* 18 (7): 583-89.

Jackson, Andrew. 2004. *Gender Inequality and Precarious Work: Exploring the Impact of Unions through the Gender and Work Database*. Proceedings of the "Gender and Work: Knowledge Production in Practice" conference. Toronto: York University.

Julvez, Jordi, and Philippe Grandjean. 2009. "Neurodevelopmental Toxicity Risks Due to Occupational Exposure to Industrial Chemicals during Pregnancy." *Industrial Health* 47 (5): 459-68.

Kloc, Joe. 2011. *The Illustrated Guide to Epigenetics*. Mother Jones. http://motherjones.com/environment/2011/02/illustrated-guide-epigenetics.

Kortenkamp, Andreas. 2008. "Low Dose Mixture Effects of Endocrine Disruptors: Implications for Risk Assessment and Epidemiology." *International Journal of Andrology* 31 (2): 233-40, doi:10.1111/j.1365-2605.2007.00862.x.

Kortenkamp, Andreas, Olwenn Martin, Michael Faust et al. 2011. *State of the Art Assessment of Endocrine Disrupters: Final Report.* European Commission. http://ec.europa.eu/environment/chemicals/endocrine/pdf/sota_edc_final_report.pdf.

Lewchuk, Wayne, Marlea Clarke, and Alice de Wolff. 2008. "Working without Commitments: Precarious Employment and Health." *Work Employment and Society* 22 (3): 387-406.

Lewis, Sarah. 2011. *Sex, Gender and Chemicals: Factoring Women into Canada's Chemicals Management Plan.* Toronto: National Network on Environments and Women's Health. http://www.nnewh.org/images/upload/attach/NNEWH_chemicals_report_for_web.pdf.

MacGregor, Sherilyn. 2010. "'Gender and Climate Change': From Impacts to Discourses." *Journal of the Indian Ocean Region* 6 (2): 223-38, doi:10.1080/19480881.2010.536669.

MacKendrick, Norah A. 2010. "Media Framing of Body Burdens: Precautionary Consumption and the Individualization of Risk." *Sociological Inquiry* 80 (1): 126-49, doi:10.1111/j.1475-682X.2009.00319.x.

–. 2011. "The Individualization of Risk as Responsibility and Citizenship: A Case Study of Chemical Body Burdens." PhD dissertation, University of Toronto.

Mergler, Donna. 2012. "Neurotoxic Exposures and Effects: Sex and Gender Matter! Hänninen Lecture 2011." *Neurotoxicology* 33 (4): 644-51. http://dx.doi.org/10.1016/j.neuro.2012.05.009; http://www.cinbiose.uqam.ca/upload/files/Mergler_2012_Neuro tox_gender__sex.pdf.

Messing, Karen, and Jeanne Mager Stellman. 2006. "Sex, Gender and Women's Occupational Health: The Importance of Considering Mechanism." *Environmental Research* 101 (2): 149-62, doi:10.1002/ajim.10225.

Messing, Karen, Laura Punnett, Meg Bond, et al. 2003. "Be the Fairest of Them All: Challenges and Recommendations for the Treatment of Gender in Occupational Health Research." *American Journal of Industrial Medicine* 43 (6): 618-29, doi:10.1016/j.envres.2005.03.015.

MIREC (Maternal-Infant Research on Environmental Chemicals). 2013. "About MIREC." http://www.mirec-canada.ca/index.php?option=com_content&view=article&id=2&Itemid=11&lang=en.

Moss, Pamela, and Katherine Teghtsoonian. 2008. *Contesting Illness: Processes and Practice.* Toronto: University of Toronto Press.

NAS/NRC (National Academy of Sciences/National Research Council). 2007. *Toxicity Testing in the 21st Century: A Vision and a Strategy.* NAS/NRC Committee on Toxicity Testing and Assessment of Environmental Agents. Washington, DC: National Academies Press.

Noack, Andrea M., and Leah F. Vosko. 2011. *Precarious Jobs in Ontario: Mapping Dimensions of Labour Market Insecurity by Workers' Social Location and Context.* Toronto: Law Commission of Ontario. http://www.lco-cdo.org/en/vulnerable-workers-call-for-papers-noack-vosko.

Orsini, Michael. 2014. "Health Social Movements: The Next Wave in Contentious Politics" in *Group Politics and Social Movements in Canada,* edited by Miriam Smith. Toronto: University of Toronto Press, 333-54.

Picchio, Antonella. 1992. *Social Reproduction: The Political Economy of the Labour Market.* Cambridge: Cambridge University Press.

Pigg, Susan. 2011. "Research Controversy: Male Mice Used to Study Diseases that Affect Women." *Toronto Star,* 29 March.

Program on Reproductive Health and the Environment. 2008. *Shaping Our Legacy: Reproductive Health and the Environment*. San Francisco: University of California at San Francisco.

Public Health Agency of Canada. 2009. "Sex and Gender Based Analysis Quick Reference Manual." Internal policy document, Section 1.1.

–. 2011. "What Determines Health?" http://www.phac-aspc.gc.ca/index-eng.php.

Quinn, Margaret M. 2011. "Why Do Women and Men Have Different Occupational Exposures?" *Occupational and Environmental Medicine* 68: 861-62, doi:10.1136/oemed-2011-100257.

Rahder, Barbara. 2009. "Invisible Sisters: Women and Environmental Justice in Canada." In *Speaking for Ourselves: Environmental Justice in Canada*, edited by J. Agyeman, P. Cole, R. Haluza-DeLay, and P. O'Riley, 81-96. Vancouver: UBC Press.

Reuben, Suzanne H. 2009. *Reducing Environmental Cancer Risk: What We Can Do Now*. 2008/9 Annual Report, President's Cancer Panel, US Department of Health and Human Services, National Institutes of Health, National Cancer Institute. Bethesda, MD: President's Cancer Panel.

Schwartz, David A., and Kenneth S. Korach. 2007. "Emerging Research on Endocrine Disruptors." *Environmental Health Perspectives* 115 (1): A13, doi:10.1289/ehp.115-a13.

Schwarzman, Megan, and Sarah Janssen. 2010. *Pathways to Breast Cancer: A Case Study for Innovation in Chemical Safety Evaluation*. San Francisco: Regents of the University of California.

Scott, Dayna Nadine. 2008. "Confronting Chronic Pollution: A Socio-Legal Analysis of Risk and Precaution." *Osgoode Hall Law Journal* 46 (2): 293-343. http://www.ohlj.ca/english/documents/OHLJ46-2_Scott_ConfrontingChronicPollution.pdf.

–. 2009. "Testing Toxicity: Proof and Precaution in Canada's Chemicals Management Plan." *Review of European Community and International Environmental Law (RECIEL)* 18 (1): 59-76, doi:10.1111/j.1467-9388.2009.00621.x.

Scott, Dayna Nadine, and Alexandra Stiver. 2009. "Methyl Mercury Exposure and Women's Bodies." In *Rising to the Challenge: Sex- and Gender-Based Analysis for Health Planning, Policy and Research in Canada*, edited by B. Clow, A. Pederson, M. Haworth-Brockman, and J. Bernier, 60-64. Halifax: Atlantic Centre of Excellence for Women's Health.

Smith, Rick, and Bruce Lourie. 2009. *Slow Death by Rubber Duck: How the Toxic Chemistry of Everyday Life Affects Our Health*. Toronto: Knopf Canada.

Source for Women's Health, The. 2013. "Unpaid Work." http://www.womenshealthdata.ca/category.aspx?catid=83.

Statistics Canada. 2012. "Canadian Health Measures Survey (CHMS): Detailed Information for August 2009 to November 2011 (Cycle 2)." http://www23.statcan.gc.ca/imdb/p2SV.pl?Function=getSurvey&SurvId=129548&InstaId=62444&SDDS=5071.

Sweeney, Ellen. 2013. "Preventing Breast Cancer: An Analysis of Canada's Regulatory Regime for Chemicals." PhD dissertation, Faculty of Environmental Studies, York University.

TEDX (The Endocrine Disruption Exchange). 2013. "Endocrine Disruption: Introduction: Overview." http://endocrinedisruption.org/endocrine-disruption/introduction/overview.

Thompson, Deborah, David Kriebel, Margaret M. Quinn, et al. 2005. "Occupational Exposure to Metalworking Fluids and Risk of Breast Cancer among Female Autoworkers." *American Journal of Industrial Medicine* 47 (2): 153-60, doi:10.1002/ajim.20132.

Thornton, Joe. 2000. *Pandora's Poison: Chlorine, Health and a New Environmental Strategy*. Cambridge, MA: MIT Press.

Tilman, Anna. 2010a. *Submission on Batch 8 of the Chemical Management Plan: Final Screening Assessments and Proposed Risk Management Approach Documents, Where Applicable*. Aurora, ON: International Institute of Concern for Public Health. http://iicph.org/files/iicph-comments-on-batch-8.pdf.

–. 2010b. *Submission on Batch 8 of the Chemical Management Plan: Revised Draft Screening Assessment for Phosphonic Acid, [[3,5-bis(1,1-dimethylethyl)-4-hydroxyphenyl]methyl]–, monoethyl ester, calcium salt (2:1), Referred to as PADMEC (CAS# 65140-91-2)*. Aurora, ON: International Institute of Concern for Public Health. Copy in possession of author.

Tilman, Anna, Sandra Madray, Mary Richardson, and Fe de Leon. 2010. *ENGO Evaluation Project: Challenge Program, Chemicals Management Plan*. Copy in possession of author.

United Nations. 1992. "Rio Declaration on Environment and Development." United Nations Conference on Environment and Development, Rio de Janeiro, Brazil, June.

US EPA (United States Environmental Protection Agency). 2013. "Endocrine Disruptor Screening Program (EDSP)." Office of Science Coordination and Policy. http://www.epa.gov/endo/.

US FDA. 2013. "Endocrine Disruptor Knowledge Base." National Center for Toxicological Research (NCTR), US Department of Health and Human Services. http://www.fda.gov/scienceresearch/bioinformaticstools/endocrinedisruptorknowledgebase/default.htm.

Vandenberg, Laura N., Theo Colborn, Tyrone B. Hayes, et al. 2012. "Hormones and Endocrine-Disrupting Chemicals: Low-Dose Effects and Nonmonotonic Dose Responses." *Endocrine Reviews* 33 (3): 378-455, doi:10.1210/er.2011-1050.

Vom Saal, Frederick S., and Daniel M. Sheehan. 1998. "Challenging Risk Assessment." *Forum for Applied Research and Public Policy* 13 (3): 11-18. http://endocrinedisruptors.missouri.edu/pdfarticles/Vomsaal1998-2.pdf.

Wald, Chelsea, and Corinna Wu. 2010. "Of Mice and Women: The Bias in Animal Models." *Science* 237 (5973): 1571-72, doi:10.11126/science.327.5973.1571.

Part 2: Routes of Women's Exposures

Chapter 4 Trace Chemicals on Tap: The Potential for Gendered Health Effects of Chronic Exposures via Drinking Water

Jyoti Phartiyal

Compared with many other countries, Canada has a high standard of drinking water quality. Yet, outbreaks of water-borne disease in Walkerton, Ontario, the Kaschechewan First Nation, and North Battleford, Saskatchewan, are tragic reminders that drinking water remains a source of human exposure to environmental contaminants. Microbiological contaminants such as *E. coli* and *Cryptosporidium* pose a serious threat, but contamination of drinking water by chemical and radiological substances can also cause harm to human health and contribute to a range of chronic health effects. Evidence is accumulating that adverse health effects can occur even at very low levels of exposure when sustained over the long term (Hogg 2007; Navas-Acien et al. 2005; Nawrot and Stassen 2006; Noonan et al. 2002; Vandenberg et al. 2012; vom Saal and Hughes 2005; Welshons, Nagel, and vom Saal 2006; Weyer et al. 2001; Zahir et al. 2005). Low-dose impacts are well established for lead and for endocrine-disrupting compounds (EDCs), and data are emerging for other contaminants, such as nitrate, arsenic, cadmium, and manganese (Bouchard et al. 2007, 2011; Järup 2003; Payne 2008).

This new knowledge base calls into question the levels of chemical exposure previously considered safe by government regulators. It is becoming increasingly clear that for some chemicals, no safe level of exposure exists, particularly during key windows of vulnerability in human development that are highly dependent on sex and gender, as also discussed in Chapters 1, 2, 3, 7, and 11. From this perspective, chronic, low-level exposures have the potential to disproportionately affect women's health. As trace amounts of a range of chemicals are routinely found in Canadian drinking water, it is important that governments assess the adverse health impacts of chemicals at low doses, and the particular gendered effects of exposure, based on the latest in environmental health research.

Why Gender?

A gender perspective is largely missing from research on water resources and related policy decisions in Canada. The majority of work on women and water comes from international development studies; very few studies focus on women and water in developed countries. Further, as mentioned in Chapters 1, 2, 3, 11, and others, environmental health research on low-dose exposures to chemical contaminants increasingly points to critical windows of vulnerability during development that are tied to sex and that influence an individual's susceptibility and absorption of certain contaminants in gendered ways. By highlighting the gendered effects of chronic low-dose exposures to contaminants in drinking water, this chapter makes a unique contribution to the emerging literature on gender and chemical exposure, and aims to inform a policy agenda that protects the health of all Canadians.

Overview

This chapter is divided into four parts. Part 1 provides a brief overview of drinking water guidelines in Canada. Part 2 describes the role of health risk assessments in the setting of drinking water guidelines and contaminant thresholds, and the various shortcomings of this approach. Part 3 reviews the environmental health research on chronic, low-level exposures for three contaminants of concern in drinking water: lead, trichloroethylene (TCE), and nitrate. Each review includes a case study of a particular Canadian community as a way of highlighting some of the current challenges in ensuring access to safe drinking water in Canada. The case studies and supporting examples were deliberately chosen from among those where health risks were a concern and data were available. Taken together, they illuminate the ubiquitous nature of toxic chemicals in drinking water and, along with research findings on gender and exposure, reinforce the need for a more gender-sensitive analysis of water contamination. Part 4 offers specific recommendations for moving forward.

Part 1: Drinking Water Guidelines in Canada

Responsibility for ensuring safe drinking water in Canada is shared by all levels of government. The provinces and territories have primary responsibility for the provision and day-to-day management of drinking water, including constructing and maintaining water systems and treatment plants, setting water standards, and monitoring and testing water samples. Drinking water on First Nations reserves and federally owned land and property (such as military bases, national parks, and nuclear facilities), and in relation to the crossing of jurisdictional boundaries, is the responsibility of the federal government. Water is supplied municipally for 87 percent of Canadian households, with the remainder of Canadian households accessing non-municipal water supplies such as private wells (10 percent) and surface sources (1 percent) (Statistics Canada 2011b). Private drinking supplies, including wells and cisterns, are largely unregulated, with water safety and management resting with the property owner. Approximately 30 percent of Canadians rely on groundwater for domestic use. Of this 30 percent, approximately two-thirds live in rural areas. The remaining one-third of users live primarily in smaller municipalities where groundwater is the primary source for their water supply systems (Environment Canada 2011).

Our discussion here will focus on the voluntary Guidelines for Canadian Drinking Water Quality, which attempt to synchronize drinking water quality regimes across the provinces and territories, and the Safe Drinking Water for First Nations Act[1] passed in June 2013.

Guidelines

The Guidelines for Canadian Drinking Water Quality are a key resource in the provision and governance of drinking water across Canada (Health Canada 2012). Developed by Health Canada and the Federal-Provincial-Territorial Committee on Drinking Water (CDW), the current edition of the Guidelines (August 2012) sets out maximum acceptable concentrations (MACs) for approximately 165 microbiological, chemical, and radiological substances found in drinking water. The CDW establishes guidelines specifically for contaminants that meet all of the following criteria:

- Exposure to the contaminant could lead to adverse health effects in humans.

- The contaminant is frequently detected, or could be expected to be found, in a large number of drinking water supplies throughout Canada.
- The contaminant is detected, or could be expected to be detected, in drinking water at a level that is of possible human health significance.

If a contaminant of interest does not meet all these criteria, the CDW may choose not to establish a numerical guideline or develop a Guideline Technical Document. In that case, a Guidance Document may be developed (Health Canada 2012).

Chemical limit values are established and updated by the CDW using health risk assessments, and represent levels below which exposure to a chemical is seen to cause minimal harm to human health and the environment (Health Canada 2012). The Guidelines also set out aesthetic standards for taste, colour, and odour, as well as operational standards for temperature and pH levels. Guidelines and MACs are subject to scientific peer-review, public consultation, and final approval by the Federal-Provincial-Territorial Committee on Health and the Environment.

Differences in Drinking Water across Canada

Because the Guidelines result from cooperation between provincial, territorial, and federal agencies, there is an element of negotiation and compromise in their development, resulting in "a mix of political and technical solutions" (Peterson and Fricker 2008, 14–15). For example, during the CDW's review of trihalomethanes, a group of compounds that can form from disinfection of drinking water with chlorine, it was concluded that the recommended exposure level for these substances should be decreased from 350 to 50 micrograms/litre (µg/L). However, some provinces disagreed with this enhanced level of safety, and a guideline of 100 µg/L was ultimately selected (Peterson and Fricker 2008). In some cases, provinces are able to retain a less stringent guideline by arguing that they require more time to update treatment plants in order to meet new limits.

In addition, whereas the United States and Europe follow federally binding drinking water standards, the Guidelines in Canada represent voluntary targets that may or may not be adopted at the provincial or territorial level (European Parliament and Council 2008; US EPA 2009). A number of reports emphasize that this lack of enforceability is a serious weakness in Canada's regulatory structure for drinking water and have recommended a legally binding regime (Boyd 2006; Christensen 2011). In particular, recent research shows that the current regulatory structure leads to uneven drinking water quality standards and protections across the country, as well as differences in the frequency of water testing and of the follow-up required for adverse test results. Indeed, while some provinces and municipalities have taken steps to make the Guidelines legally binding by incorporating them into their drinking water regulations, other provinces continue to lack this legal protection (Christensen 2011).

Inconsistencies in drinking water quality do not arise from interprovincial differences alone. Discrepancies in drinking water standards and monitoring also reflect a rural/urban gap that has resulted in "a two tier system of water supply" (Hrudey 2008, 975). Larger communities and cities often have the infrastructure and technology to test above and beyond Guideline levels, whereas treatment facilities in smaller and more remote communities, such as First Nations reserves, frequently lack appropriate resources and capacity, testing instead for a small subset of parameters within the Guidelines (Hrudey 2008; Peterson 2008). This is also the case for many private wells that are managed, tested, and monitored by the property owner. As a result, securing safe drinking water requires not only meeting chemical MACs but also sufficient education and capacity to oversee and operate water systems from source to tap (Rizak and Hrudey 2007).

Finally, although Health Canada's stated aim is to develop five to seven new water quality guidelines per year, following a three-year timeline for review and approval, there have been numerous cases in which these timelines have not been met (Federal-Provincial-Territorial Committee on Drinking Water 2002; Office of the Auditor General of Canada 2005). In 2005, Canada's Commissioner of the Environment and Sustainable Development criticized the delays in guideline development, stating that at such a pace it would take the CDW at least ten years to clear the backlog of approximately fifty guidelines that required verification and updating (Office of the Auditor General of Canada 2005). These delays result in guidelines that cover only a small fraction of new chemicals released into the environment every year. Moreover, even the most current guidelines may be out of date in light of evolving scientific knowledge and research.

The Guidelines, although not immune to criticism, do provide at least some reassurance to all who drink water provided by the provinces and territories. Aboriginal people living on reserves, however, do not fall under provincial or territorial jurisdiction. The federal government is responsible for making regulations and providing services for "Indians and Lands Reserved for Indians" under the Constitution Act, 1982 (s. 91[24]).[2] Until 2013, there was no legally enforceable federal regulatory regime for protecting drinking water on reserves (Aboriginal Affairs and Northern Development Canada 2013). This changed in June 2013 when the Safe Drinking Water for First Nations Act was enacted.

The Safe Drinking Water for First Nations Act

Data from the National Household Survey (NHS) conducted by Statistics Canada show that 1,400,685 people had an Aboriginal (Métis, Indian, or Inuit) identity in 2011, representing 4.3 percent of the total Canadian population. Of the 637,660 First Nations people who reported being Registered Indians, nearly one-half (49.3 percent) lived on an Indian reserve or Indian settlement (Statistics Canada 2011a). The federal government is therefore responsible for providing potable drinking water to approximately 315,000 people living on reserves, many of which are extremely remote.

In February 2012, the federal government introduced a bill in the Senate to ensure the provision of safe drinking water, source water protection, and good waste water management for First Nations. On 19 June 2013, the bill was passed into law. The Safe Drinking Water for First Nations Act empowers the Minister of Aboriginal Affairs and Northern Development to make regulations regarding, among other things, drinking water systems; waste water systems; the training and certification of operators of drinking water systems and waste water systems; source water protection; the distribution of drinking water by truck; the monitoring, sampling, and testing of waste water and reporting of test results; and the collection and treatment of waste water (s. 4[1]). The Minister of Health may make recommendations to the Governor-in-Council respecting standards for the quality of drinking water on First Nations lands (s. 4[2]). Importantly, the regulations may establish offences punishable on summary conviction for contraventions of the regulations and set fines or terms of imprisonment or both for such offences. The limitations of liability are to be the same as those applied in the provinces unless otherwise specified in the regulations (s. 11[2][a]).

Part 2: Limitations in Approaches to Drinking Water Safety in Canada

Safe or "tolerable" limits of exposure continue to underpin Canada's regulatory regime for contaminants in the environment. Maximum acceptable concentrations for individual contaminants are determined using government health risk assessments, which estimate risks of exposure to a given contaminant based on available scientific research (Health Canada 2009). Assessments also take into account operational considerations, such as feasibility, cost, and availability of treatment technologies. A safety factor, typically ranging from 1 to 10, is applied to each MAC calculation to account for the uncertainties involved in translating chemical effects measured in laboratory settings into estimated effects on humans (Health Canada 2009). Additional safety margins are added for known carcinogens and radionuclides. It is argued that employment of these margins means that precaution is inherent in the final MAC recommendations.

The risk assessment process also establishes a "tolerable daily intake" for each contaminant, which dictates the safe daily limit of chemical exposure from various pathways, including air, food, and drinking water (Health Canada 2009). These intake levels are set with the aim of protecting the public from acute poisoning or chronic exposure to contaminants over a lifetime. MACs in drinking water represent a proportion of an individual's tolerable daily intake for a given contaminant.

Despite these steps taken by government to assess and establish "safe" concentrations and exposure levels for chemicals in drinking water, new research on chemical exposure in relation to vulnerability, gender, chemical mixtures, methodological shortcomings, and the health risks from low-dose exposures gives rise to concerns about whether these thresholds can truly be considered precautionary for everyone. These various considerations are explored next.

Critical Windows of Vulnerability

As discussed in Chapters 1, 2, 3, and 11, a growing body of evidence reveals that chemical exposure during critical windows of vulnerability can result in greater impacts on an individual's health and the health of her offspring (Endocrine Society 2012; Selevan, Kimmel, and Mendola 2000). The most vulnerable stages are those of fetal and early postnatal development, where greater exposure patterns and unique behavioural characteristics result in infants' being more sensitive to chemicals in drinking water (World Health Organization Regional Office for Europe and European Environment Agency 2002). For example, a study in Quebec found that two-month-old formula-fed babies are at particular risk of lead exposure in drinking water, even at low doses, because they drink more water by body weight than older children and adults (Levallois et al. 2008). Water used to make formula for infants has been found to contribute to 40-60 percent of an infant's overall lead intake (US EPA 2012). At present, MACs in Canada do not take into consideration the unique sensitivities of infants, and many chemical assessments fail to acknowledge critical windows of vulnerability or sensitive populations. Even in the few cases where MAC calculations do consider children, such as with lead and nitrates, these calculations are based on the average daily intake amount for a two-year-old child, thereby neglecting exposures in infancy (Health Canada 2007).

Vulnerabilities to chemical exposure in drinking water are also influenced by social, economic, and cultural factors, with race and poverty making some groups more vulnerable to environmental contamination (Gosine and Teelucksingh 2008). Research into the genomic effects of contaminants is also just now emerging, and much remains to be learned about how contaminants affect genes, and the role of genetics in the absorption, uptake, and accumulation of contaminants in the body (Eyles et al. 2011; TEDX 2011). Further research on critical windows of vulnerability and the influence of social location and genetic makeup on vulnerability is urgently needed in order to enhance our understanding of how chemical exposures from drinking water, particularly at low levels, affect human health at different life stages, and to ensure that MACs offer suitable standards to protect the health of all Canadians.

Gender Differences

As mentioned in Chapter 3, conventional health risk assessments are often based on calculations that use a standard adult male as the prototype, based on a one-size-fits-all approach, and a majority of laboratory experiments and human-centred studies are based on male adults (Vahter et al. 2007). In these cases, it is assumed that the effects of exposure for the standard male are applicable to other population groups. Women are seldom considered in health-related investigations, even though evidence points to significant gender differences in the susceptibility, absorption, and toxicity of chemicals. Recent studies have illustrated gender differences in the exposure to, and toxic effects of, cadmium, nickel, lead, mercury, arsenic, and manganese in drinking water (Roels et al. 2012; Vahter 2008). These are important factors to take into account when setting MACs for chemicals. Gendered effects are particularly significant when the possible effects of low-dose exposures to a contaminant are examined (Mergler 2012).

Chemical Mixtures

The current approach to managing chemical exposure in drinking water is based on the assessment of individual contaminants, despite evidence that humans are exposed to a mixture of chemicals in the environment on a daily basis, and often at low doses (Chapters 1, 3, 9, and 10). These mixtures can result in cumulative, additive, and synergistic effects on health, regardless of whether each contaminant is below its MAC. The cumulative effects of chemical mixtures are currently not accounted for in the setting of MACs, which represents a fundamental weakness in the development of drinking water standards (Carpenter, Arcaro, and Spink 2002; Kortenkamp et al. 2007; Timoney 2007; Zeliger 2003).

Sampling and Causation

Limitations in epidemiological (human health) studies and laboratory experiments used to inform health risk assessments can undermine current understanding of the impact of contaminant exposures on human health. Laboratory studies can involve experimentation on a variety of animals, leading to discrepancies in the impacts of exposure depending on the species, and complications in applying laboratory results to assess the adverse human health effects associated with real-world potential exposures (Cooper and Vanderlinden 2009). Available epidemiological research often involves small sample sizes and differing exposure criteria, making it difficult to draw reliable conclusions, and few studies have explored the influence of low-level chemical exposures on human health (Manassaram, Backer, and Moll 2006; Wigle et al. 2008). Finally, it is difficult to establish chemical exposure as the cause of particular health impacts when there are a number of other potential contributing factors based on environmental context and individual behaviour (Manassaram, Backer, and Moll 2006; Wigle et al. 2008).

In addition, health risk assessments have not been updated to include emerging techniques for measuring chemical concentrations in the body. While contaminant levels are traditionally determined using blood, urine, semen, hair, or fingernail samples, new and potentially more rigorous methods are under development. New detection practices are now sensitive enough to measure chemical concentrations in an individual at low doses (Welshons, Nagel, and vom Saal 2006). For example, although lead is typically measured through blood tests, bone analysis offers an approach for assessing long-term exposure, as lead can reside in bone for up to ten years (Payne 2008). Advances in biomonitoring also mean that it is now possible to measure a person's total body burden (Brown 2007; Chapter 2). Genetic research is continually advancing, carrying with it the potential to uncover the effects of chronic low-level exposures in the environment and drinking water at the molecular level. For example, a study of the health effects of four metals at low doses found that exposure to each metal had an observable influence on gene expression (Andrew et al. 2003).

Low-Dose Exposures

Connecting all of the foregoing considerations is the current trend in environmental health research on low-dose exposures to contaminants in the environment. A number of studies are revealing that not all contaminants follow the traditional linear dose-response relationship established in toxicology, which assumes that the greater the dose of exposure to a chemical, the greater the harm to human health (Myers, Zoeller, and vom Saal 2009; Vandenberg et al. 2012). Rather, some contaminants, such as endocrine disruptors, appear to cause greater adverse health effects at low doses (Vandenberg et al. 2012; Welshons, Nagel, and vom Saal 2006). Bisphenol A (BPA) is an example of a contaminant for which health effects at low levels of exposure are widely recognized (Welshons, Nagel, and vom Saal 2006). Widespread exposure to low doses of chemicals may be causing harm that is serious and in some cases irreversible, especially for women and other vulnerable populations.

Such evidence presents problems for the methodology used in many health risk assessment studies, which often test chemicals at high doses and gradually reduce the dose until no adverse health effects are observed

(Myers, Zoeller, and vom Saal 2009). Under this model, it is assumed that no health effects could occur at doses below this level (Birnbaum 2012). This methodology may result in greater health risks for chemical exposures that have the potential to cause more harm at lower doses. Assessments also fail to explore latency periods between low-dose exposure and subsequent health outcomes, where health effects may arise after prolonged periods of exposure or as a result of multiple exposures over time, manifesting themselves after many years or in successive generations (Wigle et al. 2008). In disregarding possible low-dose exposures and how these exposures interact with other complex environmental and social factors such as sex and gender, health risk assessments and the resulting MACs may fail to adequately safeguard human health from chemical exposures from drinking water (Vahter 2008).

Part 3: Three Contaminants of Concern in Drinking Water

We now examine three contaminants that have been found in different sources of Canadian drinking water. For each contaminant, we explore the contaminant's uses and presence in the environment and its connection to adverse and gender-specific health effects, and present a specific case study of a community that has experienced contamination resulting from the release of the contaminant into water sources. Although the case studies are not gender-specific, they illustrate the ever-present nature of low-dose contaminants in drinking water and their impacts on human health. By pairing these incidents with a review of the gendered effects of exposure for each chemical, we hope to demonstrate the need for gendered analyses of exposure that more thoughtfully and effectively address human health concerns in relation to low-level drinking water contamination at all life stages.

Lead

Lead has long been an ingredient in a range of consumer products, including gasoline, paint, children's toys, and cosmetics, as well as a key material in plumbing. Although a number of regulations around the world have reduced overall lead exposure by removing it from many products, the continued presence of lead at low levels in the body suggests the need for further action to minimize harm (Wigle et al. 2008).

Lead in Drinking Water

Low levels of lead enter Canadian drinking water primarily through older sections of community water distribution systems that continue to use lead pipes and lead solder. Homes built before 1950 often contain lead plumbing, and those built as recently as 1990 can contain lead solder. The amount of lead that leaches into the water depends on a range of factors, including the age of the system, the chemistry and temperature of the water, the season, and the length of time the water might be sitting in the pipes (Health Canada 2007). For example, in warmer weather, lead dissolves faster in water, resulting in elevated levels of the contaminant in drinking water sources. Lead concentrations in water, and an individual's exposure to the substance, may therefore differ significantly over time (Ryan, Huet, and MacIntosh 2000).

Health Effects

Although lead poisoning is now rare, chronic low-dose exposures to lead have been found to adversely affect the human central nervous, renal (urinary), cardiovascular, reproductive, and hematological (blood-related) systems (ATSDR 2007). Emerging research has shown that tap water consumption is linked to an increase in blood lead levels in exposed young children (Levallois et al. 2013), making this a salient environmental health issue.

Gender Differences

Lead exposure at low doses has been shown to have unique effects on women's bodies during critical windows of vulnerability (Vahter et al. 2007). For example, over 90 percent of lead accumulates in bone, with an average half-life of about ten years (Payne 2008). During pregnancy and lactation, which are periods of increased blood turnover, lead stored in bone is released into the body, thereby increasing maternal and fetal exposure to the contaminant. Accelerated bone loss during menopause can also lead to increased exposures in women (Vahter et al. 2007). These impacts may be heightened in marginalized or racialized communities that are already disproportionately exposed to lead as a result of their status and living conditions (Levin et al. 2008; Theppeang et al. 2008).

Reproductive and Developmental Effects

Lead exposure is most serious for the fetus and young children, as their bodies absorb more chemicals than adults, and their exposure may be heightened by unique behaviours, such as hand-to-mouth activity. A fetus' exposure to lead can take place as early as the twelfth week of pregnancy, and can continue throughout its development (Health Canada 2007). Low-level lead exposure has been linked to adverse affects on a child's brain and nervous system, including behavioural and cognitive impairment, decreased IQ, attention deficit disorders, and poor social abilities (ATSDR 2007). In particular, studies have shown that lower levels of lead exposure appear to have a greater incremental impact on IQ than higher levels (Wigle et al. 2008). Table 4.1 summarizes the findings from a review that surveyed the level of epidemiological evidence for relationships between adverse reproductive and childhood health outcomes and preconceptual, prenatal, and childhood exposures to a range of chemical contaminants, including low-level lead exposure from drinking water (Wigle et al. 2008). Levels of evidence for causal relationships were based on pre-established criteria, defined as:

1 Sufficient – where at least one expert group reviewed the available evidence and published a peer-reviewed report indicating a consensus view that there is a causal relationship
2 Limited – where the evidence is suggestive of an association but is limited because chance or bias cannot be ruled out with confidence
3 Inadequate – where the available studies are of insufficient quality to permit a conclusion about the presence or absence of a causal relationship. (Wigle et al. 2008, 379)

It is important to note that although many associations are characterized by "limited" evidence, this denotes a lack of epidemiological evidence, not that no harm exists. The review authors conclude that there is "no lead exposure threshold for neurotoxicity in children," and suggest the need for public health policies aimed at virtual elimination of lead in consumer products and drinking water (Wigle et al. 2008, 486).

Case Study: Low-Level Lead Exposure in London, Ontario

Concern about low levels of lead in Ontario drinking water grew in 2007 after a study in London revealed that 54 percent of homes built before 1952 had tap water samples exceeding the provincial standard of 10 μg/L (City of London 2008). Elevated levels were observed even after the water lines had been flushed. These findings prompted

TABLE 4.1 Adverse fetal and infantile health effects associated with low-level exposure to contaminants in drinking water

Contaminant	Adverse health outcomes by strength of evidence		
	Sufficient	Limited	Inadequate
Lead	*Neuropsychological deficits* • Cognitive function in children 3 or older (high- and low-level prenatal exposures) "Recent reviews concluded that 1) lead impairs behavioral and cognitive development of children at blood levels below 10μg/dl 2) no blood lead threshold for effects has been demonstrated 3) there appears to be a steeper slope for the inverse association between IQ and blood lead below 10μg/l compared to higher levels."[1] *Renal tubular damage* • (Low- to moderate-level childhood exposures)	*Adverse pregnancy outcomes* • Spontaneous abortion (maternal and paternal exposures) • Preterm birth (maternal exposure) • Fetal growth deficit (maternal exposure) *Neuropsychological deficits* • Cognitive function in children aged 0-2 (low-level prenatal and childhood exposures) • Cognitive function in children 3 or older (low-level prenatal exposure) • Problem behaviours (childhood exposure) • Motor function in children 3 or older (childhood exposure) • Sensory function (maternal exposure) • Auditory function (low- to moderate-level childhood exposures) *Childhood cancer* • Neuroblastoma (paternal occupational exposure)	*Adverse pregnancy outcomes* • Delayed conception • Stillbirth (maternal and paternal exposures) *Birth defects* • Neural tube defects (maternal and paternal exposures) • Cardiac birth defects (maternal and paternal exposures) • Orofacial clefts (maternal and paternal exposures) *Neuropsychological deficits* • Developmental milestones • Motor function in children aged 0-2 (prenatal and early childhood exposure) • Motor function in children 3 or older (maternal exposure) *Childhood cancer* • Neuroblastoma (maternal exposure)

	Other health outcomes • Postnatal growth in height (childhood exposure) • Adolescent reproductive development: age at menarche (childhood exposure) • Adolescent reproductive development: pubic hair in girls (childhood exposure) • Dental caries (childhood exposure)	*Other health outcomes* • Postnatal growth in height (maternal exposure) • Age at breast development (childhood exposure)
Nitrate	*Adverse pregnancy outcomes* • Fetal growth deficit (maternal exposure in drinking water) *Birth defects* • Neural tube defects	*Adverse pregnancy outcomes* • Stillbirth (maternal exposure in drinking water) • Preterm birth (maternal exposure in drinking water)[2] *Birth defects* • Cardiac birth defects[3] • Musculoskeletal birth defects[4] *Cancers* • Leukemia (prenatal or childhood exposure) • Childhood brain cancer (maternal exposure) *Adult cancer* • Testicular cancer (childhood exposure)

1 Wigle et a . 2008, 431.
2 One study cited; further investigation needed.
3 Two studies quoted with different results.
4 Only one study available that showed association.
Source: Adapted from Wigle et al. 2008.

Ontario's Chief Drinking Water Inspector at the time, Jim Smith, to organize drinking water tests for older homes in thirty-six communities across the province, in order to protect human health and prevent a drinking water health hazard (Ontario Ministry of the Environment 2007a). Results obtained from these tests revealed that half of the communities had lead levels above the provincial standard; in some communities, the provincial standard was exceeded in over 40 percent of samples (Ontario Ministry of the Environment 2007b). A number of test results were twice the MAC for lead. The data revealed the extent to which lead pipes affected drinking water quality throughout Ontario.

These findings brought about sweeping legislative changes to lead testing in the province. An amendment to the Safe Drinking Water Act in 2007 requires all Ontario municipalities to test tap water through a community lead testing program.[3] The program takes samples twice a year from a specified number of households, non-residential facilities, and sections of the municipal distribution system. Where tests reveal elevated lead levels, corrective action is taken as directed by the local medical officer of health. Since the introduction of the legislation, two rounds of testing have taken place. While approximately 26 percent of municipalities (150 cities and towns) have at least one residence with elevated levels of lead in drinking water, 98 percent of all plumbing samples taken have been within the 10 μg/L standard (Stager 2009; Sun Media 2009).

The legislative changes in Ontario are among the most robust in the country and offer a model for other jurisdictions to follow. In some jurisdictions, such as the Northwest Territories and certain areas in Saskatchewan, lead testing may be conducted only once a year, and may be restricted to the water treatment plant alone (Nickel 2008; Northwest Territories Health and Social Services 2007). Indeed, the presence of lead at low levels in drinking water continues to be a nationwide issue. In 2001, 50 homes from five cities across Canada were tested for lead as part of a CBC *Marketplace* report (2001). Tests were conducted in 25 homes built before 1970 and 25 homes built after 1970. Thirty percent of the homes had lead levels above the MAC for their province. In the 25 homes built before 1970, 11 had high lead levels before the tap was run for three minutes. St. John, New Brunswick, was found to have the greatest number of homes with high lead levels (CBC *Marketplace* 2001). In another example, high levels of lead in Edmonton homes prompted Alberta Health Services to enact a province-wide public health advisory on drinking water. The advisory recommended that water be tested in any homes built before 1960 or occupied by young children or pregnant women (Aspen Regional Health 2008).

Trichloroethylene (TCE)

Trichloroethylene (TCE) is an industrial solvent used extensively in the automotive and metal industries to degrease and clean metal parts. Other applications include industrial dry cleaning, printing, and the production of printing ink. Consumer products containing TCE include paint removers and strippers, adhesives, stain removers, and cleaning fluids. There are no known natural sources of TCE.

TCE in Drinking Water

TCE contamination in drinking water is largely the result of poor handling and improper disposal of the substance during industrial activities. Although the chemical rapidly evaporates in air, TCE persists in low levels in subsurface soils and leaches into groundwater, where it can remain for many years. The highest concentrations of TCE in groundwater occur in wells located near waste disposal sites or industrialized areas (Canadian Council of Ministers of the Environment 2007). There is no current national estimate on the extent of TCE-contaminated groundwater.

In 2005, the CDW reduced the MAC for TCE from 50 μg/L to 5 μg/L (Federal-Provincial-Territorial Committee on Drinking Water 2005). This change was prompted by the discovery of low levels of TCE in the drinking water of two provinces, a review by a standing committee of the House of Commons of TCE in drinking water, and the development of regulations to reduce the release of TCE into the environment. Although it was noted that the change in MAC would have negligible effects on Canadian drinking water supplies, the subcommittee asserted that it would have a positive impact on private wells through the cleanup of contaminated sites (Federal-Provincial-Territorial Committee on Drinking Water 2004).

Health Effects

TCE exposure at low doses is associated with neurotoxicity, immunotoxicity, developmental toxicity, liver and kidney impairment, endocrine effects, and several forms of cancer (US EPA 2001). Epidemiological evidence shows strong connections between low-level TCE exposure and liver and kidney cancer, and lymphomas (Scott and Chiu 2006).

Although increased risk of cervical cancer has been found in some studies, it is confounded by socio-economic variables (Raaschou-Nielsen et al. 2003).

Gender Differences

Data on the gendered effects of low-level TCE exposure from drinking water are limited. Gendered differences in metabolism may influence susceptibility to exposure by altering how TCE interacts in the body or by providing different targets for TCE's toxicity (US EPA 2001). Additionally, the highest concentrations of TCE occur in fatty tissues (Pastino, Yap, and Carroquino 2000). Since women tend to possess more fatty tissue than men, they may experience a greater buildup of TCE in their bodies, and the chemical may remain in their bodies for longer periods of time (Clow et al. 2009; US EPA 2011).

Reproductive and Developmental Effects

Whereas epidemiological studies of TCE exposure provide evidence of reproductive effects in men, there are virtually no studies that explore the effects of TCE on reproductive function in women (Scott and Chiu 2006). Research on the relationship between maternal exposure to low-level TCE in drinking water and the development of embryos is at an early stage, and more research is required to understand the specific forms and modes of action of TCE influencing fetal health (Pastino, Yap, and Carroquino 2000; Bove, Shim, and Zeitz 2002). Existing evidence suggests that TCE can diffuse rapidly across the placenta during pregnancy, and build up in the fetus (ATSDR 2011; Johnson et al. 2003; Scott and Chiu 2006). This process can lead to elevated risks of birth anomalies and impaired fetal growth (Shaw et al. 2003; US NRC 2006). Some animal studies have also suggested a link between congenital heart defects and low-level exposure, but epidemiological data for congenital defects is not conclusive (Pastino, Yap, and Carroquino 2000; Watson et al. 2006; US TCE Review Panel 2002). In addition, TCE has been detected in low doses in human breast milk, a potential route of exposure for nursing infants that requires further study (Pastino, Yap, and Carroquino 2000; Scott and Chiu 2006).

Case Study: Low-Level TCE exposure in Shannon, Quebec

In December 2000, SNC TEC, a company operating near the municipality of Shannon Quebec, conducted a water quality test on the residential well of an employee. The test revealed that the municipality's groundwater had been contaminated with TCE (Lafrance 2004). Local public health authorities issued a warning, instructing residents to stop drinking water from their wells and to ventilate their houses when using tap water. Shannon began testing other residential wells for TCE, and by June 2001, thirty-five wells were found to contain TCE at levels over 50 µg/L, with one well measuring TCE levels of 955 µg/L (Asselin 2001a, 2001b). By 2004, the contamination had affected 400 residences and two schools (Lafrance 2004). Residents have requested that an epidemiological study be carried out to measure TCE exposure levels in the local population.

Shannon is connected to the water distribution system of the nearby federally owned Valcartier military base. TCE contamination originated from the base in the 1950s. Maximum concentrations in groundwater near the source have ranged from 800 to 71,000 µg/L, and the resultant toxic plume, 4 kilometres long and 650 metres wide, contained over an estimated 1,500 kilograms of dissolved TCE and reached as far as the Jacques Cartier River (Lefebvre et al. 2003). Evidence suggests that both the Department of National Defence and SNC TEC were aware of the contamination as early as 1997, with a report commissioned in 1999 stressing the vulnerability of the aquifer and suggesting that nearby property owners be informed (House of Commons 2012). This information was never shared with the municipality.

In 2009, residents of Shannon launched a class action lawsuit against the Department of National Defence for TCE contamination of their drinking water wells. The residents claimed that the TCE contamination was responsible for over 500 cases of cancer, including almost 300 deaths (Laffont and Godbout 2012). Élaine Michaud, NDP member of Parliament for Portneuf-Jacques-Cartier, where the municipality of Shannon is located, filed a Private Member's Bill asking the government to formally acknowledge its responsibility and compensate the victims (Laffont and Godbout 2012). The proposal was defeated even though it was supported by both the NDP and the Liberal Party, and the legal battle continued. In 2012, the Superior Court judge acknowledged that the government did contaminate the water but said that the plaintiffs in the case failed to prove a causal link between TCE detected in the town's water supply and the number of cases of cancer reported in the community (CBC News 2012). Some residents were compensated for the inconvenience of each month that they had no domestic water supply, up to $12,000 per person.

Since the decision, the Quebec City regional public health agency has set up a committee of international experts to determine whether there is a causal link between the cancers seen in Shannon and TCE (Séguin 2013). The agency has explained that more information is available now as they can consider the data brought forward during the class action lawsuit, not just the administrative data and provincial averages that were considered earlier. The class members also began the appeal process in early 2013 (Séguin 2013).

The Ministry of National Defence had invested in the construction of a new water system, which was extended in 2004 to a good part of Shannon, and in 2009, the Shannon water system became independent from the Valcartier military base. However, the risk of exposure by groundwater remains, as the decontamination process will begin only in 2015 and is expected to take fifteen years to complete (Laffont and Godbout 2012).

The case of Shannon highlights the need for improved monitoring and surveillance of drinking water quality, especially in areas located near contaminated land. Many other municipalities have had similar experiences. For example, TCE has been found in the water of a number of Ontario cities at low doses (Mittelstaedt 2001). In 2000, the contaminant was detected in private wells in Beckwith as a result of a nine-kilometre-long plume extending from an abandoned landfill. The residents were unable to receive government assistance because their TCE contaminant levels did not exceed the MAC guideline of 50 μg/L at the time (Cheadle 2000). More recently, over 400 homes in Cambridge were affected by TCE vapour that was being emitted from contaminated groundwater and escaping through cracks in the foundation (Hancock 2008).

Shannon also provides some insights into contamination on federally owned lands. In a 2002 report, the federal Commissioner of the Environment and Sustainable Development raised concerns about the environmental and health risks associated with federally owned contaminated sites, remarking on the government's limited knowledge about and research on these sites, and its failure to enact proper legislative instruments and actions towards effective cleanup (Office of the Auditor General of Canada 2002). The Federal Contaminated Sites Action Plan, initiated in 2005, has sought to address these gaps (Government of Canada 2012). Given the events in Shannon, however, it is questionable whether the status of ground and surface water in many federally contaminated sites is known. Certain communities may therefore continue to be at risk of low-level drinking water contamination.

Nitrate

Nitrate is the most common chemical contaminant in groundwater, with concentrations increasing around the world due to the widespread use of nitrogenous fertilizers and animal manure in modern agricultural production (Canadian Council of Ministers of the Environment 2002; Environment Canada 2011; European Environment Agency 2003). Septic systems and leaking sewers are also major contributors to nitrate contamination, as are airborne compounds emitted by industry that precipitate into water sources. The MAC for nitrate, which came under review in 2009 as part of an update of existing guidelines, sits at 45 mg/L (equivalent to 10 mg/L measured as nitrate-nitrogen) (Federal-Provincial-Territorial Committee on Drinking Water 2012).

Nitrate in Drinking Water

Low-level nitrate contamination is of particular concern to households relying on private water supplies that draw from groundwater. Over 4 million Canadians depend on private sources for drinking water, many of which are private wells (Corkal, Schutzman, and Hilliard 2004). Evidence suggests that testing and treatment of private water supplies by owners are done infrequently, and that "most rural water users do not know whether their drinking water sources are safe or suitable for any given use" (Corkal, Schutzman, and Hilliard 2004, 1621). This lack of awareness is confirmed by data suggesting that nitrate levels in private wells often surpass regional drinking water standards. It is estimated that between 20 and 40 percent of all rural wells have nitrate concentrations in excess of drinking water guidelines, with up to 60 percent of wells containing unsafe levels in regions with intensive agriculture or animal farming operations (van der Kamp and Grove 2001). Indeed, an estimated 45 percent of all waterborne bacterial outbreaks in Canada involve non-municipal systems, largely in rural or remote areas (Corkal, Schutzman, and Hilliard 2004). Private wells have been found to contain a number of other chemicals at low levels, including atrazine, arsenic, copper, lead, mercury, radium, and volatile organic compounds that could exacerbate contamination and subsequent public health and safety outcomes (Health Canada 2006; Weir 2005; Wilson, Schreier, and Brown 2008).

Health Effects

Recent studies have linked continuous low-dose nitrate exposure to chronic health outcomes (Ward et al. 2005). Nitrate is a precursor in the formation of a class of genotoxic compounds that are known to cause cancer in animals and that are potentially cancerous in humans (Canadian Council of Ministers of the Environment 2009). Low levels of exposure are also associated with type 1 (childhood) diabetes, increased blood pressure, and acute respiratory tract infections (Ward et al. 2005).

Reproductive and Developmental Effects

The connection between nitrate contamination in drinking water and methoglobemia, or "blue baby syndrome," has been well recognized for over fifty years. Nitrate/nitrite combines with hemoglobin in the blood of infants and hinders the transfer of oxygen to the rest of the body (Knobeloch et al. 2000). Formula-fed infants younger than six months are particularly susceptible, and can experience life-threatening health effects if the condition is not recognized and treated appropriately (Knobeloch et al. 2000). The Canadian drinking water standard for nitrate has been set to protect against this condition, with few suspected cases reported in Canada (Federal-Provincial-Territorial Committee on Drinking Water 1987), but methoglobemia remains a serious concern, especially in relation to low-level chemical and microbacterial contamination of drinking water in private wells (Powlson et al. 2008; van Grinsven et al. 2006).

The body of evidence in relation to nitrate exposure and reproductive health is limited, with varied experimental results. These inconclusive findings may result from the time periods over which exposure was assessed, the levels of nitrate used across studies, and the ways in which the contaminant interacted with other environmental or biological factors (Manassaram, Backer, and Moll 2006).

Scientific reviews of nitrate in drinking water show limited evidence of connections between low-level exposure and fetal growth deficit or neural tube defects in infants (Ward et al. 2005; Wigle et al. 2008).

These results point to a lack of epidemiological research in this area. See Table 4.1 on page 120 for the findings of the review conducted by Wigle and colleagues (2008) on the level of epidemiological evidence for a relationship between low-level nitrate exposure in drinking water and adverse health effects on the fetus and infant.

Case Study: Low-Level Nitrate Exposure in Abbotsford, British Columbia

Abbotsford, located in the Fraser Valley, is the fifth-largest municipality in British Columbia (Tourism British Columbia n.d.). It is described as "a city in the country," and the area outside the metropolitan core is predominantly farmland (City of Abbotsford 2012). Although most residents in the municipality get their drinking water from a local creek, at times of the year when there is high turbidity, water is taken from the Abbotsford aquifer, one of the largest in the province (BC Provincial Health Officer 2001). The aquifer is also the sole source of drinking water for several private suppliers and thousands of residents who draw groundwater from private wells (City of Abbotsford 2012). In total, the aquifer provides groundwater to over 100,000 people in the city and surrounding region (Environment Canada 2008). Due to its largely unconfined nature and highly permeable sands and gravels, it has been identified as "highly vulnerable" to contamination, with groundwater pollution in the area acknowledged by the BC government (BC Ministry of Water, Land and Air Protection 2002, 5). Intensive agricultural practices in the region, which rely heavily on the use of fertilizers, have led to widespread low-level nitrate contamination of the aquifer.

The state of the Abbotsford aquifer has received considerable scientific attention. Environment Canada has run ongoing groundwater monitoring using a network of monitoring wells since the early 1970s. The data have shown that since 1992 there has been a rise in nitrate contamination in local groundwater over time, with 71 percent of samples frequently exceeding the nitrate MAC of 10 mg/L, and concentrations as high as 91.9 mg/L (Wassenaar, Hendry, and Harrington 2006). A study comparing nitrate levels in the aquifer in 1993 and 2004 found that concentrations for both years exceeded drinking water guidelines by over 15 percent, and that contamination appeared to be worsening (Hendry 2006). It is expected that widespread nitrate contamination of the aquifer will continue, unless agricultural practices are dramatically altered to reduce the amount of fertilizer applied to the land (Wassenaar, Hendry, and Harrington 2006).

Abbotsford is not alone in its experience: low-level nitrate contamination of groundwater wells remains a reality for many other areas across the country. In Nova Scotia, an estimated 22 percent of wells exceed the nitrate drinking water guideline; in Prince Edward Island, an estimated 6 to 11 percent; and in Ontario, 15 percent (Corkal, Schutzman, and Hilliard 2004; Nova Scotia Environment 2011; PEI Commission on Nitrates in Groundwater 2008). These examples point to weaknesses in the current water governance system in Canada, especially insufficient water testing and monitoring practices.

Part 4: Moving Forward

The foregoing analysis raises many questions about the current regulatory approach to managing chemical contaminants found in Canadian drinking water, and points to shortcomings in policy and research to address the potential health harms associated with chronic, low-level contaminant exposures. This is of particular importance when considering how women and children might be uniquely vulnerable to exposures during key windows in development and as a result of distinct physical traits and susceptibilities to illness. Each case study represents an example of low-dose chemical contamination in drinking water. In conjunction with research findings of associations between each chemical and gendered or developmental health effects, these instances point to the need for programs, policies, and studies that consider gendered exposures at low doses. The following recommendations highlight specific actions that governments can take to address chronic, low-level exposure to chemicals in our drinking water.

Strengthening the Guidelines for Canadian Drinking Water Quality

Federal, provincial, and municipal governments should work to ensure that all Canadians are able to enjoy the same drinking water quality, regardless of where they live, by adopting the Guidelines as mandatory minimum standards. The Guidelines should be health-based, long-term objectives that are sensitive to gender concerns and that protect vulnerable groups, such as children and pregnant women. Additionally, the MACs set out in the Guidelines should be equivalent to or more stringent than those found in other industrialized nations. Finally, the process for developing the Guidelines should be flexible and participatory, involving regular reviews to update existing standards based on emerging scientific evidence and research, such as the risks associated with low-level exposure, and to ensure the timely development of new standards for contaminants where there is currently no guideline. National standards should support regional adaptation initiatives that take into account local drinking water conditions and specific testing and monitoring needs to maintain drinking water quality.

Provinces and territories should strive to adopt drinking water guidelines that are equivalent to or stricter than those provided in the Guidelines. For all chemicals listed in the Guidelines, testing should be mandatory and occur at regular intervals, taking into consideration seasonal patterns and regional effects that may impact results. Where there is a high probability of chronic low or high-dose contamination, testing frequency should be increased. If a contaminant is detected at levels exceeding its MAC, a protocol for corrective action should be implemented immediately, and the public should be notified.

Reviewing Public Health Education Programs and Information

To ensure that steps are being taken to protect all Canadians, a review of educational initiatives related to low-level exposures to chemicals and their impacts on women should be implemented by the Public Health Agency of Canada. A review should also be conducted of programs that inform Canadians, and specifically private well owners, on proper well monitoring and maintenance, and the risks to human health from low-level well water contamination. Where needed, further funding and resources should be directed towards making these programs more widespread and accessible, with special guidance given to pregnant women and parents with infants.

Re-envisioning Health Risk Assessments

In determining drinking water guidelines, health risk assessments should take into account gendered critical windows of vulnerability. Sufficient safety margins should be included in an assessment where health risk data are missing or incomplete. Assessment methodologies should be updated in order to effectively evaluate non-linear dose-response curves for certain contaminants and examine scenarios where low dose may lead to greater harms to human health (Endocrine Society 2012). Current risk assessment models should also recognize the limitations of conducting evaluations of health risk using a chemical-by-chemical approach, and should design assessment processes that take into account the health effects of exposure to chemical mixtures from multiple sources (Endocrine Society 2012).

Continuing Research on Low-Level Exposures to Chemicals

There is a need for further research on the effects of low-level exposures to chemicals in drinking water on human health, especially at key stages of development, and the synergistic effects of multiple low-level exposures using gender/sex-sensitive methodologies. Populations that are especially vulnerable to low-dose exposures should be identified, and standards that take these populations into account should be set. Research should also be conducted to explore groundwater contamination and the health effects associated with chronic, low-level exposures to contaminants in drinking water from private wells. Such research will help fill important gaps in the literature and reduce uncertainty in the assessment process.

Carrying out Biomonitoring Studies

Longitudinal biomonitoring studies should be carried out on vulnerable populations, such as pregnant women, beginning before conception and continuing into the child's adulthood. The purpose of these studies would be to identify the health outcomes of various low-level chemical exposures on the fetus and into infancy, childhood, adolescence, and adulthood (Wigle et al. 2008), and to compare those data with biomonitoring results for Canadians as a whole. At the same time, waiting for results from such studies should not be used by government to justify the postponement of precautionary regulatory action.

Making Groundwater Protection and Source Water Protection a Priority

Ensuring the safety of drinking water over the long term depends on the protection of source waters. In order to safeguard the health of all populations across the country, it is critical that governments and communities work together to prevent chemical contaminants from entering the environment and Canadian waterways, even at low concentrations.

Improving Access to Drinking Water Data

Obtaining municipal drinking water data can be difficult and time consuming. All Canadians should have equal access to drinking water test results. Water suppliers should be required to provide drinking water and system approval data to the public through news releases and public announcements, and on the Internet.

Enacting the Precautionary Principle in All Decisions Concerning the Management of Contaminants in Canadian Drinking Water

An emerging understanding of non-linear dose-response curves and low-dose exposures, chemical mixtures, gendered and developmental health effects, critical windows of vulnerability, and pervasive gaps in research point to the need for a truly precautionary approach in the management and regulation of drinking water, a practice discussed in a number of chapters in this volume. Such an approach mandates action despite scientific uncertainty, in order to prevent potential harm to human health or the environment. Such a practice would support policy positions that protect the Canadian population from exposures to chemicals that may be harmful to human health, especially with respect to vulnerable groups such as women and children, and would involve a more sensitive health risk assessment and standard-setting process for water contaminants.

Conclusions

Although many questions remain, the foregoing analysis points to a number of shortcomings in Canada's current regulatory systems and programs meant to protect drinking water at the local and national level. It is clear that even in the absence of conclusive scientific proof of harm associated with certain chemicals, there is sufficient evidence to support a strengthening of standards for certain contaminants or an assertive policy position ensuring that current standards are being met in communities across the country. This is of particular importance in light of possible low-level, gendered exposures and subsequent health effects.

A sole focus on standards is insufficient, however. It is important for governments and the public to consider whether existing standards can be met, and where standards are not being met, what specific actions can be taken to address the risks of particular chemicals in their relevant contexts. There is a need for targeted programs, based on precaution, that reduce the exposure of vulnerable communities or populations that may be disproportionately affected by chronic, low-dose drinking water contamination. Such programs must examine the nature of risk, how contaminants might be entering drinking water, and viable solutions that fully protect the health of all individuals.

Acknowledgment

This chapter is based on the report by Susanne Hamm, *The Gendered Health Effects of Chronic Low-Dose Exposures to Chemicals in Drinking Water,* produced by the National Network on Environments and Women's Health in 2009. An earlier co-authored draft of this chapter is available on SSRN.

Notes

1 *Safe Drinking Water for First Nations Act,* S.C. 2013, c. 21 (http://laws-lois.justice.gc.ca/eng/acts/S-1.04/index.html).
2 See "Constitution Acts, 1867 to 1982," http://laws-lois.justice.gc.ca/eng/Const/index.html.
3 O. Reg. 170/03, *Safe Drinking Water Act,* 2002, S.O. 2002, c. 32, 2003 (http://www.e-laws.gov.on.ca/html/regs/english/elaws_regs_030170_e.htm).

References

Aboriginal Affairs and Northern Development Canada. 2013. "Backgrounder: Safe Drinking Water for First Nations Act." http://www.aadnc-aandc.gc.ca/eng/1330529331921/1330529392602.

Andrew, Angeline S., Amy J. Warren, Aaron Barchowsky, et al. 2003. "Genomic and Proteomic Profiling of Responses to Toxic Metals in Human Lung Cells." *Environmental Health Perspectives* 111 (6): 825–37, doi:10.1289/ehp.6249.

Aspen Regional Health. 2008. *Lead in Drinking Water: Public Health Advisory.* Edmonton: Alberta Health Services. http://www9.albertahealthservices.ca/default.aspx/Default.aspx?cid=330&lang=1.

Asselin, Pierre. 2001a. "La Contamination s'étend." *Le Soleil,* 13 June, A4.

–. 2001b. "La Doute qui ronge." *Le Soleil,* 8 February, A1.

ATSDR (Agency for Toxic Substances and Disease Registry). 2007. *Toxicological Profile for Lead.* Atlanta: Department of Health and Human Services, Public Health Service. http://www.atsdr.cdc.gov/toxprofiles/tp13.pdf.

–. 2011. *Medical Management Guidelines for Trichloroethylene.* Atlanta: Division of Toxicology and Human Health Sciences. http://www.atsdr.cdc.gov/mmg/mmg.asp?id=168&tid=30.

BC Ministry of Water, Land and Air Protection. 2002. *Environmental Indicator: Groundwater in British Columbia*. http://www.env.gov.bc.ca/soe/archive/reports/et02/technical_documents/Groundwater_2002.pdf.

BC Provincial Health Officer. 2001. *Drinking Water Quality in British Columbia: The Public Health Perspective*. Provincial Health Officer's Annual Report 2000. Victoria: Ministry of Health Planning. http://www.health.gov.bc.ca/pho/presentations/annual_report_2000.pdf.

Birnbaum, Linda S. 2012. "Environmental Chemicals: Evaluating Low-Dose Effects." *Environmental Health Perspectives* 120 (4): 143-44, doi:10.1289/ehp.1205179.

Bouchard, Maryse F., François Laforest, Louise Vandelac, et al. 2007. "Hair Manganese and Hyperactive Behaviors: Pilot Study of School-Age Children Exposed through Tap Water." *Environmental Health Perspectives* 115 (1): 122-27.

Bouchard, Maryse F., Sébastien Sauvé, Benoit Barbeau, et al. 2011. "Intellectual Impairment in School-Age Children Exposed to Manganese from Drinking Water." *Environmental Health Perspectives* 119 (1): 138-43.

Bove, Frank, Youn Shim, and Perri Zeitz. 2002. "Drinking Water Contaminants and Adverse Pregnancy Outcomes: A Review." *Environmental Health Perspectives* 110 (Supp. 1): 61-74, doi:10.1289/ehp.02110s1161.

Boyd, David Richard. 2006. *The Water We Drink: An International Comparison of Drinking Water Quality Standards and Guidelines*. Healthy Environment, Healthy Canadians Series. Vancouver: David Suzuki Foundation. http://www.davidsuzuki.org/publications/downloads/2006/DSF-HEHC-water-web.pdf.

Brown, Phil. 2007. *Toxic Exposures: Contested Illnesses and the Environmental Health Movement*. New York: Columbia University Press.

Canadian Council of Ministers of the Environment. 2002. *Effects of Agricultural Activities on Water Quality*. CCME Linking Water Science to Policy Workshop Series, Report 1, Winnipeg. http://www.ec.gc.ca/inre-nwri/Default.asp?lang=En&n=795D939F-0%23groundwater#effects.

–. 2007. *Canadian Soil Quality Guidelines: Trichloroethylene Environmental and Human Health Effects*. PN 1393. http://www.ccme.ca/assets/pdf/tce_ssd_1393.pdf.

–. 2009. "Source to Tap: Nitrate and Nitrite." http://www.ccme.ca/sourcetotap/nitrates.html.

Carpenter, David O., Kathleen Arcaro, and David C. Spink. 2002. "Understanding the Human Health Effects of Chemical Mixtures." *Environmental Health Perspectives* 110 (Supp. 1): 25-42, doi:10.1289/ehp.02110s125.

CBC *Marketplace*. 2001. "Results of Marketplace Tests of Lead in Tap Water." http://www.cbc.ca/marketplace/pre-2007/files/health/leadwater/tests.html.

CBC News. 2012. "Judge Rules No Cancer Link in Shannon, Que. Water Contamination Case." http://www.cbc.ca/news/canada/montreal/judge-rules-no-cancer-link-in-shannon-que-water-contamination-case-1.1174150.

Cheadle, Bruce. 2000. "Residents Want Action on Well Water." *Chronicle Herald*, 12 October, A18.

Christensen, Randy. 2011. *Water Proof 3: Canada's Drinking Water Report Card*. Vancouver: Ecojustice. http://www.ecojustice.ca/publications/reports/waterproof-3-canadas-drinking-water-report-card/attachment.

City of Abbotsford. 2012. "Groundwater: Economic Development and Planning Services." http://www.abbotsford.ca/.

City of London. 2008. "Lead Mitigation Program and Community Lead Testing." Report to the Environment and Transportation Committee, 7 April. http://council.london.ca/meetings/Archives/Agendas/Environment%20and%20Transportation%20

Committee%20Agendas/ETC%20Agendas%202008/2008-04-07%20Agenda/item%202.pdf.

Clow, Barbara, Ann Pederson, Margaret Haworth-Brockman, et al., eds. 2009. *Rising to The Challenge: Sex- and Gender-Based Analysis for Health Planning, Policy and Research in Canada.* Halifax: Atlantic Centre of Excellence for Women's Health.

Cooper, Kathleen, and Lorne Vanderlinden. 2009. "Pollution, Chemicals and Children's Health: The Need for Precautionary Policy in Canada." In *Environmental Challenges and Opportunities: Local-Global Perspectives on Canadian Issues,* edited by Christopher D. Gore and Peter J. Stoett, 183-224. Toronto: Emond Montgomery.

Corkal, Darrell R., William C. Schutzman, and Clint R. Hilliard. 2004. "Rural Water Safety from the Source to the On-Farm Tap." *Journal of Toxicology and Environmental Health Part A* 67 (20-22): 1619-42, doi:10.1080/15287390490491918.

Endocrine Society. 2012. *Experts Say Protocols for Identifying Endocrine-Disrupting Chemicals Inadequate.* News Room: 2012 News Releases. http://www.endo-society.org/media/press/2012/Experts-Say-Protocols-for-Identifying-Endocrine-Disrupting-Chemicals-Inadequate.cfm.

Environment Canada. 2008. "Nitrate Levels in the Abbotsford Aquifer: An Indicator of Groundwater Contamination in the Lower Fraser Valley." http://www.ecoinfo.ec.gc.ca/env_ind/region/nitrate/nitrate_e.cfm.

–. 2011. "Groundwater." http://www.ec.gc.ca/inre-nwri/default.asp?lang=En&n=1EDF83E1-1.

European Environment Agency. 2003. "Nitrate in Groundwater." http://www.eea.europa.eu/data-and-maps/indicators/nitrate-in-groundwater.

European Parliament and Council. "Directive 2008/105/EC ... on Environmental Quality Standards in the Field of Water Policy ..." http://eur-lex.europa.eu/LexUriServ/LexUriServ.do?uri=OJ:L:2008:348:0084:0097:EN:PDF.

Eyles, John, K. Bruce Newbold, Anita Toth, et al. 2011. *Chemicals of Concern in Ontario and the Great Lakes Basin: Update 2011 – Emerging Issues.* Hamilton: McMaster Institute of Environment and Health.

Federal-Provincial-Territorial Committee on Drinking Water. 1987. "Nitrate-Nitrite." Guidelines for Canadian Drinking Water Quality: Supporting Documentation. Ottawa: Health Canada, Water Quality and Health Bureau, Healthy Environments and Consumer Safety Branch. http://www.hc-sc.gc.ca/ewh-semt/alt_formats/hecs-sesc/pdf/pubs/water-eau/nitrate_nitrite/nitrate_nitrite-eng.pdf.

–. 2002. "Minutes of 31st Meeting, Ottawa, Ontario (October 24-25, 2002)."

–. 2004. "Minutes of the 35th Meeting, Ottawa, Ontario (October 27-29, 2004)."

–. 2005. "Trichloroethylene." Guidelines for Canadian Drinking Water Quality: Supporting Documentation. Ottawa: Health Canada, Water Quality and Health Bureau, Healthy Environments and Consumer Safety Branch.

–. 2012. "Guidelines for Canadian Drinking Water Quality Summary Table." Prepared by the Federal-Provincial-Territorial Committee on Drinking Water of the Federal-Provincial-Territorial Committee on Health and the Environment, August.

Gosine, Andil, and Cheryl Teelucksingh. 2008. *Environmental Justice and Racism in Canada: An Introduction.* Toronto: Emond Montgomery.

Government of Canada. 2012. *Federal Contaminated Sites: The Action Plan.* http://www.ec.gc.ca/default.asp?lang=En&n=D87FA775-1&news=54541450-0E90-4539-9E84-AFFE1CF64BD3.

Hancock, Melissa. 2008. "Resident Tired of Filter System on Her Property." *Cambridge Times,* 30 December. http://www.cambridgetimes.ca/news/article/157554.

Health Canada. 2006. *Guidelines for Canadian Drinking Water Quality: Guideline Technical Document – Arsenic.* Ottawa: Health Canada. http://www.hc-sc.gc.ca/ewh-semt/alt_formats/hecs-sesc/pdf/pubs/water-eau/arsenic/arsenic-eng.pdf.

–. 2007. *Water Talk: Drinking Water Quality in Canada.* Ottawa: Health Canada. http://www.hc-sc.gc.ca/ewh-semt/pubs/water-eau/drink-potab-eng.php.

–. 2009. "Part 1: Approach to the Derivation of Drinking Water Guidelines." http://www.hc-sc.gc.ca.

–. 2012. *Guidelines for Canadian Drinking Water Quality: Summary Table.* Ottawa: Health Canada, Water, Air and Climate Change Bureau, Healthy Environments and Consumer Safety Branch.http://www.hc-sc.gc.ca/ewh-semt/pubs/water-eau/2012-sum_guide-res_recom/index-eng.php.

Hendry, James M. 2006. "Comments on Future Challenges in Canada's Groundwater Resource: Quality Issues." Presentation for the Water Policy in Canada National Workshop Series. Workshop 2: Emerging Challenges to Sustainable Water Policy in Canada. Lethbridge, AB, 15-16 March 2006.

Hogg, Matthew. 2007. "Chemicals Harmful to Health in Low as Well as High Doses." The Environmental Illness Resource, http://www.ei-resource.org/news/general-environmental-health-news/chemicals-harmful-to-health-in-low-as-well-as-high-doses/.

Hrudey, Steve E. 2008. "Safe Water? Depends on Where You Live!" *Canadian Medical Association Journal* 178 (8): 975, doi:10.1503/cmaj.080374.

House of Commons. 2012. "Groundwater Contamination." Motion of Élaine Michaud. *House of Commons Debates* 146 (104), 2 April. 41st Parliament, 1st Session. http://www.parl.gc.ca/HousePublications/Publication.aspx?DocId=5493945&Language=E&Mode=1.

Järup, Lars. 2003. "Hazards of Heavy Metals Contamination." *British Medical Bulletin* 68 (1): 167-82, doi:10.1093/bmb/ldg032.

Johnson, Paula D., Stanley J. Goldberg, Mary Z. Mays, et al. 2003. "Threshold of Trichloroethylene Contamination in Maternal Drinking Waters Affecting Fetal Heart Development in the Rat." *Environmental Health Perspectives* 111 (3): 289-92, doi:10.1289/ehp.5125.

Knobeloch, Lynda, Barbara Salna, Adam Hogan, et al. 2000. "Blue Babies and Nitrate-Contaminated Well Water." *Environmental Health Perspectives* 108 (7): 675-78, doi: 10.1289/ehp.00108675.

Kortenkamp, Andreas, Michael Faust, Martin Scholze, et al. 2007. "Low-Level Exposure to Multiple Chemicals: Reason for Human Health Concerns?" *Environmental Health Perspectives* 115 (Suppl. 1): 106-14, doi:10.1289/ehp.9358.

Laffont, Nicholas, and Nicolas Godbout. 2012. "TCE à Shannon: le NPD envisage le dépôt d'un projet de loi." Le Huffington Post Québec, http://quebec.huffingtonpost.ca/2012/10/15/tce-a-shannon-npd-projet-de-loi_n_1966388.html.

Lafrance, Annie. 2004. "Contamination de l'eau au TCE à Shannon: Le problème touche les écoles primaires." *Cyberpresse, Le Soleil,* 7 February, A10.

Lefebvre, René, Alexandre Boutin, Richard Martel, et al. 2003. *Caratérisation et modélisation numérique de l'écoulement et de la migration de la contamination en TCE dans l'eau souterraine du secteur Valcartier, Québec, Canada.* Institut National de la Recherche Scientifique, Eau, Terre et Environnment, Université du Québec. Rapport de recherche R-631 soumis à la Garnison Valcartier et à RDDC Valcartier.

Levallois, Patrick, Suzanne Gingras, Madeleine Caron, et al. 2008. "Drinking Water Intake in Infants Living in Rural Quebec (Canada)." *Science of the Total Environment* 397 (1-3): 82-85, doi:j.scitotenv.2008.02.026.

Levallois, Patrick, Julie St-Laurent, Denis Gauvin, et al. 2013. "The Impact of Drinking Water, Indoor Dust and Paint on Blood Lead Levels of Children Aged 1-5 Years in Montréal (Québec, Canada)." *Journal of Exposure Science and Environmental Epidemiology,* doi:10.1038/jes.2012.129.

Levin, Ronnie, Mary Jean Brown, Michael E. Kashtock, et al. 2008. "Lead Exposures in US Children, 2008: Implications for Prevention." *Environmental Health Perspectives* 116 (10): 1285-93, doi:10.1289/ehp.11241.

Manassaram, Deana M., Lorraine C. Backer, and Deborah M. Moll. 2006. "A Review of Nitrates in Drinking Water: Maternal Exposure and Adverse Reproductive Health and Developmental Outcomes." *Environmental Health Perspectives* 114 (3): 320-27, doi:10.1289/ehp.8407.

Mergler, Donna. 2012. "Neurotoxic Exposures and Effects: Gender and Sex Matter!" *Neurotoxicology* 33 (4): 644-51.

Mittelstaedt, Martin. 2001. "Hazardous Solvent Found in Ontario Groundwater." *Globe and Mail,* 21 March, A1.

Myers, John Peterson, R. Thomas Zoeller, and Frederick S. vom Saal. 2009. "A Clash of Old and New Scientific Concepts in Toxicity, with Important Implications for Public Health." *Environmental Health Perspectives* 117 (11): 1652-55, doi:10.1289/ehp.0900887.

Navas-Acien, Ana, A. Richey Sharrett, Ellen K. Silbergeld, et al. 2005. "Arsenic Exposure and Cardiovascular Disease: A Systematic Review of the Epidemiologic Evidence." *American Journal of Epidemiology* 162 (11): 1037-49, doi:10.1093/aje/kwi330.

Nawrot, Tim S., and Jan A. Stassen. 2006. "Low-Level Environmental Exposure to Lead Unmasked as Silent Killer." *Circulation* 114 (13): 1347-49, doi:10.1161/CIRCULATION AHA.106.650440.

Nickel, Rod. 2008. "Province to Order More Water Testing." *Star Phoenix,* 28 November, B6.

Noonan, Curtis W., Sara M. Sarasua, Dave Campagna, et al. 2002. "Effects of Exposure to Low Levels of Environmental Cadmium on Renal Biomarkers." *Environmental Health Perspectives* 110 (2): 151-55, doi:10.1289/ehp.02110151.

Northwest Territories Health and Social Services. 2007. "Northwest Territories Drinking Water Sampling and Testing Requirements." http://www.hlthss.gov.nt.ca/.

Nova Scotia Environment. 2011. *Well Water Nitrate Monitoring Program: 2011 Report.* Halifax: Province of Nova Scotia. http://www.gov.ns.ca/nse/groundwater/docs/WellWaterNitrateMonitoringProgram-2011Report.pdf.

Office of the Auditor General of Canada. 2002. "Chapter 2: The Legacy of Federal Contaminated Sites." In *Report of the Commissioner of the Environment and Sustainable Development to the House of Commons.* Ottawa: Minister of Public Works and Government Services Canada. http://www.oag-bvg.gc.ca/.

–. 2005. "Chapter 4: Safety of Drinking Water – Federal Responsibilities." In *Report of the Commissioner of the Environment and Sustainable Development to the House of Commons.* Ottawa: Minister of Public Works and Government Services Canada. http://www.oag-bvg.gc.ca/internet/docs/c20050904ce.pdf.

Ontario Ministry of the Environment. 2007a. "Provincial Officer's Order: Chief Drinking Water Inspector Orders Municipalities to Do Lead Testing." On file with author.

–. 2007b. *Summary of Lead Sampling Results for Municipalities Required to Sample under "Ontario Tap Water Order."* Toronto: Queen's Printer for Ontario. http://www.ontla.on.ca/library/repository/mon/17000/274264.pdf.

Pastino, Gina M., Wendy Y. Yap, and Maria Carroquino. 2000. "Human Variability and Susceptibility to Trichloroethylene." *Environmental Health Perspectives* 108 (Suppl. 2): 201-14, doi:10.1289/ehp.00108s2201.

Payne, Mark. 2008. "Lead in Drinking Water." *Canadian Medical Association Journal* 179 (3): 253-54, doi:10.1503/cmaj.071483.

PEI Commission on Nitrates in Groundwater. 2008. *The Report of the Commission on Nitrates in Groundwater.* Report for the Government of Prince Edward Island.

Charlottetown: Commission on Nitrates in Groundwater. http://www.gov.pe.ca/photos/original/cofNitrates.pdf.

Peterson, Hans. 2008. "Safe Drinking Water in Jeopardy: Rural Canadians Get Contaminated Drinking Water from Livestock and Human Waste Run-Off." *Global Research,* 14 June. http://www.globalresearch.ca/index.php?context=va&aid=9323.

Peterson Hans, and Colin Fricker. 2008. "A Framework for Safe Drinking Water: Using Science over Politics in the Search for Safe Drinking Water Solutions." *Canadian Water Treatment* (November/December): 14-15. http://www.safewater.org/PDFS/waternewsmagazines/cwtnovdecframeworkforsafedrinkingwater.pdf.

Powlson, David S., Tom M. Addiscott, Nigel Benjamin, et al. 2008. "When Does Nitrate Become a Risk for Humans?" *Journal of Environmental Quality* 37 (2): 291-95, doi:10.2134/jeq2007.0177.

Raaschou-Nielsen, Ole, Johnni Hansen, Joseph K. McLaughlin, et al. 2003. "Cancer Risk among Workers at Danish Companies using TCE: A Cohort Study." *American Journal of Epidemiology* 158 (12): 1182-92, doi:10.1093/aje/kwg282.

Rizak, Samantha, and Steve E. Hrudey. 2007. "Achieving Safe Drinking Water: Risk Management Based on Experience and Reality." *Environmental Reviews* 15 (1): 169-74, doi:10.1139/A07-005.

Roels, H.A., R.M. Bowler, Y. Kim, et al. 2012. "Manganese Exposure and Cognitive Deficits: A Growing Concern for Manganese Neurotoxicity." *Neurotoxicology* 33 (4): 872-80.

Ryan, Barry P., Natalie Huet, and David L. MacIntosh. 2000. "Longitudinal Investigation of Exposure to Arsenic, Cadmium, and Lead in Drinking Water." *Environmental Health Perspectives* 108 (8): 731-35, doi:10.1289/ehp.00108731.

Scott, Cheryl Siegel, and Weihsueh A. Chiu. 2006. "Trichloroethylene Cancer Epidemiology: A Consideration of Select Issues." *Environmental Health Perspectives* 114 (9): 1471-78, doi:10.1289/ehp.8949.

Selevan, Sherry G., Carole A. Kimmel, and Pauline Mendola. 2000. "Identifying Critical Windows of Exposure for Child's Health." *Environmental Health Perspectives* 108 (Suppl. 3): 451-55, doi:10.1289/ehp.00108s3451.

Séguin, Rhéal. 2013. "Quebec Area Where Cancer Rates Are 80 Times Higher to Be Studied." *Globe and Mail,* 20 May. http://www.theglobeandmail.com/life/health-and-fitness/health/quebec-area-where-cancer-rates-are-80-times-higher-to-be-studied/article12033398/.

Shaw, Gary M., Verne Nelson, David M. Iovannisci, et al. 2003. "Maternal Occupational Chemical Exposures and Biotransformation Genotypes as Risk Factors for Selected Congenital Anomalies." *American Journal of Epidemiology* 157 (6): 475-84, doi:10.1093/aje/kwg013.

Stager, John. 2009. "Re: Release of Community Lead Testing Program – Round 1 (December 2007 – April 2008) Results." Memo to Owners of Municipal and Non-Municipal Year-Round Residential Drinking Water Systems. Toronto: Ministry of the Environment, Drinking Water Management Division. http://www.town.minto.on.ca/content/water/Community_Lead_Testing_Results.pdf.

Statistics Canada. 2011a. *Aboriginal Peoples in Canada: First Nations People, Métis, and Inuit.* Ottawa: Statistics Canada. http://www12.statcan.gc.ca/nhs-enm/2011/as-sa/99-011-x/99-011-x2011001-eng.cfm.

–. 2011b. *Households and the Environment.* Ottawa: Statistics Canada. http://www5.statcan.gc.ca/bsolc/olc-cel/olc-cel?lang=eng&catno=11-526-XWE.

Sun Media. 2009. "Lead in Drinking Water Still a Problem." *The Observer,* 9 March. http://www.theobserver.ca/ArticleDisplay.aspx?e=1469816.

TEDX (The Endocrine Disruption Exchange). 2013. "Endocrine Disruption: Introduction: Overview." http://endocrinedisruption.org/endocrine-disruption/introduction/overview.

Theppeang, Keson, Thomas A. Glass, Karen Bandeen-Roche, et al. 2008. "Gender and Race/Ethnicity Differences in Lead Dose Biomarkers." *American Journal of Public Health* 98 (7): 1248-55, doi:10.2105/AJPH.2007.118505.

Timoney, Kevin P. 2007. *A Study of Water and Sediment Quality as Related to Public Health Issues, Fort Chipewyan, Alberta*. Fort Chipewyan, AB: Nunee Health Board Society. http://www.borealbirds.org/resources/timoney-fortchipwater-111107.pdf.

Tourism British Columbia. n.d. "Abbotsford." http://www.hellobc.com/abbotsford.aspx.

US EPA (Environmental Protection Agency). 2001. *Health Assessment Document for Trichloroethylene Synthesis and Characterization (External Review Draft)*. EPA/600/P-01/002A. Washington, DC: US EPA, Office of Research and Development, National Center for Environmental Assessment. http://ofmpub.epa.gov/eims/eimscomm.getfile?p_download_id=4580.

–. 2009. "Drinking Water Contaminants: National Primary Drinking Water Regulations." http://www.epa.gov/safewater/mcl.html#mcls.

–. 2011. "Toxicokinetics." Chapter 3 in *Toxicological Review of Trichloroethylene* (CAS 79-01-6). EPA/635/R-09/011F. Washington, DC: US EPA. http://www.epa.gov/iris/toxreviews/0199tr/Chapter3_0199tr.pdf.

–. 2012. "Water: Lead – Basic Information." http://www.epa.gov/safewater/lead/basicinformation.html#healtheffects.

US NRC (US National Research Council, Committee on Human Health Risks of Trichloroethylene). 2006. *Assessing the Human Health Risks of Trichloroethylene: Key Scientific Issues*. Washington, DC: National Academies Press. http://www.nap.edu/openbook.php?record_id=11707&page=R1.

US TCE Review Panel (TCE Review Panel of the Environmental Health Committee of the US Environmental Protection Agency Science Advisory Board). 2002. *Review of Draft Trichloroethylene Heath Risk Assessment Synthesis and Characterization: An EPA Science Advisory Board Report*. EPA-SAB-EHC-03-002. Washington, DC: US EPA. http://www.epa.gov/sab/pdf/ehc03002.pdf.

Vahter, M. 2008. "Health Effects of Early Life Exposure to Arsenic." *Basic and Clinical Pharmacology and Toxicology* 102 (2): 204-11, doi:10.1111/j.1742-7843.2007.00168.x.

Vahter, Marie, Agneta Åkesson, Carola Lidén, et al. 2007. "Gender Differences in the Disposition and Toxicity of Metals." *Environmental Research* 104 (1): 85-95, doi:10.1016/j.envres.2006.08.003.

van der Kamp, Garth, and Gary Grove. 2001. "Well Water Quality in Canada: An Overview." In *An Earth Odyssey: Proceedings of the 54th Canadian Geotechnical Conference and 2nd Joint IAH-CNC and CGS Groundwater Specialty Conference*, edited by M. Mahmoud, R. van Everdingen, and J. Carss, 39-41. Calgary: Canadian Geotechnical Society.

Van Grinsven, Hans J.M., Mary H. Ward, Nigel Benjamin, et al. 2006. "Does the Evidence about Health Risks Associated with Nitrate Ingestion Warrant an Increase of the Nitrate Standard for Drinking Water?" *Environmental Health: A Global Access Science Source* 5 (26): 1-6, doi:10.1186/1476-069X-5-26.

Vandenberg, Laura N., Theo Colborn, Tyrone B. Hayes, et al. 2012. "Hormones and Endocrine-Disrupting Chemicals: Low-Dose Effects and Nonmonotonic Dose Responses." *Endocrine Reviews* 33 (3): 378-455, doi:10.1210/er.2011-1050.

Vom Saal, Frederick S., and Claude Hughes. 2005. "An Extensive New Literature Concerning Low-Dose Effects of Bisphenol A Shows the Need for a New Risk Assessment." *Environmental Health Perspectives* 113 (8): 926-33, doi:10.1289/ehp.7713.

Ward, Mary H., Theo M. deKok, Patrick Levallois, et al. 2005. "Workgroup Report: Drinking-Water Nitrate and Health – Recent Findings and Research Needs." *Environmental Health Perspectives* 113 (11): 1607-14, doi:10.1289/ehp.8043.

Wassenaar, Leonard I., M. Jim Hendry, and Nikki Harrington. 2006. "Decadal Geochemical and Isotopic Trends for Nitrate in a Transboundary Aquifer and Implications for Agricultural Beneficial Management Practices." *Environmental Science and Technology* 40 (15): 4626-32, doi:10.1021/es060724w.

Watson, Rebecca E., Catherine F. Jacobson, Amy Lavin Williams, W. Brian Howard, and John M. DeSesso. 2006. "Trichloroethylene-Contaminated Drinking Water and Congenital Heart Defects: A Critical Analysis of the Literature." *Reproductive Toxicology* 21 (2): 117-47, doi:10.1016/j.reprotox.2005.07.013.

Weir, Erica. 2005. "Well-Water Maintenance." *Canadian Medical Association Journal* 172 (11): 1438, doi:10.1503/cmaj.050377.

Welshons, Wade V., Susan C. Nagel, and Frederick S. vom Saal. 2006. "Large Effects from Small Exposures. III. Endocrine Mechanisms Mediating Effects of Bisphenol A at Levels of Human Exposure." *Endocrinology* 147 (6): s56-s69, doi:10.1210/en.2005-1159.

Weyer, Peter J., James R. Cerhan, Burton C. Kross, et al. 2001. "Municipal Drinking Water Nitrate Level and Cancer Risk in Older Women: The Iowa Women's Health Study." *Epidemiology* 12 (3): 327-38.

Wigle, Donald T., Tye E. Arbuckle, Michelle C. Turner, et al. 2008. "Epidemiologic Evidence of Relationships between Reproductive and Child Health Outcomes and Environmental Chemical Contaminants." *Journal of Toxicology and Environmental Health, Part B: Critical Reviews* 11 (5-6): 373-517, doi:10.1080/10937400801921320.

Wilson, Julie, Hans Schreier, and Sandra Brown. 2008. *Arsenic in Groundwater in the Surrey-Langley Area.* Vancouver: Institute for Resources and Environment, University of British Columbia, for Fraser Health Authority and BC Ministry of the Environment. http://www.fraserhealth.ca/media/ArsenicReportSurreyLangley.pdf.

World Health Organization Regional Office for Europe and European Environment Agency. 2002. *Children's Health and Environment: A Review of Evidence.* Environmental Issue Report 29. Copenhagen: European Environment Agency. http://www.eea.europa.eu/publications/environmental_issue_report_2002_29/at_download/file.

Zahir, Farhana, Shamim J. Rizwi, Soghra K. Haq, et al. 2005. "Low Dose Mercury Toxicity and Human Health." *Environmental Toxicology and Pharmacology* 20 (2): 351-60, doi:10.1016/j.etap.2005.03.007.

Zeliger, Harold I. 2003. "Toxic Effects of Chemical Mixtures." *Archives of Environmental and Occupational Health* 58 (1): 23-29, doi:10.3200/AEOH.58.1.23-29.

Chapter 5 Consuming "DNA as Chemicals" and Chemicals as Food

Bita Amani

Edible: good to eat, and wholesome to digest,
as a worm to a toad, a toad to a snake, a snake to
a pig, a pig to a man, and a man to a worm.

– *Ambrose (Gwinett) Bierce,* The Devil's Dictionary

There are risks associated with consuming food. In fact, food itself can be one of our most intimately administered noxious substances. Food is essential to life and therefore constitutes a fundamental medium of exposure to *risk.* Food controversies demonstrate the sensitive, politicized, and sometimes fatal relationship between people and their consumption choices. Health risks from food arise from contaminants, additives, toxins, and other disease-causing agents, and call into question the effectiveness of regulatory standards and surveillance to ensure that food safety is maintained in production, processing, management, storage, and trade. Health Canada scientists are responsible for assessing the risks to human health from exposure to food-borne chemical contaminants, the presence of natural toxins or chemical pesticide residue. The risk of consuming food is rarely understood in terms of food itself *as the chemical.*

Advances in molecular genetics have significantly shifted food production away from traditional cross-breeding methods towards engineered products. Governments now strive to allay mounting public concerns over the health effects and environmental safety of genetically modified organisms (GMOs) forming "novel foods," GM ingredients, and animal feed.

Modern food risks come not only from the unknown risks of consuming recombinant DNA but also from continued industrial reliance on and chronic exposure to chemicals, either as "add-ons" (such as pesticides) or increasingly as a manufactured component in the crop itself (such as Bt crops genetically engineered to produce a naturally occurring toxin found in the soil bacterium *Bacillus thuringiensis*). As a result, many non-governmental organizations (NGOs) and opponents of GMOs herald the arrival of the precautionary principle in public debates, citing the unknowns related to the complexity of the interaction between genes and their environments (Tokar 2001; see also Chapters 1-4, and 8 for other discussions of the precautionary principle). Public interest advocates are also working to reveal the political processes through which technocratic frames of governance, centred on the mantra of "assess, predict, and manage," gain regulatory dominance over novel foods (Scott 2007, 16) and are demanding greater citizen participation in the governance of issues so fundamental to our health. They strive to dispel the mythology that scientific processes are themselves neutral or void of moral, cultural, and political subjectivity.

This chapter assesses the current limitations of the Canadian regulatory framework for novel foods and argues for a broader conceptual approach to "DNA as chemicals" and chemicals as food. Although health consequences are of general concern, a gender-based analysis of risk arising from GMOs and novel food consumption is necessary to address their differential impact on women. "Consuming DNA as Chemicals" will examine the bases of a conceptual framework for DNA as chemicals. Once adopted in the regulatory approval processes for novel foods, a chemical understanding of DNA already endorsed by the courts in the biopatent field would ensure cross-policy coherence between the regulated incentive mechanisms for agrochemical and novel food innovation and their commercialization. "Consuming Chemicals as Food" argues that novel food consumption is in fact chemical consumption. "Canada's Novel Food Regulations" outlines Canada's novel food framework and the principle of *substantial equivalence* in assessing food safety in order to demonstrate the potential deficiency of such an approach in capturing the chemical consumption risks of novel foods. "Chemical Consumption and the Implications for Women's Health" further examines some of the scientific evidence and argues that, as with other chemicals, the gendered impact of novel food consumption on women's health and their environments merits specific analysis. This is followed by "Recommendations for Law and Policy Reforms," and the chapter concludes with a reminder of Canada's obligations under the United Nations Convention on the Elimination of All Forms of Discrimination against Women (CEDAW),[1] which demand that we move beyond merely conducting gender-based analyses of novel food safety towards gender-based regulatory impact assessments of Health Canada's novel food regulations.

Novel food consumption is chemical consumption. Understanding this helps dispel the need for more (or absolutes in) scientific knowledge as a precondition for precaution. It shifts the focus to consumers' right to know the chemicals they consume every day through food, and reinforces the demand for mandatory labelling. Canadians are passionately invested in the regulation of GM food because of its potential to cause harm to human health. Mandatory labelling is imperative because, "while scientists may be able to tell the public what the risks are, they have no particular expertise in decisions as to whether risks are worth taking" (Epps 2008b, 362).

Consuming DNA as Chemicals

> Recombinant DNA [rDNA] technology faces our society with problems unprecedented not only in the history of science, but of life on Earth. It places in human hands the capacity to redesign living organisms, the products of three billion years of evolution. Such intervention must not be confused with previous intrusions upon the natural order of living organisms: animal and plant breeding ... All the earlier procedures worked within single or closely related species ... Our morality up to now has been to go ahead without restriction to learn all that we can about nature. Restructuring nature was not part of the bargain ... this direction may be not only unwise, but dangerous. Potentially, it could breed new animal and plant diseases, new sources of cancer, novel epidemics. (Wald 1979, 127-28)

Deoxyribonucleic acid (DNA) provides the genetic code for all life and makes possible new applications for a bioeconomy: "Each new gene is a potential target for drug development – to fix it when broken, to shut it down, to attenuate or amplify its expression, or to change its product, usually a protein" (Cook-Deegan 1994).

The *Canadian Oxford Dictionary* defines "chemical" as "a compound or substance that has been purified or prepared, esp. artificially." Enzymes are used in DNA sequencing to cut the DNA chain into fragments. While some enzyme mechanisms "terminate" the DNA chain (cutting the sequence in a way that blocks any further addition of nucleotides) (Alberts et al. 2002), other enzymes leave staggered ends:

> Once the cuts have been made in the DNA backbones, the resulting fragments are held together only by the relatively weak hydrogen bonds which hold complementary base pairs together in the DNA double helix. Since these bonds are not strong enough to hold the fragments together for long at 37°C (98°F), the fragments soon separate from one another, leaving behind protruding 5' ends composed of unpaired bases. (Peters 1993, 67)

These "sticky ends" allow for recombinant DNA molecules to be formed and novel traits to be introduced through a change in the organism's genetic code.

The proliferation of GM food innovation correlates with a shift in patent law to recognize DNA sequences as inventions rather than as a product of nature or discovery. Developments in agrochemicals and GM products are spurred by the promise of a patent – an instrumental institutional tool conferring exclusive rights created by statute for a limited term (twenty years in Canada) to encourage the development of new, useful, and not obvious inventions – and the accepted legal practice of gene patenting that occurred quietly through patent office decisions and was endorsed by the judiciary. In an early landmark decision, *Re Howard Florey Institute/Relaxin* (1995), a patent claim over the hormone human H2-relaxin found naturally in the bodies of pregnant women was made based on the Institute's determination of the hormone's chemical structure (WIPO 2006). The Opposition Division of the European Patent Office (EPO) found that "until a cDNA encoding human H2-relaxin and its precursors was isolated ... the existence of this form of relaxin was unknown." Further, the decision noted that "it is established patent practice to recognize novelty for a natural substance which has been isolated for the first time and which had no previously recognized existence" (*Howard Florey*, para. 4.3.1). The charge that DNA encoding the human H2-relaxin hormone was a *discovery* and thus not patentable subject matter under the European Patent Convention (EPC, Art. 52[2]) was defeated (*Howard Florey*, paras. 5.1-5.2), as was the assertion that the patent over the DNA fragment encoding the human H2-relaxin was offensive to morality under Art. 53(a) of the EPC. The decision found that "no woman is affected in any way by the present patent" (para 6.3.3).[2] And, according to the Opposition Division, "it is worth pointing out that DNA is not 'life' but a *chemical substance* which carries genetic information and can be used as an intermediate in the production of proteins" (*Howard Florey,* 541-42). The conceptual shift from DNA as code to DNA as chemicals, purified and isolated, reinforced if not spurred the growth of related industries (Amani 2009, 2014).

The Supreme Court of Canada, in *Harvard College v. Canada (Commissioner of Patents)* (2002), contrary to other jurisdictions, denied the eligibility of the claim to non-human higher life as a patentable product. The court found that, "in this case, we are dealing with a transgenic mouse, all of whose cells contain a foreign oncogene which was assembled and artificially incorporated into the genome of the mouse as a result of human intervention" (*Harvard,* para. 130); but Harvard did not invent, nor could it control, the reproduction of the whole mouse and so was not entitled to the product patent over it (see discussion in Amani 2009, 2014). The majority confirmed, however, the patentability of transgenic material.

In the United States, the issue of patent rights over the BRCA1 and BRCA2 genes linked to increased risk of breast and ovarian cancer in women came before the US Court of Appeals for the Federal Circuit in 2011. In *AMP (Association for Molecular Pathology) v. Myriad Genetics* (2012), the court found:

> Chemically, the human genome is composed of deoxyribonucleic acid ("DNA"). Each DNA molecule is made up of repeating units of four nucleotide bases – adenine ("A"), thymine ("T"), cytosine ("C"), and guanine ("G") – which are covalently linked, or bonded, together via a sugar-phosphate, or phosphodiester, backbone. DNA generally exists as two DNA strands intertwined as a double helix in which each base on a strand pairs, or hybridizes, with a complementary base on the other strand: A pairs with T, and C with G.

In a 2-1 vote, the Court of Appeals overturned the decision of the US District Court for New York and found in favour of the firm Myriad Genetics Inc.:

> The distinction ... between a product of nature and a human-made invention ... turns on a change in the claimed composition's identity compared with what exists in nature ... Applying this test to the isolated DNA in this case, we conclude that the challenged claims are drawn to patentable subject matter because the *claims cover molecules that are markedly different – have a distinctive chemical identity and nature – from molecules that exist in nature.*
>
> *It is undisputed that Myriad's claimed isolated DNAs exist in a distinctive chemical form – as distinctive chemical molecules – from DNAs in the human body, i.e.,* native DNA. Native DNA exists in the body as one of forty-six large, contiguous DNA molecules. Each DNA molecule is itself an integral part of a larger structural complex, a chromosome. Isolated DNA, in contrast, is a free-standing portion of a native DNA molecule, frequently a single gene.
>
> Isolated DNA has been cleaved (i.e., had covalent bonds in its backbone chemically severed) or synthesized to consist of just a fraction of a naturally occurring DNA molecule. For example, the *BRCA1* gene in its native state resides on chromosome 17, a DNA molecule of around eighty million nucleotides. (*AMP* 2012, 1351; emphasis added)

The decision views a covalent bond as separating one chemical species from another (1353) and treats DNA as a "distinct chemical entity"(1352). The dissenting appeals judge sided with the lower court decision in finding that to allow a gene patent for mere isolation would be like allowing a claim to a leaf snapped from a tree. The analogy was dismissed by the majority: "Snapping a leaf from a tree is a physical separation, not one creating a new chemical entity" (1354); they also found that their decision "comports with the long-standing practice of the USPTO" such that any change should come from Congress and not the court (1354). On further appeal, the US Supreme Court found, in *Association for Molecular Pathology et al. v. Myriad Genetics, Inc., et al.* (2013), that isolated naturally occurring DNA segments, in this case human genes, are products of nature and thus not patent-eligible compositions. The decision does not disturb the established chemical view of DNA: "Myriad's [patent] claims are simply not expressed in terms of chemical composition, nor do they rely in any way on the chemical changes that result from the isolation of a particular section of DNA" (14). Rather, Myriad's "claim is concerned primarily with the information contained in the genetic *sequence,* not with the specific chemical composition of a particular molecule" (15); such a characterization was decisive in the court's finding of patent ineligibility.

The chemical view of DNA opens the door to the appreciation and investigation of new forms of risk, such as GM contamination of farmers' fields and transgenic pollution. Regulating novel food ought to be consistent with current approaches to chemicals legislation that work to curtail their harmful effects and are sensitive to dosage and/or long-term exposures as determinants of toxicity. A chemical approach to DNA expands the scope of regulatory oversight for GMOs as *chemical* products. As chemical labelling, mandatory labelling of novel foods becomes imperative.

Consuming Chemicals as Food

As prepared substances, food crops can be genetically engineered (GE) with rDNA to slow the ripening of fruit, resist harsher temperatures, produce chemicals that act as pesticides or create chemical resistance to chemical applications, eliminate fertility, foster sterility through "terminator" and "suicide" seeds, or respond only to specific chemicals in order to germinate. Traditional chemical assessments of food have looked at the mitigation of chemical pesticide residues and chemicals found in food. What happens when food is the chemical residue? GM food consumption may lead to chemical risks from:

- new chemicals in GM food, including: (1) GM varieties, such as animal-plant-human transgenic combinations, or (2) built-in chemicals, such as insecticides
- new uses of agrochemicals (herbicides) applied to GM "ready" varieties.

To assess the routes of chemical exposure, we need more data to determine whether the GM food itself can be considered a chemical risk, in terms of either the digestibility of novel DNA or the chemical consequences of built-in agrochemicals in GMOs. Major public concern regarding GM foods relates to the molecular mixing of genes from *unrelated* species to form various transgenic animal-plant-human (chimera) combinations. Molecular manipulation allows for "the expression of novel protein or changes in the levels of endogenous proteins" (Perr 2002, 475). Examples of genetic mixing include tomato (with a flounder), potato (with silk moth), corn (with firefly), and rice (with bacteria) (Babula 1999, 128). DNA of a mouse, combined with human tumour fragments, has been inserted into tobacco DNA to create a vaccine against non-Hodgkin's lymphoma in the tobacco plant (MacDonald Glenn 2004).

With such examples, we can appreciate how novel foods qualify as "prepared chemical substances," but as Crouch (2001, 31) persuasively argues, "organisms are not machines nor are they merely chemicals. They grow, reproduce, interact, and change. Scientists accelerate the rate of change by moving genes between different species, shaking up long-standing relationships in unforeseen ways. In nature, novelty is disruptive." Even without manipulated changes in the genetic code, DNA is more like a script that, when read by cells, can result in different productions and life histories (Carey 2012, 2). With molecular intervention, DNA retains its informational capacity, raising added issues of unintended consequences. Jürgen Habermas (2003, 47), in characterizing biotechnology as intervention, cites Hans Jonas: "In dealing with organisms, activity is confronted with activity: biotechnology is collaborative with the auto-activity of the active material, the biological system in its natural functioning into which a new determinant has to be incorporated." From this perspective, DNA may be better understood as "a communication medium in a complex biological and biochemical systems network with other actants"; it operates discursively as *biomedia* (Amani 2014).

Those opposed to GM practices argue that the release of GMOs into our food and environment results in the treatment of consumers as a human laboratory (Burrows 2001), implicated in the long-term determination and assessment of food and biosafety. GM production gives rise to an inherent challenge in regulation: once GMOs are introduced into the environment and our bodies, *there is no recall* (Crouch 2001). By way of example, Crouch offers the case of the GM *Klebsiella planticola* bacterium, a common soil microbe engineered with the genes of another soil bacterium, *Xanthomonas,* in order to produce ethanol from crop waste for industrial use. Grass farmers would collect plant residues, ferment the material with the engineered bacteria, and sell the resulting alcohol as a fuel additive, helping to reduce air pollution. The residue from fermentation would also be used as fertilizer. The GM bacteria were to be field-tested before release for commercial use, but the US Environmental Protection Agency (US EPA) performed safety experiments and discovered that "when wheat seedlings were planted in the soil containing the engineered bacteria, the plants died in about a week"

(Crouch 2001, 30-31). The conventional bacteria lived at the roots of plants and these new genes "used secretions produced by the plants and organic matter in the soil to make alcohol, and thus poisoned the seedlings" (30-31). Field trials would have proven dire:

> Pathogens flourished. Beneficial soil organisms died, including the ones whose presence is essential for the uptake of nutrients by plants ... [and] the genetically engineered *Klebsiella* proved to be highly competitive, persisting in the soil, causing changes in ecosystem structure and functions, and having substantial long-term effects. (Burrows 2001, 71)

Crouch (2001, 30-31) argues that such evidence suggests it may be perilous to depend exclusively on scientific evidence gathered through conventional safety assessment methods: "The EPA could easily have missed this danger if they had used their standard tests. A soil ecologist ... just happened to suggest a different experimental design, which caught the problem before the bacteria were let out of the laboratory."

GM animal applications have existed for some time in Canada. The University of Guelph's Enviropig™ is genetically engineered with mouse DNA and bacterial protein in order to "produce an enzyme to help it better digest plant phosphorous [sic], a vital nutrient in their feed ... which means there is less phosphorous [sic] – up to 60 percent less than ordinary pigs – in their waste" (Ogilvie 2008, A19). Cecil Forsberg, one of the engineering scientists, says that the Enviropig is designed to decrease the environmental footprint of industrial pig farms. Reducing phosphorus in pig waste means that there would be subsequently less phosphorus leaching "from pig manure, a major fertilizer source for farmers, into freshwater lakes and streams where it can trigger vast algal blooms and kill fish" (Ogilvie 2008, A19). The projected benefits of this endeavour are claimed to be significant, considering that in 2007 "Ontario farmers raised 3.9 million hogs, each producing a possible 450 kilograms of waste every six months" (Ogilvie 2008, A19). Scientists believe that the evidence suggests that the Enviropig is safe for human food consumption, "since the chemical analysis has shown the animal's tissue composition is the same as an ordinary Yorkshire pig, and the introduced bacterial protein is not found in any major food tissues, such as the ham, loin, heart and skin" (Ogilvie 2008, A19). Consumers may remain concerned, however, as body parts other than "major food tissues" are often found in processed food and animal feed, and what constitutes "major food tissues" may be culturally dependent. The novel food application is still pending, although the pigs have since been euthanized due to funding issues (Schmidt 2012).

Other developments occurring in the field of animal biotechnology include the production of biomedical products in animal blood or milk, and transgenic livestock or fish with "improved" production traits or genetic disease models (Wheeler 2007). Salmon, for example, have been genetically modified, using a combination of Chinook salmon and ocean pout DNA, to overexpress a growth hormone that accelerates the farming period for market size (sixteen to eighteen months instead of three years) (Schmidt 2011). The Massachusetts-based engineering firm AquaBounty Technologies claims that the GM salmon will grow year-round and "will make fish farming more efficient, a boon to producers and to consumers, who can continue to buy cheap salmon" (Ogilvie 2008, A19). CEO Ronald Stotish contends that "it is an opportunity that we have to take if we want to maintain our current quality of life" (Ogilvie 2008, A19). It is unclear what this "quality of life" is or why it merits preserving, but it is evident that such understandings in science and governance are based on normative values of economic progress. As Tokar (2001, 7) suggests, "in each instance, biotechnology helps perpetuate the myth that the inherent ecological limitations of a thoroughly nature-denying economic and social system can simply be engineered out of existence." Beth Burrows (2001, 73) asks: "Do we care if this technology feeds need or feeds greed as long as we can be in on the investment?"

In September 2008, the US Food and Drug Administration released draft guidelines on the regulation of transgenic animals for human consumption. The guidelines drew criticism for enabling the full-scale introduction of transgenic animal products, such as GM salmon, into supermarkets without sufficient procedural transparency or scientific assessment of environmental impact. In January 2009, final guidance was given to producers of genetically engineered animals to "help them meet their obligations and responsibilities under the law" (US FDA 2009b).[3] The FDA position is that there are no biologically relevant differences between GM and conventional salmon species (CBC 2010). Two 2012 FDA reports find the salmon safe for consumption (Moore 2013); it is believed that the GM salmon, after almost two decades, may finally be approved by the FDA (Ledford 2013).

AquaBounty also awaits regulatory approval in Canada. AquaBounty's position in both jurisdictions is that its product presents little risk to health or the environment. Sarah Schmidt reports, however, that federal scientists were in fact aware of the risk of GM fish, including contamination of wild species and harms that could befall Canadian fish stocks if the world's first genetically engineered salmon were approved for commercialization (Schmidt 2011). Schmidt uncovered risk information after obtaining internal records and correspondence, including a journal article by leading fisheries scientist Dr. Robert Devlin, attached to email correspondence acquired through an access to information request. Devlin's paper assessing the risks of GE fish, writes Schmidt, reported that "dispersal behaviour has been affected by introducing an outside gene into a fish, so GE fish may venture into habitat previously not used by wild fish" (Schmidt 2011). The scientific concerns regarding risk and regulation found in these records, she contends, contrast with media releases that proclaim the efficacy of the "regulatory framework for protecting the environment from potential risks of GE fish" (Schmidt 2011). Schmidt notes that, according to Michael Hanson, a scientist at the New York-based Consumers Union, "the real issue here is (DFO) [the Department of Fisheries and Oceans] are raising credible scientific issues because, frankly, the assessment that FDA did was scientifically completely inadequate." Or, as Lucy Sharratt, co-coordinator of the Canadian Biotechnology Action Network, suggests, "this could be a case of good scientists inside departments constrained by regulations" (Schmidt 2011).

AquaBounty relies on technology developed by scientists from Memorial University of Newfoundland (Schmidt 2011) and the University of Toronto to develop sterile, all-female eggs (Doucette 2013). Health Canada, along with DFO and Environment Canada, voiced concerns during early consultations about GM fish migrating back to affect Canadian fish stocks (Schmidt 2011). In 2013, however, Environment Canada approved AquaBounty's GM salmon under the Canadian Environmental Protection Act (CEPA), finding that the eggs were not harmful to the environment or human health as long as farming was contained (Doucette 2013). AquaBounty has a hatchery for egg production in Atlantic Canada and a second farm in Panama, where eggs are shipped to be grown into fish and processed for export (Doucette 2013; Moore 2013). Critics are concerned that the environmental safety regulatory determination was made without public consultation (Doucette 2013); the Canadian federal government now faces a lawsuit by various environmental groups for the decision (CBC 2014). In the interim, the eggs and fish will not be available for sale or public consumption until the novel food application for GM salmon has been approved by Health Canada and the US FDA (Doucette 2013). Although GM animal and fish products have yet to be approved, as of 2011 a total of 127 novel foods have been approved for sale in Canada; of these, 26 are products with "no history of safe use" (e.g., *Camelina sativa* oil); 5 are products made with novel processes (e.g., UV-treated apple cider), and 96 are GM products (e.g., Roundup Ready corn, Clearfield rice) (Bean 2011).

Canada's Novel Food Regulations

To be approved for the Canadian market, novel food must meet national safety standards. Health Canada (Health Products and Food Branch) is responsible for undertaking risk assessments of product safety under the Food and Drugs Act[4] and the 1999 Novel Food Regulations (Division 28 of the Food and Drug Regulations). Novel foods are defined by Health Canada as products that result from processes previously not used for food, that are without a history of safe use as food, or that have been genetically modified.

Part B, Division 28 (Novel Food Regulations), of the Canadian Food and Drug Regulations provides several definitions. To "genetically modify" means "to change the heritable traits of plant, animal or microorganisms by means of intentional manipulation." A "major change" in food is defined as:

> a change in the food that, based on the manufacturer's experience or generally accepted nutritional or food science theory, places the modified food outside the accepted limits of natural variations for that food with regard to (a) the composition, structure or nutritional quality of the food or its generally recognized physiological effects; (b) the manner in which the food is metabolized in the body; or (c) the microbiological safety, the chemical safety or the safe use of the food. (B.28.001)

Safety is assessed in each individual case through various prescribed steps before a GM crop can be sold in Canada as food, feed, or seed, and includes pre-market notification and fee payments to be made by firms to the Health Products and Food Branch of Health Canada. The regulations mandate that the firm or product developer submit a safety assessment data package to the Novel Foods Section to demonstrate the safety of the food. The information provided is to include a description of the particular genetic change; microbiological, toxicological, and chemical data on food safety, nutritional information, and the food's major and minor constituents; and any allergenicity considerations as well as unintended secondary effects (Health Canada 2006). Health Canada carries out a scientific assessment based on the submitted data to determine whether the food's safety can be verified. Firms are obliged to collect and provide the initial data because of its exorbitant costs, but Health Canada advises that the verification process, which includes chemical safety, nutritional, and microbial hazards assessment, provides adequate due diligence before novel foods are approved. GM food is considered safe if it demonstrates the same characteristics (such as nutrient combinations) as its conventional unmodified parent crop on the traits compared for safety assessment (University of Guelph Food Science Network 2012).

The substantial equivalence principle has been adopted by Health Canada as well as by regulators in the United States as a regulatory tool for determining GM food safety. Many GM products are intended for export, making convergence of national standards among trading partners attractive and compliance with international standards necessary. On the scientific evidence, GM products are considered safe under standards established by a number of authoritative international scientific bodies, such as the Codex Alimentarius Commission (Health Canada 2006b). Codex provides guidelines on conducting food safety assessments on foods derived from recombinant DNA plants and recombinant DNA animals. It operates on probabilities, however, as the guidelines indicate (Codex Alimentarius Commission 2003, para. 17):

> A variety of data and information are necessary to assess unintended effects because *no individual test can detect all possible unintended effects or identify, with certainty, those relevant to human health.* These data and information, when considered in total, provide *assurance* that the food is *unlikely* to have an adverse effect on human health. [emphasis added]

There may be no conclusive evidence of the toxicity of novel foods if the right metrics are omitted or when risk is assessed on a case-by-case basis. In terms of determinants of safety, "the choice of the one-in-a-million goal is a policy choice that cannot be determined by science"

(Epps 2008a, 158). Science may well be manipulated by various stakeholders and is subject to potential capture by vested interests (Epps 2008b, 379). The aggregate compounded effect of GM product consumption could escape detection in the Health Canada regulatory approval process, a process that some view as highly undemocratic and "an incrementalist approach that essentially removes the power to influence the *direction* of policy development from the political community" (Scott 2007, 18). Instead, we have "a discourse of cold calculation – of mundane, routine technical procedures for the assessment of risk and its weighing against an unstated economic payback" (Scott 2007, 19). The burden is on firms to make the case for safety, but "the comparison of the levels of key nutrients and toxicants in a GM variety with that of a traditional counterpart is limited to the presence of known components. This precludes the identification of unknown components of toxicological significance which might emerge in the modified food" (Health Canada 2002a). Health Canada's website indicates that no applications for genetically modified foods have ever been denied; applications that do not meet the criteria for approval are voluntarily withdrawn (Health Canada 2012b). This practice may be susceptible to criticism for skewing the optics of safety and safety assessment.

The precautionary principle, on the other hand, is commonly invoked in environmental discourse and is the prevalent approach used in the European Union, which has been able to enact more stringent regulations than Canada (see Chapters 1-4, and 8). Precaution with GMOs makes good scientific sense: a 2006 Central Science Laboratories report revealed various areas of potential long-term risks and benefits associated with novel foods (Henry 2006). The report stressed the need for longitudinal and statistically meaningful scientific studies on safety.

The regulatory safety assessment of novel foods in Canada is shared and coordinated with the Canadian Food Inspection Agency (CFIA). Whereas Health Canada is tasked with regulations, guidelines, and assessment of novel foods, the CFIA is charged with the assessment of associated environmental risks relating to GM seeds, feeds, and the cultivation of plants with novel traits (Health Canada 2006), import permissions, and registration of different GM products in Canada. CFIA, together with Environment Canada and Health Canada, commissioned the Royal Society of Canada (RSC) to investigate a series of questions relating to novel foods developed through GM techniques. The RSC

Expert Panel reported a lack of consistency in scientific findings regarding novel food safety. The panel found that "substantial equivalence," as a decision threshold to exempt GM products from risk assessment, is scientifically unjustifiable and inconsistent with the precautionary principle. As an "unambiguous" and "reasonably conservative" standard of safety, the substantial equivalence principle, it was found, is consistent with the precautionary standard (Royal Society of Canada 2001). Some critics posit that the Canadian government falls considerably short of implementing the panel's fifty-eight recommendations, including its emphasis on the importance of transparency and peer review of decisions, both of which could be given effect by making public the scientific and experimental data on which GM food and crop approval decisions are made (Andree and Sharratt 2004).

A literature review of risk by Epps (2008b) reveals more personalized accounts than the quantitative, scientific assessments of product safety used by governments. Epps notes (368):

> While consumers do not understand "risk" in the same way as scientists, their non-statistical, more qualitative, and heuristic-based risk perceptions are often arguably quite sensible. Pointing to the case of genetically modified foods, they suggest that when there are familiar and relatively risk-free alternatives to GM foods available, who is to say that consumer avoidance of GM foods is illogical?

Others are of the view that although "biotechnological innovations may entail risks … as with most such concerns, what is needed is more scientific knowledge (rather than ideological reckoning)" (Moschini 2008, 6).

If scientific knowledge means absolute evidence of actual risks, it is much more difficult to come by when the available data prove to be controversial (see, for example, the Seralini study discussed below), contested, or inconclusive and long-term data are not available. Scientists may be loath to come forward with research that contradicts the conventional wisdom on GM safety for fear of sharp criticism or loss of funding (Freedman 2013). Critiques may go beyond the particulars of a study (Arjo et al. 2013) to the integrity of the scientists, with even attempts to edit the scientific record with journals captured by commercial interests (Robinson and Latham 2013).

Political relations and trade agreements often prevail over true determinants and determinations of risk. The introduction of mandatory labelling and traceability standards in Canada, for example, may create significant practical, political, and economic challenges for trade and result in additional costs for consumers, farmers, and producers. The economic value of the biotech industry[5] makes the long-term consequences of genetic modification of food easy to ignore, particularly when risks are narrowly construed and difficult to assess or evidence. Eliminating risk, according to Scott, "is not an option – conventional approaches simply seek to 'manage, regulate and distribute risk'" (2008, 30).

Chemical Consumption and the Implications for Women's Health

A chemical understanding of DNA will help meet the demand for broadened conceptions of risk and transparency through mandatory labelling and will allow the public's qualitative perceptions, and not just the quantitative evidence espoused by scientific experts (Epps 2008; Scott 2005), to inform food consumption choices. The unknown risks of consuming DNA as chemicals and chemicals as (novel) foods affect women's health. Indeed, human health and welfare may be adversely affected by a loss of feelings of security (Chang 2004). As noted by Robert Howse (2000, 2337):

> If citizens believe they need a certain regulation, however "deluded" such a belief is, their utility will be reduced if they do not get it, in the sense that they will believe themselves exposed to a risk they believe to be significant ... part of the problem in these cases may be an absence of trust in the information and judgments that expert regulator/bureaucrats feed into the regulatory process, and part of the solution to distrust is "openness in government."

The mounting evidence of risk, moreover, validates the use of the precautionary principle as effective and sound science in light of the unknown risks of biotechnology, potential for allergenicity, and the interactive and transformative nature of DNA.

In a comprehensive review of the literature, Hilary Perr (2002, 482) identifies three basic health questions that arise from GM food consumption: (1) Does the insertion of novel genes into plants or animals alter the expression of other proteins, thereby increasing toxicity or decreasing nutritional value in the body? (2) Can GM plants transfer genes to bacteria in soil, animal digestive tracts, or the human gut that may, in turn, contribute to toxicity or enhanced disease potential in an individual? Finally, (3) Are there health concerns associated with the potential absorption and incorporation of GM DNA into the body, such that an individual might experience increased risk of disease? Increased risk to women stems in part from the focus on women as targeted consumers.

Rice, for example, has been genetically enriched with vitamin A and iron to help prevent illness and disease related to vitamin deficiencies. Vitamin A deficiency is responsible for blindness, particularly in children in developing countries. Iron deficiency affects 5-10 percent of women of childbearing age, and the resulting "low birth weight infants of iron-deficient mothers are at increased risk of hemorrhage, sepsis [illness that develops from reaction to bacteria], and death during childbirth," while "postnatal iron deficiency ... impairs subsequent cognitive development and immune function" (Perr 2002, 476). Iron enrichment strategies focus on optimizing the uptake of iron by increasing its content and improving its absorption in the body (477). Various foreign genes have been introduced into the rice plant to enhance vitamin A and iron uptake, and are in the process of being transferred to commercial varieties.[6] Perr (2002) posits that the rice will be internationally distributed once possible risks to human health and the environment are assessed. By current accounts, however, "no country has definite plans to grow Golden Rice" (Freedman 2013, 82). Cellular biologist David Williams specializes in vision and warns against the impulse to naïvely forge ahead: "Anyone in the field knows the genome is not a static environment. Inserted genes can be transformed by several different means, and it can happen generations later" (Freedman 2013, 80). It is far from clear how genetic manipulation affects plant metabolism and other gene functions, how hybrid products are metabolized in humans, or how human health is impacted in the long term.

Perr's review (2002) examines the use of GM foods to address human morbidity or mortality through edible vaccines in development, including: "(1) food vaccines ingested by mothers to passively immunize the fetus via the placenta or via breast milk, (2) oral delivery of 'autoantigen' to suppress automimmune diseases, and (3) oral vaccines for *H. pylori*" (482). In each instance, human proteins are incorporated into common fruits or vegetables, such as bananas, potatoes, and tomatoes, to achieve the intended effect (Macdonald Glenn 2004).

Beyond the use of women as conduits of fetal "inoculation" through such vaccines, GM foods under investigation seek to directly target the particular nutritional needs of pregnant women. The "biofortification" of tomatoes, for example, promises to increase folate levels to twenty-five times that of regular tomatoes. Promotional materials emphasize folate deficiency in early pregnancy as linked to increased risk of neural tube defects, such as spina bifida (the incomplete development of the spinal cord) and anencephaly (the incomplete development of the brain or skull), in infants (Daniells 2007):

> Tomatoes that expressed both genes were found to accumulate about 840 micrograms of folate per 100 grams of fruit, enough to provide the RDA for a pregnant women [sic] *"in less than one standard serving,"* said the researchers. *"To our knowledge, the folate levels we achieved are the highest reported for plants; our tomatoes accumulate up to seven times more folates than green leafy vegetables, which are considered rich folate sources,"* they said. The researchers also note that the concentrations of the folate precursors *are likely at safe* consumption levels, but stated that more research is required to fully assess the potential and safety of accumulated pteridines in GM fruit and plants.

Over a decade ago, before the mass proliferation of agri-biotech foods, Millstone, Brunner, and Mayer (1999) proposed that an obvious solution would be

> for legislators to have treated GM foods in the same way as novel chemical compounds, such as pharmaceuticals, pesticides and food additives, and to have required companies to conduct a range of toxicological tests, the evidence from which could be used to set "acceptable daily intakes" (ADIs). Regulations could have been introduced to ensure that ADIs are never, or rarely exceeded.

This chapter renews the call to treat novel foods as novel chemical compounds and substances. New conceptual understandings are helpful and probative of the gendered and long-term impact of novel food consumption as chemical consumption. Long-term health concerns have also been expressed over the presence of antibiotic resistance in some GM plants (Conway 2000, 8),[7] because there is the potential for rapid transfer of genetic resistance to other life as well as the potential to contribute to the growing trend of antibiotic resistance in humans (Whitney, Maltby, and Carr 2004, 264).[8] Assessing the risks related to the transfer and interactivity of a variety of foreign genes is difficult but may be significant, as the cauliflower mosaic virus (CaMV) example suggests. CaMV is a vital component in the gene transfer of herbicides and promotes disease resistance in modified plant sources. As a close relative of HIV, human leukemia virus, and hepatitis B, CaMV recombines with existing plant viruses in insects and creates new and more virulent viruses (Whitney, Maltby, and Carr 2004, 263). Between 1991 and 1996, four studies revealed that plant viruses can acquire a variety of viral genes from transgenic plants (Wintermantel and Schoelz 1996). CaMV recombines with the transgenic host plant to acquire copies of such genes. The African cassava mosaic virus, which has caused the devastation of cassava crops in East Africa, has also developed from natural recombination. These examples demonstrate that "plants that are modified to contain genes from viral pathogens of crops might exchange these genes with other viral pathogens, creating entirely new viral strains with unknown properties" (Conway 2000, 6). The risks of new biological characteristics arising from "gene mixing" (Tepfer 1993) are unknown, and cannot be measured or managed.

There is evidence that a number of autoimmune diseases may be enhanced by foreign DNA fragments not fully digested in the human stomach and intestines (Strauss 2006). Women are generally at higher risk of developing autoimmune diseases than men. They have more vigorous immune responses to infections and are thus more "pro-inflammatory"; there may also be immune modulating effects of sex hormones or important differences in sex chromosomes (McCarthy 2000, 1088). Of the 8.5 million people with autoimmune diseases in the United States, women are reported to make up 80 percent (McCarthy 2000).

Perr (2002) admits that safety assessment criteria and procedures are continually being adapted to identify unintentional consequences of genetic modification, particularly given the uncertainty of where a foreign gene will integrate into a host plant genome, or its possible disturbance to the functionality of other genes in the host. Genetic instability may result from changes to gene expression and activation, or from the development of antinutritionals, "acknowledged as essentially compounds that result in suboptimal nutrient utilization" (Perr

2002). To answer the question of whether novel genes increase toxicity or decrease nutritional value, Perr adopts the position put forward by an expert panel: "Taking into account the natural variations of DNA sequences, the present use of recombinant techniques in the food chain does not introduce changes in the chemical characteristics of the DNA." In the biopatent field, such a view is disputed by the same stakeholders who would rely on it for the purpose of asserting safety. If DNA is chemical and simple severance of a bond sufficient chemical change in the sequence to warrant a patent, introducing a foreign gene via recombinant molecular manipulation should be understood as a chemical change of relevance for risk, however negligible the shift in biology.

In terms of whether gene transfer might occur in an organism and cause toxicity or pathogenicity, Perr (2002, 483) posits that "the transfer and functional integration of ingested plant DNA into gut microflora or human cells appears unlikely, but is still under investigation." And, on whether there are health concerns related to absorbing DNA in food, she concludes from a review of data on the human ingestion of GM plants that there is no "increased risk of human disease or risk of incorporation of novel genes into the human genome"(483). In particular, Perr refers to a comprehensive position paper, the "Novel Foods Task Force ILSI Europe Workshop on Safety Considerations of DNA in Food," which reviews the structure, function, presence, and stability of nucleic acids in food, and the safety of foreign DNA in food. Taking these findings into account, she endorses novel foods on the basis that "no human hazard has yet been identified," but also points to the need to "identify and monitor beneficial and deleterious proteins in novel food plants, regardless of technology," particularly for their long-term effects on children, as it is they who will have the greatest exposure and thus have the most to benefit (483–84).

Health Canada (2012a) acknowledges that there are risks to novel food consumption:

> Potential hazards remain those associated with toxic or allergenic compounds which are inherently present already in the food supply. However, unlike traditional breeding, techniques such as recombinant DNA technology permit the transfer of genetic material form [sic] unrelated species and this is precisely why a safety assessment is considered to be necessary. Similarly, a gene may be transferred from an organism expressing a protein that has

> no history of use as a food. Safety assessment provides assurances that toxic and or [sic] allergenic compounds are not transferred along with the desired trait when new DNA is introduced into an organism.

Steve Taylor and Susan Hefle (2001) claim that GM foods will not be allergenic as there will be no transfer of genes from known allergens. This position is a practical one rather than a scientific determination of GMO allergenicity per se. Their reasoning is that scientists will practise due diligence in identifying allergenic sources and guarding against their transfer, a view also held by Health Canada. They therefore believe that the allergenic potential of GMOs will be successfully contained, if not eliminated. According to Health Canada, the strategy has been successful. In one case, DNA coding for storage protein was transferred from a Brazil nut to soybean to improve the quality and durability of soybean meal as animal feed (Health Canada 2012a):

> Since the Brazil nut is a known cause of allergic reaction for a small number of sensitive individuals, laboratory tests using sera from Brazil nut-sensitive individuals were conducted in order to determine whether an allergenic protein had been transferred to the soybean. The results of the laboratory tests showed that the gene obtained from the Brazil nut likely encoded the major Brazil nut allergen and research on this product discontinued.

Health Canada (2012a) states that "the product was never commercially developed and soybeans containing a Brazil nut protein were not available on the market."

Health Canada's approach of assessing food safety based on known risks and substantial equivalence may be criticized, as it leaves open the question of whether the transfer of genes from unknown allergenic sources and/or recombinant DNA may induce allergenic or toxic properties in resulting GMOs. It also leaves a significant margin for error in the identification and elimination process. Scientists consider there to be an inherent challenge in identifying these allergens and individuals sensitive to them, stating that "our challenge scientifically is how to assess novel proteins that have little to no exposure in the general population and thus no readily available tools for the prediction of exposures" (Bernstein et al. 2003, 1120). Subsequent determination of toxicity or allergenicity based on good scientific evidence may prove too late, as evidenced by the fatalities from Eosinophilia-Myalgia Syndrome (EMS). EMS, a fatal flu-like condition linked to a tryptophan food supplement derived from GM bacteria designed to overproduce tryptophan, was contracted by some 5,000 people in the United States. Those affected suffered painful swollen joints and muscles from toxic accumulation; 37 died and over 1,500 were left permanently disabled (Whitney, Maltby, and Carr 2004, 263). Health Canada observes that "in addition to changing the production organism there were also changes made in the recovery and purification steps; specifically those steps involved in removing impurities," and notes that "the toxic metabolite was also produced by natural or non-genetically modified strains of bacteria" to support the conclusion that "it was most likely the change in the purification step that allowed the toxic metabolite to contaminate the tryptophan" (Health Canada 2012a). Consumers may be less comfortable than regulators, however, with informed speculation of risks when the outcome is severe or fatal.

Health and environmental risks have also been associated with unknown consequences of newly developed chemicals, such as Bt insecticide-producing potatoes, corn, and tomatoes. Bt crops have generated significant public rally, as demonstrated by the StarLink™ corn example, which highlights the limitations of post-market release surveillance (Bernstein et al. 2003, 1120). Perr explains the source of controversy (2002, 481):

> Corn borers are insect larvae which cost farmers nearly one billion dollars annually in destroyed corn crops. Starlink corn contains a foreign gene derived from the soil bacterium, *Bacillus thuringiensis.* The gene encodes a Bt toxin called cryc9c, which confers insect resistance. Bt toxin naturally exists in hundreds of forms with varying insecticidal properties. Bt toxin is harmless to humans and has been used in sprays and powders for decades. Safety assessment of cryc9c did not reveal allergenicity of the gene source and there was no homology with known allergens. However, the protein was resistant to digestion. While characteristic of some allergens, digestive stability is not fully predictive of allergenicity. Nonetheless, approval was restricted to use in animals and delayed for human consumption. Limited planting ensued from 1998 to 2000. Faulty segregation of harvested corn led to the presence of StarLink corn in human food products such as taco shells, which were subsequently recalled.

A joint report by Indigenous communities, peasant farmers, and other interested parties including Canada's Action Group on Erosion, Technology and Concentration (ETC Group), on transgene contamination in Mexico describes a number of government studies performed over a two-year period to determine whether transgenes were present in traditional corn crops. Government officials and the scientific community have since acknowledged the impossibility of segregating traditional crops from GM varieties, despite a government ban on GM seed planting (ETC 2003):

> Of 2,000 maize plants tested, samples from 33 communities in nine Mexican states tested positive for contamination. In some cases as many as four GM traits, all patented by multinational Gene Giants, were found in a single plant. Traces of insecticidal toxin (Cry9c), an engineered trait found in StarLink maize formerly sold by Aventis CropScience, were also detected causing alarm because the US government had never approved StarLink for human consumption due to possibilities of increased allergic reaction.

Despite the recall of StarLink products from the US market, the broader issue of GM contamination remains a global problem that has, according to the ETC Group (2003), affected cotton in Greece, canola (rapeseed) in Canada, soy in Italy, and papaya in Hawaii. StarLink™ was approved in the United States as animal feed but not for human consumption; it has "never been approved for use as human food, animal feed or environmental release in Canada" (Health Canada 2012a).

Health Canada considers Bt potatoes safe for human consumption. "The introduction of genetic information (DNA) into these potatoes to make them resistant to CPB [Colorado Potato Beetle] does not result in any differences in the composition or nutritional quality of the potatoes," and therefore these are "as safe and nutritious as other commercially available potato varieties" (Health Canada 1997). Health Canada's position is that even though Bt proteins in plants are toxic to certain species of insects, they are digestible like other proteins in the human digestive system and so remain safe for human consumption, rendering the Bt crop nutritionally and substantially equivalent to their unmodified counterparts (Health Canada 2012a). Health Canada also says that it is "unaware" of any reports linking adverse human health effects to consumption of Bt proteins, and that Bt has been used safely "for more than 30 years to control insect pests by home gardeners, organic growers and other farmers" as part of active insecticidal agents in certain commercial pest control products. All of this is, apparently, good evidence of safety.

Compare this with the account set out in F. William Engdahl's exposé *Seeds of Destruction: The Hidden Agenda of Genetic Manipulation* (2007), which outlines the risks associated with the invasion of the common food supply by GM seeds. Engdahl describes the experience of Gottfried Glockner, a German farmer who planted Bt176 Syngenta corn as feed for his cows. Glockner gradually increased the amount of GM feed from 10 to 100 percent. Side effects appeared only after the full switch to GM feed: the cattle suffered "gluey-white feces and violent diarrhea." (231) Some cattle stopped producing milk, others produced blood-contaminated milk, and nearly all seventy cows eventually died. Glockner obtained independent scientific analysis that verified, apparently contrary to Syngenta's claim, that the toxins found in Glockner's feed sample were "active" and "extremely stable" in form (231-32). A recent study of rats fed 30 percent GM insect-resistant corn (produced by incorporating Monsanto's borer-resistant trait) found "several changes in organs/body weight and serum biochemistry," indicating "potential adverse health/toxic effects of GM corn and further investigations still needed" (Gab-Alla et al. 2012, 1122). Bt crops are designed for their chemical property to kill pests; toxicity to mammals may well prove to be dose and exposure-dependent.

Chemical Add-Ons to Chemical Foods

Health Canada's list of approved novel foods contains many "ready" varieties. These are GM plants that have been designed to survive chemical sprays designed to kill. Roundup Ready® Canola (RRC), for example, is engineered to express a particular gene that makes the plant resistant to Monsanto's glyphosate-based herbicide Roundup (the chemical "add-on"). Roundup, when sprayed, will kill all other plants but the "ready" variety, making it easier to control weeds. In exchange, farmers agree not to save seeds from one year's crop for future replanting or inventory, and not to share (sell or give) the seed to third parties. "Sudden Death Syndrome" has been observed in fields with substantial glyphosate use, because this chemical not only functions as a herbicide but is also a biocide:

> One of the biocides linked to NHL [non-Hodgkin's lymphoma] by the Hardell study is Glyphosate. A previous study in 1998 had implicated Glyphosate to hairy cell leukemia. Several animal studies have shown that Glyphosate can cause gene mutations and chromosomal aberrations. Denmark banned Glyphosate in September 2003 because it was so persistent that it polluted most of the water table. The response of many regulatory authorities is to ensure that use of Glyphosate is increased substantially around the world with the approval of "Roundup Ready" genetically modified crops. (Leu 2011)

The Canola Council of Canada (CCC) reports that there are three main groups of herbicide-resistant canola: Roundup Ready and Liberty Link varieties, produced using genetic modification, and Clearfield varieties, developed using mutagenesis, a traditional plant breeding technique. "GM or transgenic canola," it advises, "burst on the scene in 1995 and the acreage rose rapidly. In 2004, transgenic Roundup Ready and Liberty Link varieties were grown on 75 percent of the acres, while Clearfield varieties were on 18 percent and conventional on seven percent of the acres" (Canola Council of Canada 2005). The CCC (2005) claims that farmers benefit substantially from growing GM crops, with higher price per yield, lower transaction costs, and less chemical use:

> Transgenic canola growers used less fuel because of fewer field operations, including tillage, harrowing, fertilizing and less summerfallow. Fuel saved by transgenic canola growers totalled 3.12 million litres in 2000, a savings of over $13 million. Transgenic growers also used less herbicide than conventional growers. Total herbicide use was cut by 6,000 tonnes of total product in 2000.

Canada is one of four countries (with the United States, Brazil, and Argentina) responsible for 90 percent of the global GM crop production (Freedman 2013). The CCC (2005) touts the safety of its products for consumption:

> It is important to note that the transgenic gene inserted into the canola plant to produce Roundup Ready and Liberty Link InVigor herbicide resistance is a protein. All protein is removed

> from canola oil during processing. Therefore, canola oil contains no GM material and is identical to canola oil from a non-GM canola plant.[9]

Ferrara and Dorsey (2001, 61) observe, however, that "plants can take up glyphosate into parts used for food." Despite precautions taken by companies, glyphosate has been found in "strawberries, wild blueberries and raspberries, lettuce, carrots and barley after crops or wild areas have been sprayed. The herbicide has been found in food plants a year after fields have been sprayed" (61). The impacts may be far more serious than industry admits:

> When glyphosate is used before harvest on grain crops to dry out the grain, it can leave residues in the grain. Glyphosate can contaminate surface water when the soil particles it binds to are washed into rivers, streams, lakes and ponds. The herbicide kills beneficial insects and worms, and has indirect hazardous effects on birds and small mammals ... [and is] acutely toxic to fish. (Ferrara and Dorsey 2001, 61)

In their 2004 article "'This Food May Contain ...' What Nurses Should Know about Genetically Engineered Foods," Whitney, Maltby, and Carr draw attention to Monsanto products. They claim that elevated levels of phytoestrogens, a type of estrogen normally found in minute quantities in plants, have been discovered in Roundup Ready Soybeans (RRS). Increased consumption of estrogen has been associated with, among other things, breast cancer in women. Although Monsanto asserts that its products are safe, Ferrara and Dorsey (2001, 61) report, more generally, that

> studies show that glyphosate causes long-term toxic and reproductive effects, and glyphosate-containing products cause genetic damage. Though the EPA has deemed it to be a non-carcinogen, the Oregon-based Northwest Coalition for Alternatives to Pesticides has analysed studies for glyphosate and determined that carcinogenicity ... remains unknown pending more long-term studies.

The authors note the evidence of acute toxicity derived from a 1993 Berkeley study, which found that "glyphosate was the most common cause of pesticide-related illness among landscape maintenance workers in California, and the third leading cause among agricultural workers" (61).

Another high-profile and controversial study (Séralini et al. 2012) looked at the health effects of a Roundup-tolerant GM maize (from 11 percent of the diet), cultivated with or without Roundup, and Roundup in rats over a two-year period. GM-fed rats had higher cancer rates. Among females, all treated groups had two to three times greater mortality than controls, with death occurring more rapidly (4,223). Females were more sensitive to Roundup in the water supply. Mammary tumours, disabled pituitary organs, and sex hormonal balance were affected by GMO and Roundup treatments. The authors note that "currently, no regulatory authority requests mandatory chronic animal feeding studies to be performed for edible GMOs and formulated pesticides" (4,221). Some scientists discredit the findings of this study, taking issue with the "many errors and inaccuracies" as well as the methodology (Arjo et al. 2013), while others caution against the quick dismissal of such remarkable results (Freedman 2013, 85). The position of Health Canada and the CFIA (Health Canada 2012b) is that the methodology was inadequately described, the data were neither fully nor transparently presented, and the conclusions regarding the long-term safety of Roundup Ready maize and glyphosate were unsupported by their analysis. While the debate over the safety of "ready" varieties rages on, "more than 90 percent of US soybean farms use Monsanto seeds" (Holland 2013). Soybeans, of course, make up a good portion of go-to health choices, such as tofu and soymilk.

Many proponents of GM foods assert that engineered foods hold the promise of improving the world's food supply and promoting food security. The most widely commercialized are not the enriched GM seeds, however, but those with built-in chemicals and resistant varieties (Lappé and Lappé 2013). The economic motivations are transparent: the expansion of intellectual property rights has served to consolidate food, drug, and agro-chemical markets (Drahos and Braithwaite 2002; Tansey and Rajotte 2008):

> With recurring waves of new mergers in the pharmaceutical and agricultural biotech industries, the concentrations of corporate power in these areas has grown to truly staggering proportions. By 1999, five companies – AstraZeneca, Dupont (owner of Pioneer Hi-Bred, the world's largest seed company), Monsanto,

> Novartis and Aventis – controlled 60 percent of the global pesticide market, 23 percent of the commercial seed market and nearly all of the world's genetically modified seeds. Aventis – formed from the merger of the chemical giants Hoechst and Rhone [sic] Poulenc – was also the world's largest pharmaceutical company, and Novartis and AstraZeneca were numbers four and five, respectively, in global pharmaceutical sales. (Tokar 2001, 8)

As Tokar (2001) explains, this concentration of control is "significantly driven by developments in biotechnology, while at the same time profits from the sale of herbicides and other chemicals are channelled towards the development of new genetically modified life forms" (8).

Transgenic "ready" variety crops are often defended for contributing to a reduction in traditional agrochemical use. Indeed, the global proliferation of GM crops (James 2012) would not have been possible without farmer buy-in to the benefits of agricultural innovation. Yet, as Richard Steinbrecher (2001, 81) highlights, "the driving force behind herbicide-tolerant crops is the agrochemical industry." Insofar as some crops are engineered to contain their own chemicals (e.g., Bt), they may in fact reduce farmer hospitalization from toxic poisoning due to chemical pesticide applications (CSPI 2012, 8), but any argument over the pesticide efficiency of GMOs in the aggregate may be misleading by virtue of the fact that pesticides may be built-in. Moreover, some claim, GM "ready" varieties could "potentially double or triple the residual pesticides found in food and water supplies" (Whitney, Maltby, and Carr 2004, 264), particularly since certain weeds become resistant to chemicals with time. Ryegrass *(Lolium rigidum),* Australia's most common weed, developed a resistance to Roundup and "was tolerant to nearly five times the recommended spraying dose" (Steinbrecher 2001, 80). There is evidence also of farmer non-compliance with planting restrictions and refuge requirements established by the US Environmental Protection Agency that will in time contribute to insect resistance and loss of efficacy of Bt crops and Bt microbial insecticides (CSPI 2012, 11). Post-market surveillance for health consequences, moreover, are moot if effective crop segregation cannot or is not maintained. The profitability and proliferation of GM products and the corporate resistance to mandatory labelling suggest that more energy is being invested in research and development of agrochemicals and agrofoods, with a clear interest in promoting greater reliance on both.

Recommendations for Law and Policy Reforms

According to a 2006 Farm Women and Canadian Agricultural Policy Report, initiated by the National Farmers Union and funded by Status of Women Canada, women have thus far been left out of agricultural policy, and will need to better assert their position as producers and consumers of food in order to participate more fully in constructing regulatory regimes responsive to their specific needs and health concerns. Much evidence has accumulated demonstrating the gendered effects of biotechnology. Around the world, women farmers disproportionately suffer the negative health outcomes associated with chemical exposure and pesticide use (Chapter 11).

A report for the Working Group on Women, Health and the New Genetics (2000) asserts that in order "to sustain the pressure towards maintaining public health and safety as the highest standard, lobbying efforts must be continuous at all levels of government involved in this issue" (Ford 2000, 10). The report draws on past experience with Monsanto's recombinant bovine growth hormone (rBGH), which was used to increase milk production in cows. Proponents of rBGH argued that cows injected with the hormone produce 10-25 percent more milk (Ford 2000, 11). Monsanto has consistently asserted that rBGH is safe (Powell and Leiss 1997, 151) and the US Food and Drug Administration agrees (FDA 2009c). Opponents, on the other hand, have drawn attention to scientific evidence that demonstrates adverse health effects in both humans and cows from rBGH consumption: cows given rBGH more commonly suffered inflammation in their udders, which required antibiotics treatment (Ford 2000, 11). These antibiotics are in turn transferable to milk, and milk consumption by humans could lead, with exposure, to cases of antibiotic resistance. Women and children would be particularly adversely affected as they tend to drink more milk and are targets of consumer marketing campaigns that focus on bone strengtening, osteoporosis prevention, and improved/increased lactation (Ford 2000, 11). Monsanto continued to market rBGH despite scientists' warnings that rBGH elevated insulin-like growth factor (IGF-1) protein levels in humans, which was linked to an increase in the risk of developing breast, prostate, and colon cancer (Powell and Leiss 1997, Chapter 6). In Canada, rBGH has not been approved for use, a decision that was said to follow rather controversial and extensive "whistleblowing" action by six high-profile senior scientists at Health Canada (Ford 2000, 11). The scientists accused senior bureaucrats at

Health Canada of showing "unusual favour" to Monsanto (Ford 2000, 11). In response to allegations of bribes offered to Health Canada that were raised in media, "Monsanto Canada insisted there was 'never an offer of money in exchange for product approval ... we proposed to carry out additional meaningful research to complement existing data and to add to the worldwide body of knowledge on BST [bovine somatotropin, or bovine growth hormone]'" (Powell and Leiss 1997, 139). Although Health Canada approved rBGH as safe for human consumption as early as 1986 (Powell and Leiss 1997, 137), after the Canadian Senate's agriculture committee investigated the issue, Health Canada chose to ban rBGH (Ford 2000). Health Canada's position is that concern over animal health was the basis of its decision (Health Canada 2012c). For some, the outcome was more political and showed poor risk communication as well as a greater need for transparency in regulatory decision making (Powell and Leiss 1997, Chapter 6); it also showed that "many consumers have a much broader definition of risk" than Monsanto (Powell and Leiss 1997, 151). Indeed, "the exact pattern of consumer outrage in response to not being explicitly told a product was in the food supply would be repeated six years later after field-test crops of Monsanto's New Leaf potato, genetically engineered with Bt-gene ... was released without notice" (Powell and Leiss 1997, 135).

One might expect that as health-associated risks of GMOs appear, the credibility and authority of experts within scientific institutions and regulatory bodies will become further eroded (Hiskes 1998). There appears to be a fundamental difference between Bt found in plants or as part of chemicals applied to food that may be washed, scrubbed, or peeled, and Bt engineered into food. As for transparency, Health Canada is launching a pilot project under the Regulatory Transparency and Openness Framework to make summaries of novel food safety assessments available online – but will this be enough?

Women have always played a significant role in food production and family consumption choices, as caregivers, health workers, and providers of nutrition (Howard 2003; Chapter 2). Women have also been integral to domestic family farming and have made significant contributions to the economy (Boserup 1970; Brophy et al. 2002; Chapter 11), even as their role as contributors to the political economy – and engagement as constituents – has been limited because of gender inequality. Perhaps for this reason, a report prepared for the Working Group on Women, Health and the New Genetics (2001) found that

> polls have shown that women are more suspicious of biotechnology than men (Environics 2000). Yet, women are treated as a group to be managed through public relations, rather than major stakeholders to be consulted. The federal government's awareness that *gender matters* in biotechnology policy has not been directed to evaluating the different impacts which biotechnology and the CBS might have on women and men, as directed by federal policies.

Women's relationship with food has a long history and reveals the unique service women provide their families in managing nutrition and health needs through food. Women also have a long-standing role as traditional breeders of new plant varieties, which they have adapted to meet particular cultural needs and the demands of local agricultural and environmental conditions (Howard 2003). Women continue to act as stewards of biodiversity in a relationship of care that extends from their families to the sustainable use of crops. Anne Rochon Ford (2000, 4) notes that "part of women's role as gatekeepers of health in the home involves growing, selecting, purchasing and preparing food. Therefore, genetically engineered foods, their labelling and patenting are issues of particular concern to women, as are the safety of the seeds from which these foods derive."

While consuming chemicals may be generally harmful, health impacts are gendered due to a woman's unique body composition (See Introduction and Chapters 1, 3, 4, 7, and 8). GM products are metabolized differently and can produce chemical reactions (DNA mutations, gene expression, etc.) in women's bodies that are compounded by women's greater exposure, as a result of their being targeted consumers. The substantial equivalence metrics for assessing safety is flawed. Like Snow White's apple,[10] GM products may look, taste, and even be nutritionally equivalent to traditional GM-free counterparts, but, as we all know, the apple was fatally toxic. Law and policy reform should reflect the view that with GM foods, the chemical difference is meaningful; it is what makes the difference.

The scientific evidence currently available is enough to support the adoption of the precautionary principle in novel food regulation. The definition of "genetically modify" in the Novel Food Regulations should be amended to adopt the conceptual understanding of DNA as chemicals and to include any recombinant molecular manipulation that results in chemical change. Our understanding of what is "chemical" should be expanded to include genetic changes in novel food and crop events. The Novel Food Regulations should adopt some of the measures available in intellectual property governance regimes and include a mechanism for third-party filings in relation to pending applications. Provision should be made for filing of a "protest" or relevant evidence – as allowed for prior art in patent applications – for consideration in the approval process. Formal opposition proceedings should also be provided for; they are allowed for trademark registration applications, so why not for novel food? There should be freedom to access research and research results (CWHN 2001); this would facilitate both the filing and opposition processes. Approval should be subject to subsequent regulatory re-examination on request by third parties (as is the case on filing a request with the Patent Commissioner) and impeachment before a court by "any interested person."

Labelling should be mandatory. If women are the primary consumers of food in the market, labelling would enable them to make more informed choices. It would also enhance transparency and raise awareness about the risks and benefits of biotechnology (Baker and Burnham 2001). Although there is some provision for labelling where the GM food is not substantially equivalent,[11] mandatory labelling has to do with overcoming knowledge asymmetries regarding what the novel food is: it is chemicals, and novel enough to be patentable. In Canada, labelling is mandatory for foods where health or safety is a concern, such as foods that might possess allergens, under subsection 5(1) of the Food and Drugs Act and section 7 of the Consumer Packaging and Labelling Act[12] (Canadian General Standards Board 2004). The Canadian Food Inspection Agency is responsible for protecting consumer interests; for ensuring that labelling, advertising, and packaging are neither misleading nor fraudulent; and for prescribing labelling and advertising standards – but these do not encompass toxic substances in food.

In Canada, several attempts made through Private Member's Bills to introduce mandatory GM labelling legislation in Parliament (e.g., Bills C-287, C-310, and C-517) have failed.[13] The latest attempt, Bill C-257 (which passed Introduction and First Reading in the House of Commons in 2011), was reinstated in the second session of Parliament (October 2013) and is still pending (Parliament of Canada, LegisInfo).[14] In 2004, the Standards Council of Canada adopted a voluntary standard for labelling biotech foods, but it has been criticized for having inadequate threshold limits for defining a GMO compared with more stringent international standards. Mandatory labelling should be based on fair information practices providing consumers with "honest, objective, value-free and non-disparaging information" to permit informed food choices (CSPI 2001). Extensive public consultation will help ensure that standards are clear and credible. Consumers should understand, for example, that "genetically modified" captures only "bioengineering" (CSPI 2001). Without mandatory labelling, the public's greatest capital asset – market demand – is diminished, silencing consumer voices, limiting their choices, and undermining public values – all of this, it has been argued, with little to no return in consumer benefit (Leiss 2003). Extending labelling to GM foods would not only acknowledge and validate consumer health and safety concerns but would also be consistent with Health Canada's mandate to provide oversight for food labelling in the interests of greater protection of consumer health and safety. For now, the current regime asks Canadians to trust government "experts with a highly circumscribed set of skills and knowledge," thereby "discounting and silencing ... alternative knowledges" (Scott 2008, 7).

In the United States, the FDA released a draft guidance report for voluntary labelling in 2001 (US FDA 2001) and continues to maintain the position first set out in its 1992 "Statement of Policy: Foods Derived from New Plant Varieties," namely, that there is "no basis for concluding that bioengineered foods differ from other foods in any meaningful or uniform way, or that, as a class, foods developed by the new techniques present any different or greater safety concern than foods developed by traditional plant breeding" (US FDA 2001). The FDA notes that if the food is "significantly different" from its traditional counterpart, then it would have to be reflected in the label and the food may have to be renamed (US FDA 2001). This information has been criticized for its generality and lack of definition, as there is no indication of what constitutes significant difference, nutritionally or otherwise; comments to the FDA caution against mandating GM labelling, suggesting that such labels might mislead or deceive consumers into perceiving GM foods as inferior in quality, and could ultimately lead to price gouging (rising prices) in the marketplace (CSPI 2001). On 23 May 2013, the US Senate voted 71-27 against a measure that would let states decide whether to impose mandatory labelling (McAuliff 2013). Just a year earlier, a similar GM labelling amendment proposed by the same senator to the Agriculture Reform, Food, and Jobs Act of 2012 (S. 3240) was defeated 73-26; it would have permitted individual states to require GM labelling on food, beverages, and other edible products.

On 8 May 2014, Vermont became the first state to legislate mandatory labelling. The law is set to take effect on 1 July 2016 and is expected to be challenged in court before then by food and agricultural companies (Gillam 2014b). At least twenty states are currently considering GM labelling bills (Freedman 2013, 82). Such state efforts are discouraged by the threat of litigation for pre-empting federal authority (McAuliff 2013), although some speculate that a federal standard may become attractive compared with the prospect of multi-state variation in labelling laws.

In the interim, lack of responsiveness by federal US regulators to consumer demands for mandatory GMO labelling has resulted in the creation of a citizen action movement known as the Label it Yourself Campaign. Sewell (2012) reports that "with an estimated 80 percent of processed food in the USA containing GMOs, Label It Yourself is asking: 'If there's nothing to hide, why hide it?'" Mandatory labelling is indeed consistent with values of market freedom (Leiss 2003; Pardy 2013). Food Democracy Now (FDN) is an organization that acts as educators and activists for "food democracy," working to bring public attention to the actions of agrochemical firms such as Monsanto, for transgenic pollution, soil and crop contamination, and potentially irreversible compromise to our future food supply.

There is considerable consumer demand for mandatory GM labelling. A vast majority of Americans support GM labels: in a 2010 Thompson Reuters NPR Health Care Service Poll, 93 percent of respondents said that GM food should be labelled; in some polls, the figure is as high as 96 percent (CFS 2013). The position has found renewed support among a few lawmakers and more than 200 food companies, organic farming groups, health and environmental organizations, and other interested stakeholders in the United States (Gillam 2014a). A significant divide in the food industry is emerging and has been invigorated by the well-established public debate over the demand for mandatory labelling. Growing support among well-established companies such as Ben and Jerry's, cereal maker Nature's Path, and organic yogurt maker Stonyfield Farm, along with the Consumer Federation of America (Gillam 2014a), points to the success of collective efforts to shift the discourse away from scientific risk determinations of safety and their metrics to the consumer's right to make informed food choices. Companies responsive to such demands will likely generate considerable good will, not because they are GM-free per se but because they support food democracy. On the other hand, the Grocery Manufacturer's Association, representing more than 300 food companies, has mobilized to pre-empt any state mandatory labelling laws, even pushing for the ability to adopt the label "natural" for some GMOs (Gillam 2014a). Meanwhile, there are reportedly over a hundred class action lawsuits now filed "across the US, charging major food corporations with labelling fraud for labelling or marketing GMO-tainted or chemically processed foods and cooking oils as 'natural' or 'all natural'" (Cummins 2014), misleading consumers into thinking that "unregulated, non-certified 'natural' products are 'nearly organic'[.]"

More than forty governments worldwide have adopted – or are in the process of adopting – GM labelling laws that assess products for human, animal, and environmental safety (Sewell 2012). Public concerns over potential and unknown dangers to health, safety, the environment, and biodiversity from trade in GMOs (McNelis 2000) have met with a stronger regulatory response in these other jurisdictions. Lim Li Lin (2000) remarks: "Public opposition to LMOs [living modified organisms] and their products has already forced many countries such as Japan and Australia to pass labelling laws. The European Union already has a law requiring segregation and labelling of LMOs and their products." GM crops are banned in eight EU nations; in some jurisdictions, such as India and China, "governments have yet to approve most GM crops" (Freedman 2013, 82). While critics have expressed doubt that mandatory labelling will meet its stated objectives (Carter and Gruère 2003), some food companies, confronted with renewed consumer demand for GM labels, are finally making the strategic market switch to GM-free/non-GMO products (Parker 2014).

The ability to demonstrate scientific risk may, it has been argued here, have more to do with the metrics and methodology for safety assessments than with actual risks of objective scientific harm. If we accept DNA as a chemical, we begin to appreciate its potentially harmful and gendered effects based on chemical differences rather than substantial equivalence. A chemical understanding of novel food risk and exposure should be adopted. More scientific studies are needed that analyze potential toxicity and the consequences of long-term exposures on our health and environments, with sensitivity to gender differences in risks and benefits among consumers and to any disparate impact on marginalized workers (CWHN 2001). Any animal studies, if they are to be conducted, should be on both male and female subjects.

Finally, laws that regulate chemicals, such as the 2001 Consumer Chemicals and Containers Regulations made pursuant to the Canada Consumer Product Safety Act[15] regulating consumer products – including for circulation and import, to prevent dangers to human health or safety, and governing the labelling of consumer chemical products purchased through retail distribution – are worth re-examining in order to see how they might be expanded and coordinated to integrate the chemical view of DNA for novel food regulation. The current exclusion of foods, cosmetics, pest control, and tobacco products from these regulations (in Schedule 1) merits reassessment, along with a review and reform of novel food regulations.[16]

Conclusion

A chemical approach to DNA helps capture advances in critical scholarship and advocacy that seek to reveal environmental contamination and pollution as "business" – a way of "turning risk into profit" (Scott 2008) by morphing risk into the social and, in terms of biotechnologies, ecological landscape. Biopatenting has made this business far more profitable. A growing reliance on bioengineered crops and GM seeds facilitates the mass industrialization of food production and justifies agrochemical use. Large corporations have been entitled to enjoy all the rights of proprietary regimes such as patent law without any corresponding duties. Monsanto is often cited as an example. Monsanto succeeded in aggressively enforcing patent rights against a Canadian farmer when unlicensed "ready" crops appeared in his fields. In *Monsanto Canada v. Schmeiser* (2004), the Supreme Court of Canada found the defendant, Mr. Schmeiser, liable for patent infringement despite his claim that the presence of Roundup Ready® Canola in his fields was the result of innocent contamination rather than cultivation, and that he had not derived any benefit from the GM variety as he had not sprayed his fields with Monsanto's glyphosate herbicide. In *Hoffman v. Monsanto Canada Inc.* (2007), Monsanto also succeeded in avoiding liability for GM contamination when organic farmers failed to win class action certification in a suit against Monsanto for negligence, trespass, nuisance, and strict liability.

Some commentators claim that the establishment of herbicide/pesticide "ready" varieties works to reduce the environmental and health harms of chemical use, while others claim that the "intent is for farmers to be able to increase pesticide use without harmful effects to their crops" (Whitney, Maltby, and Carr 2004, 264). Regardless, it appears that the consequence of our current regulatory approach is not a reduction in the health and environmental footprint of agricultural practice (companies tend to capitalize on innovation to increase efficiency and profitability) but rather the transformation of that footprint into less traceable and verifiable forms, thereby making invisible any resulting risk.

Women are confronted by biotechnology in complex ways: as source material for patentable inventions (for which they must pay to realize the therapeutic benefits), as targeted consumers of biotech marketing, as conduits for gestational and fetal inoculation, and as exploited labourers with limited access to processes or services to improve their disparate work conditions.

Canada became a signatory to the Convention on the Elimination of All Forms of Discrimination against Women in 1980, committing to sex and gender equality in domestic regulation. CEDAW's concept of gender equality is premised on three principles: non-discrimination, state obligation, and substantive equality. To give effect to gender equality obligations, the Canadian government has committed to the gender-based analysis of all regulatory activities. Status of Women Canada oversees capacity building within federal departments to support a gender-based analysis framework, and has been very active in pursuing its mandate, despite insufficient resources. The 2009 Report of the Auditor General of Canada found that departments were not meeting their obligations to carry out a gendered analysis (Office of the Auditor General of Canada 2009), calling into question whether the Canadian government, and departments such as Health Canada, were genuinely committed to gender equality in regulation and under CEDAW. Gender-based analyses (GBA) are integral to any regulatory impact assessment and are required to inform policy making. Health Canada, it was found, had GBA tools available for employees, but it did not have a GBA champion, the dedicated human resources for GBA were highly variable for the reviewed period, and it had failed to carry out the evaluation scheduled for 2008 in the Gender Based Analysis Implementation Strategy (12-13). GBA of regulations are needed and would reinforce the substantive recommendation that novel food and GM crop events be assessed for their impact on women consumers and farmers, on their health and fertility. The failure to do so is as much a policy failure as it is a scientific one. The consequences of GM exposure are currently externalities in the novel food regulatory regime. Regulatory impact assessments of novel foods ought to be undertaken with the cooperation of Industry, Health, and Environment Canada and the Canadian Food Inspection Agency. Public policies should be supported by complementary approaches and effective biomonitoring practices based on more appropriate metrics, premised on chemical differences, for measuring risk communication. The risks to women of modern biotechnology may be disproportionate and distinct due to their particular biological makeup and roles in society. Women have unique health outcomes and therefore specific policy interests in the determination of the safety of novel foods and associated agrochemicals.

On modern technological risk, Hiskes (1998, 6) writes: "Our present sensation of not being in complete control of modern risk adds a special feeling of helplessness in the face of a new political reality." It is hoped that the view of DNA as chemicals may spur an "upsurge in political participation by citizens who stand to be most affected by new risks" (Hiskes 1998, 7). To this end, the Government of Canada's website provides a very helpful overview of chemical regulation, outlining Canada's approach to chemical substances including those "*in* food, drugs, pesticides and products" (emphasis added), with reference to such legislation as the Pest Control Products Act, the Food and Drugs Act, the Feeds Act, and the Seeds Act.[17] This is promising for novel food regulation and demonstrates that the theorizing around DNA *as* chemicals, although novel outside the realm of patent law, does not depart too greatly from existing regulatory oversights.[18]

Acknowledgment

I would like to thank the National Network on Environments and Women's Health for commissioning this work; former research assistants Sabrina Heyde, Morgan Jarvis, and Katrina Leung, for their able contributions; the anonymous reviewers for helpful comments; and Dayna Scott, Vanessa Scanga, and Sarah Lewis for their invaluable direction.

Notes

1 *Convention on the Elimination of All Forms of Discrimination against Women*, New York, 18 December 1979, 1249 U.N.T.S. 13 (entered into force 3 September 1981).

2 "She is free to live her life as she wishes and has exactly the same right to self-determination as she had before the patent was granted." It was also found that "patents covering DNA encoding human H2-relaxin, or any other human gene do not confer on their proprietors any rights whatever to individual human beings, any more than do patents directed to other human products such as proteins including human H2-relaxin" (*Re Howard Florey Institute/Relaxin* at para 6.33).

3 United States Food and Drug Administration, Genetically Engineered Animals, http://www.fda.gov/NewsEvents/Newsroom/PressAnnouncements/2009/ucm109066.htm.
4 *Food and Drugs Act*, R.S.C., 1985, c. F-27 (http://laws-lois.justice.gc.ca/eng/acts/F-27/index.html).
5 BIOTECanada states that Canada's bioeconomy constitutes approximately 7 percent of GDP, and is worth about $86.5 billion annually (McDonell 2011).
6 "A gene encoding cellular ferritin was introduced into the rice genome from the French bean. To decrease inhibition of iron absorption by decreasing phytates contained in rice, a gene encoding phytates was introduced from *Aspergillus fumigatus*. Another gene was introduced from basmati rice which encodes a cysteine-rich protein. Cysteine improves iron resorption in the human digestive tract. The β-carotene-producing rice was subsequently crossed with the iron-enriched strain. The resulting transgenic rice plant contains seven foreign genes which offer the potential to improve intakes of vitamin A and iron" (Perr 2002, 477).
7 The authors add: "For example, transgenic tomatoes, already available in some grocery stores, carry a gene resistance to the antibiotic kanamycin. Kanamycin is an antibiotic commonly used in the treatment of tuberculosis, a disease which is increasing in incidence" (Whitney, Maltby, and Carr 2004, 264).
8 But see Health Canada's determination that a "series of highly improbable and complex events must occur in the human GI tract for such an event to occur. The hostile nature of the GI tract provides a highly unfavourable environment for the survival of DNA coding for antibiotic resistance genes if released from the cells of the modified food. Free DNA for uptake by microorganisms is continuously degraded in the GI tract ... [A] series of steps are required for the DNA to be transferred and expressed in a recipient microorganism" (Conway 2000, 8). For some, this perspective may not be entirely comforting, as it works on probabilities rather than possibilities.
9 It is worth noting the error in this statement: a gene is *not* a protein; rather, genes code for proteins.
10 Thanks to Ariel Katz for offering this fitting metaphor in a conversation in which I described this paper.
11 Any "significant nutritional or compositional" change requires labelling, but the mere fact that genetic engineering is used in the process would not, such that "oil derived from high oleic soybean lines must be listed in the food ingredients by the common name 'high oleic soybean'" (Health Canada 2012a).
12 *Consumer Packaging and Labelling Act*, R.S.C., 1985, c. C-38 (http://laws-lois.justice.gc.ca/eng/acts/C-38/index.html).
13 See e.g., Bill C-287, *An Act to Amend the Food and Drugs Act (Genetically Modified Food)*, 1st Sess., 37th Parl. (2001); Bill C-517, *An Act to Amend the Food and Drugs Act (Mandatory Labeling for Genetically Modified Foods)*, 2d Sess., 39th Parl. (2008).
14 Bill C-257, *An Act to Amend the Food and Drugs Act (Mandatory Labelling for Genetically Modified Foods)*, 1st Sess., 41st Parl. (2013), reinstated 2d Sess., 41st Parl. (2013).
15 *Consumer Chemicals and Containers Regulations, 2001*, SOR/2001-269; *Canada Consumer Product Safety Act*, S.C. 2010, c. 21.
16 For example, to consider adopting in the novel food regulations the definition of "chemical product" from the Consumer Chemicals and Containers Regulations, 2001 as "a product used by a consumer that has the properties of one or more of the following: (a) a toxic product," and adding to the definition a new paragraph: "(e) rDNA."
17 *Pest Control Products Act*, S.C. 2002, c. 28 (http://laws-lois.justice.gc.ca/eng/acts/P-9.01/index.html); *Food and Drugs Act; Feeds Act*, R.S.C., 1985, c. F-9 (http://laws-lois.justice.gc.ca/eng/acts/F-9/index.html); *Seeds Act*, R.S.C., 1985, c. S-8 (http://laws-lois.justice.gc.ca/eng/acts/S-8/index.html).

18 Canada's System for Addressing Chemicals, http://www.chemicalsubstanceschimiques.gc.ca/approach-approche/system-eng.php.

References

Alberts, Bruce, Alexander Johnson, Julian Lewis, et al. 2002. *Molecular Biology of the Cell.* 4th ed. New York: Garland Science.

Amani, Bita. 2009. *State Agency and the Patenting of Life in International Law: Merchants and Missionaries in a Global Society.* London: Ashgate.

–. 2014. "Biopatenting and Industrial Policy Discourse: Decoding the Message of Biomedia on the Limits of Agents and Audiences." In *Intellectual Property Law for the 21st Century: Interdisciplinary Approaches,* edited by B. Courtney Doagoo, Mistrale Goudreau, Madelaine Saginur, and Teresa Scassa. Toronto: Irwin Law.

Andrée, Peter, and Lucy Sharratt. 2004. *Genetically Modified Organisms and Precaution: Is the Canadian Government Implementing the Royal Society of Canada's Recommendations?* Ottawa: Polaris Institute. http://www.cban.ca/Resources/Tools/Reports/Genetically-Modified-Organisms-and-Precaution-Is-the-Canadian-Government-Implementing-the-Royal-Society-of-Canada-s-Recommendations.

Arjo, Gemma, Manuel Portero, Carme Pinol, Juan Vinas, Xavier Matias-Guiu, Teresa Capell, Andrew Bartholomaeus, Wayne Parrott, and Paul Christou. 2013. "Plurality of Opinion, Scientific Discourse and Pseudoscience: An In Depth Analysis of the Séralini et al. Study Claiming that Roundup™ Ready Corn or the Herbicide Roundup™ Cause Cancer in Rats." *Transgenic Research,* 22 February. http://www.ask-force.org/web/Seralini/Arjo-Plurality-Opinion-Scientific-Discourse-Seralini-2013.pdf.

Babula, Jared. 1999. "Transgenic Crops: A Modern Trojan Horse." *Journal of Law and Social Challenges* 3 (1): 127-37.

Baker, Gregory A., and Thomas A. Burnham. 2001. "Consumer Response to Genetically Modified Foods: Market Segment Analysis and Implications for Producers and Policy Makers." *Journal of Agriculture and Resource Economics* 26 (2): 387-403.

Bean, Jordan. 2011. "The Regulation of Novel Foods and Novel Feeds." Canadian Food Inspection Agency; Health Canada. http://www.ontariogenomics.ca/sites/default/files/Bean_1.pdf

Bernstein, Jonathan A., I. Leonard Bernstein, Luca Bucchini, Lynn R. Goldman, Robert G. Hamilton, Samuel Lehrer, Carol Rubin, and Hugh A. Sampson. 2003. "Clinical and Laboratory Investigation of Allergy to Genetically Modified Foods." *Environmental Health Perspectives* 111 (8): 1114-21, doi:10.1289/ehp.5811.

Boserup, Esther. 1970. *Women's Role In Economic Development.* London: George Allen and Unwin.

Brophy, James T., Margaret M. Keith, Kevin M. Gorey, Ethan Laukkanen, Deborah Hellyer, Andrew Watterson, Abraham Reinhartz, and Michael Gilbertson. 2002. "Occupational Histories of Cancer Patients in a Canadian Cancer Treatment Centre and the Generated Hypothesis regarding Breast Cancer and Farming." *International Journal of Occupational and Environmental Health and Safety* 8: 346-53.

Burrows, Beth. 2001. "Safety First." In *Redesigning Life: The Worldwide Challenge to Genetic Engineering,* edited by Brian Tokar, 67-74. Montreal and Kingston: McGill-Queen's University Press.

Canadian General Standards Board. 2004. *Voluntary Labeling and Advertising of Foods That Are and Are Not Products of Genetic Engineering.* CAN/CGSB-32.315. Gatineau, QC: Canadian General Standards Board.

Canola Council of Canada. 2005. "Canadian Canola Industry." http://ws373847.websoon.com/ind_overview.aspx.

Carey, Nessa. 2012. *The Epigenetics Revolution*. New York: Columbia University Press.

Carter, C.A., and G.P. Gruère. 2003. Mandatory Labeling of Genetically Modified Foods: Does It Really Provide Consumer Choice? *AgBioForum* 6 (1 and 2): 68-70. http://www.agbioforum.org.

CBC (Canadian Broadcasting Corporation). 2010. "Genetically Modified Salmon Safe, FDA Says." *CBC News Online,* 10 September. http://www.cbc.ca/news/canada/prince-edward-island/genetically-modified-salmon-safe-fda-says-1.932009.

–. 2014. "Genetically-Modified Salmon Approval Faces Lawsuit: Environment Canada Did Not Follow Its Own Rules Says Group." *CBC News Online,* 22 January. http://www.cbc.ca/news/canada/prince-edward-island/genetically-modified-salmon-approval-faces-lawsuit-1.250624.

CFS (Center for Food Safety). 2013. "US Polls on GE Food Labeling." http://www.centerforfoodsafety.org/issues/976/ge-food-labeling/us-polls-on-ge-food-labeling#.

Chang, Howard F. 2004. "Risk Regulation, Endogenous Public Concerns, and the Hormones Dispute: Nothing to Fear but Fear Itself?" *Southern California Law Review* 77: 743-76, doi:10.2139/ssrn.432220.

Codex Alimentarius Commission. 2003. "Guideline for the Conduct of Food Safety Assessment of Foods Derived from Recombinant-DNA Plants." CAC/GL 45-2003. http://www.fao.org/fileadmin/user_upload/gmfp/docs/CAC.GL_45_2003.pdf.

–. 2008. "Guideline for the Conduct of Food Safety Assessment of Foods Derived from Recombinant-DNA Animals." CAC/GL 68-2008. http://www.fao.org/fileadmin/user_upload/gmfp/resources/CXG_068e.pdf.

Conway, Gordon. 2000. "Genetically Modified Crops: Risks and Promise." *Conservation Ecology* 4 (1): 2.

Cook-Deegan, Robert Mullan. 1994. "Origins of the Human Genome Project." *RISK: Health, Safety and Environment* 5: 97-118.

Crouch, Martha L. 2001. "From Golden Rice to Terminator Technology: Agricultural Biotechnology Will Not Feed the World or Save the Environment." In *Redesigning Life: The Worldwide Challenge to Genetic Engineering,* edited by Brian Tokar, 22-39. Montreal and Kingston: McGill-Queen's University Press.

CSPI (Center for Science in the Public Interest). 2001. "Re: Comments on Docket Number 00D-1598, Guidance for Industry on 'Voluntary Labeling *Indicating Whether Foods Have or Have Not Been Developed Using Bioengineering*.'" http://www.cspinet.org/biotech/vol_labeling.html.

–. 2012. *Straight Talk on Genetically Engineered Foods: Answers to Frequently Asked Questions*. Washington, DC: Center for Science in the Public Interest. http://cspinet.org/new/pdf/biotech-faq.pdf.

Cummins, Ronnie. 2014. "GMO and the "Natural" Food Fight: The Treacherous Terrain of Food Labeling." *Global Research,* January 4. http://www.globalresearch.ca/gmo-and-the-natural-food-fight-the-treacherous-terrain-of-food-labeling/5363798.

CWHN (Canadian Women's Health Network). 2001. *If Women Mattered: A Critical View of the Canadian Biotechnology Strategy (CBS) and Alternative Visions for Community Action*. August. http://www.cwhn.ca/sites/default/files/groups/biotech/altcbs/alt-cbs.pdf.

Daniells, Stephen. 2007. "GM Tomatoes to Provide Daily Folate Needs?" *Nutra Ingredients-USA,* 6 March. http://www.nutraingredients-usa.com/Research/GM-tomatoes-to-provide-daily-folate-needs.

Doucette, Keith. 2013. "Environment Canada Approves Commercial Production of Genetically Modified Salmon." *Maclean's Online,* 25 November. http://www2.macleans.ca/2013/11/25/environment-canada-approves-commercial-production-of-genetically-modified-salmon/.

Drahos, Peter, and John Braithwaite. 2002. *Information Feudalism: Who Owns the Knowledge Economy?* London: Earthscan Publications.

Engdahl, F. William. 2007. *Seeds of Destruction: The Hidden Agenda of Genetic Manipulation.* N.p.: Global Research.

Epps, Tracey. 2008a. *International Trade and Health Protection: A Critical Assessment of the WTO's SPS Agreement.* Northampton, MA: Edward Elgar Publishing.

–. 2008b. "Reconciling Public Opinion and WTO Rules under the SPS Agreement." *World Trade Review* 7 (2): 359-92, doi:10.1017/S1474745608003819.

ETC (Action Group on Erosion, Technology and Concentration). 2003. "GM Maize Contamination in Mexico: Is the Genie Out of the Bottle?" *The Guardian,* 22 October. http://www.cpa.org.au/z-archive/g2003/1160maize.html.

European Commission. 2007. "Questions and Answers on the Regulation of GMOs in the EU." Memo/07/117. http://europa.eu/rapid/pressReleasesAction.do?reference=MEMO/07/117.

Ferrara, Jennifer, and Michael K. Dorsey. 2001. "Genetically Engineered Foods: A Minefield of Safety Hazards." In *Redesigning Life: The Worldwide Challenge to Genetic Engineering,* edited by Brian Tokar, 51-66. Montreal and Kingston: McGill-Queen's University Press.

Ford, Anne Rochon. 2000. *Biotechnology and the New Genetics: What It Means for Women's Health.* Canadian Women's Health Network, http://www.cwhn.ca/en/node/21251.

Freedman, David. 2013. "Are Engineered Foods Evil?" *Scientific American* 309 (3): 80-85.

Gab-Alla, A.A., Z.S. El-Shamei, A.A. Shatta, et al. 2012. "Morphological and Biochemical Changes in Male Rats Fed on Genetically Modified Corn (Ajeeb YG)." *Journal of American Science* 8 (9): 1117-23.

Gillam, Carey. 2014a. "Organic Food and Farm Groups Ask Obama to Require GMO Food Labels." Reuters, 16 January. http://uk.reuters.com/article/2014/01/16/us-usa-gmo-labeling-idUKBREA0F10H20140116.

–. 2014b. "Vermont Becomes First US State to Mandate GMO Labeling." Reuters, 9 May. http://www.reuters.com/article/2014/05/08/usa-gmo-labeling-idUSL2N0NU1R720140508.

Habermas, Jürgen. 2003. *The Future of Human Nature.* Cambridge: Polity Press.

Health Canada. 1997. "Safety Assessment of Potatoes Resistant to the Colorado Potato Beetle." http://www.hc-sc.gc.ca/fn-an/gmf-agm/appro/potatoes_resistant_pommes_de_terre_resistantes-eng.php.

–. 2002. "Technical Discussion on the Health and Safety of the Government of Canada Action Plan." http://www.hc-sc.gc.ca/sr-sr/pubs/gmf-agm/tech-rep-rap_04_2002-eng.php.

–. 2006. *Guidelines for the Safety Assessment of Novel Foods.* http://www.hc-sc.gc.ca/fn-an/legislation/guide-ld/nf-an/guidelines-lignesdirectrices-eng.php#a1.

–. 2012a. "Frequently Asked Questions – Biotechnology and Genetically Modified Foods." http://www.hc-sc.gc.ca/fn-an/gmf-agm/fs-if/faq_1-eng.php.

–. 2012b. "Health Canada and Canadian Food Inspection Agency Statement on the Séralini et al. (2012) Publication on a 2-Year Rodent Feeding Study with Glyphosate Formulations and GM Maize NK603." 25 October. http://www.hc-sc.gc.ca/fn-an/gmf-agm/seralini-eng.php.

–. 2012c. "Questions and Answers: Hormonal Growth Promoters." http://www.hc-sc.gc.ca/dhp-mps/vet/faq/growth_hormones_promoters_croissance_hormonaux_stimulateurs-eng.php.

Henry, Christine. 2006. *Cumulative Long-Term Effects of Genetically Modified (GM) Crops on Human/Animal Health and the Environment: Risk Assessment Methodologies.* York, UK: Central Science Laboratory. http://ec.europa.eu/food/food/biotechnology/reports_studies/docs/report_lt_effects_2006_en.pdf.

Hiskes, Richard P. 1998. *Democracy, Risk, and Community: Technological Hazards and the Evolution of Liberalism.* New York: Oxford University Press.

Holland, Jesse J. 2013. "High Court Rules for Monsanto in Patent Case." Associated Press, 13 May. http://bigstory.ap.org/article/high-court-rules-monsanto-patent-case.

Howard, Patricia, ed. 2003. *Women and Plants: Gender Relations in Biodiversity Management and Conservation.* London: ZED Books.

Howse, Robert. 2000. "Democracy, Science, and Free Trade: Risk Regulation on Trial at the World Trade Organization." *Michigan Law Review* 98: 2329-57.

James, Clive. 2012. *Global Status of Commercialized Biotech/GM Crops 2012.* ISAAA Brief 44. Ithaca, NY: International Service for the Acquisition of Agri-Biotech Applications. http://www.isaaa.org/resources/publications/briefs/44/executivesummary/pdf/Brief%2044%20-%20Executive%20Summary%20-%20English.pdf.

Kopun, Francine. 2010. "Genetically Modified Salmon Is Ready for Dinner." *Toronto Star,* 8 September.

Lappé, Frances Moor, and Anna Lappé. 2013. "Choice of Monsanto Betrays Food Prize Purpose, Say Global Leaders." http://www.huffingtonpost.com/frances-moore-lappe-and-anna-lappe/choice-of-monsanto-betray_b_3499045.html.

Ledford, Heidi. 2013. "Transgenic Salmon Nears Approval." *Nature* 497 (2 May): 17-8. http://www.nature.com/news/transgenic-salmon-nears-approval-1.12903.

Leiss, William. 2003. "The Case for Mandatory Labeling of Genetically-Modified Foods." Paper prepared for the Consumers' Association of Canada, 8 November. http://www.consumer.ca/pdfs/case_for_labelling_genetically_modified_foods.pdf.

Leu, Andre. 2011. "The Myths of Safe Pesticides." Organic Federation of Australia, http://www.ofa.org.au/papers/Mythpesticidesv2.html.

Levario-Carrillo, Margarita, Dante Amato, Patricia Ostrosky-Wegman, et al. 2004. "Relation between Pesticide Exposure and Intrauterine Growth Retardation." *Chemosphere* 55 (10): 1421-27, doi:10.1016/j.chemosphere.2003.11.027.

Lin, Lim Li. 2000. "Biosafety Talks End on Mixed Note." *Third World Network,* 31 January. http://www.twnside.org.sg/title/mixed2-cn.htm.

Lyndhurst, Brooke. 2009. *An Evidence Review of Public Attitudes to Emerging Food Technologies.* Social Science Research Unit: Food Standards Agency. http://www.food.gov.uk/multimedia/pdfs/emergingfoodtech.pdf.

MacDonald Glenn, Linda. 2004. "Ethical Issues in Genetic Engineering and Transgenics." American Institute of Biological Sciences, http://www.actionbioscience.org/biotech/glenn.html.

McAuliff, Michael. 2013. "GMO Labeling Bill Voted Down in Senate." *Huffington Post,* 23 May. http://www.huffingtonpost.com/2013/05/23/gmo-labeling-bill-genetically-modified-food_n_3325972.html.

McCarthy, Michael. 2000. "The 'Gender Gap' in Autoimmune Disease." *Lancet* 356 (9235): 1088, doi:10.1016/S0140-6736(05)74535-9.

McDonell, D'arcy. 2011. "BIOTECanada's Goal: Make Canada a Biotech World Leader by 2020." *Hill Times,* 12 September.

McNelis, Natalie. 2000. "EU Communications on the Precautionary Principle." *Journal of International Economic Law* 3 (3): 545-51, doi:10.1093/jiel/3.3.545.

Millstone, Erik, Eric Brunner, and Sue Mayer. 1999. "Beyond 'Substantial Equivalence': Showing that a Genetically Modified Food Is Chemically Similar to Its Natural Counterpart Is Not Adequate Evidence that It Is Safe for Human Consumption." *Nature* 401: 525-26.

Moore, Dene. 2013. "Canadian GM Salmon Nets FDA Thousands of Comments." *Huffington Post,* 13 February. http://www.huffingtonpost.ca/2013/02/13/-genetically-modified-salmon-fda_n_2679911.html.

Moschini, GianCarlo. 2008. "Biotechnology and the Development of Food Markets: Retrospect and Prospects." *European Review of Agricultural Economics* 35 (3): 331-55.

Office of the Auditor General. 2009. "Chapter 1: Gender-Based Analysis." In *Report of the Auditor General of Canada.* Spring. Ottawa: Minister of Public Works and Government Services Canada. http://www.oag-bvg.gc.ca/internet/docs/parl_oag_200905_01_e.pdf.

Ogilvie, Megan. 2008. "Genetically Engineered Meal Close to Your Table." *Toronto Star,* 22 November.

Pardy, Bruce. 2013. "Markets, Laws and Labels for Genetically Modified Foods." Conference paper for Yale Food Systems Symposium, New Haven, CT (on file with the author).

Parker, Laura. 2014. "The GMO Labeling Battle Is Heating Up – Here's Why." *National Geographic,* 11 January.

Parliament of Canada. 2013. LegisInfo. http://www.parl.gc.ca/Legisinfo/BillDetails.aspx?Language=E&Mode=1&billId=5100644&View=3.

Perr, Hilary A. 2002. "Children and Genetically Engineered Food: Potentials and Problems." *Journal of Pediatric Gastroenterology and Nutrition* 35 (4): 475-86.

Peters, Pamela. 1993. *Biotechnology: A Guide to Genetic Engineering.* Dubuque, IA: William C. Brown.

Powell, Douglas, and William Leiss. 1997. *Mad Cows and Mother's Milk: The Perils of Poor Risk Communication.* Montreal and Kingston: McGill-Queen's University Press.

Robinson, Claire, and Jonathan Latham. 2013. "The Goodman Affair: Monsanto Targets the Heart of Science." *Independent Science News,* 20 May. http://www.independentsciencenews.org/science-media/the-goodman-affair-monsanto-targets-the-heart-of-science/.

Royal Society of Canada. 2001. *Elements of Precaution: Recommendations for the Regulation of Food Biotechnology in Canada.* Ottawa: Royal Society of Canada. http://rsc-src.ca/sites/default/files/pdf/GMreportEN.pdf.

Schildkraut, Joellen M., Wendy Demark-Wahnefried, Emily DeVoto, Claude Hughes, John L. Laseter, and Beth Newman. 1999. "Environmental Contaminants and Body Fat Distribution." *Cancer Epidemiology, Biomarkers and Prevention* 8: 179-83.

Schmidt, Sarah. 2011. "GM Salmon Could Harm Fish Stocks: DFO Scientists." *Montreal Gazette,* 23 February.

–. 2012. "Genetically Engineered Pigs Killed after Funding Ends." *Post Media News,* 22 June.

Scott, Dayna Nadine. 2005. "Nature/Culture Clash: The Transnational Trade Debate over GMOs." Global Law Working Paper 06/05, Hauser Global Law School Program, New York.

–. 2007. "Risk as a Technique of Governance in an Era of Biotechnological Innovation: Implications for Democratic Citizenship and Strategies of Resistance." In *Risk and Trust: Including or Excluding Citizens?* edited by the Law Commission of Canada, 23-56. Halifax: Fernwood Publishing. https://apps.osgoode.yorku.ca/osgmedia.nsf/0/A438915A7BF77EA9852571DC0064E40D/$FILE/Risk-as-Techniqueof Governance.pdf.

–. 2008. "Confronting Chronic Pollution: A Socio-Legal Analysis of Risk and Precaution." *Osgoode Hall Law Journal* 46 (2): 293-343. http://www.ohlj.ca/english/documents/OHLJ46-2_Scott_ConfrontingChronicPollution.pdf.

Séralini, Gilles-Eric, Emilie Clair, Robin Mesnage, et al. 2012. "Long Term Toxicity of a Roundup Herbicide and a Roundup-Tolerant Genetically Modified Maize." *Food and Chemical Toxicology* 50 (11): 4,221-31. http://www.sciencedirect.com/science/article/pii/S0278691512005637.

Sewell, Anne. 2012. "Op-Ed: 'Label It Yourself' Campaign – Citizen Action to Label GMOs." *Digital Journal,* http://digitaljournal.com/article/321535.

Steinbrecher, Richard A. 2001. "Ecological Consequences of Genetic Engineering." In *Redesigning Life: The Worldwide Challenge to Genetic Engineering,* edited by Brian Tokar, 75-102. Montreal and Kingston: McGill-Queen's University Press.

Strauss, Debra M. 2006. "The International Regulation of Genetically Modified Organisms: Importing Caution into the US Food Supply." *Food and Drug Law Journal* 61 (2): 167-96.

Tansey, Geoff, and Tasmin Rajotte, eds. 2008. *The Future Control of Food: A Guide to International Negotiations and Rules on Intellectual Property, Biodiversity and Food Security.* London: Earthscan.

Taylor, Steve L., and Susan L. Hefle. 2001. "Will Genetically Modified Foods Be Allergenic?" *Journal of Allergy and Clinical Immunology* 107 (5): 765-71, doi:10.1067/mai.2001.114241.

Tepfer, Mark. 1993. "Viral Genes and Transgenic Plants." *Nature Biotechnology* 11: 1125-32, doi:10.1038/nbt1093-1125.

1,125?

Tokar, Brian, ed. 2001. *Redesigning Life: The Worldwide Challenge to Genetic Engineering.* Montreal and Kingston: McGill-Queen's University Press.

University of Guelph Food Science Network. 2012. "Substantial Equivalence." https://www.uoguelph.ca/foodsafetynetwork/substantial-equivalence.

US FDA (US Food and Drug Administration). 2001. "Guidance for Industry: Voluntary Labeling Indicating Whether Foods Have or Have Not Been Developed Using Bioengineering; Draft Guidance." http://www.fda.gov/food/guidanceregulation/guidancedocumentsregulatoryinformation/labelingnutrition/ucm059098.htm.

–. 2009a. "About the GRAS Notification Program." http://www.fda.gov/Food/FoodIngredientsPackaging/GenerallyRecognizedasSafeGRAS/GRASNotificationProgram/default.htm.

–. 2009b. "FDA Issues Final Guidance on Regulating Genetic Engineered Animals." 15 January. http://www.fda.gov/NewsEvents/Newsroom/PressAnnouncements/2009/ucm109066.htm.

–. 2009c. "Report on the Food and Drug Administration's Review of the Safety of Recombinant Bovine Somatotropin." 23 April. http://www.fda.gov/animalveterinary/safetyhealth/productsafetyinformation/ucm130321.htm.

–. 2013. "General QandA: The Technology." http://www.fda.gov/AnimalVeterinary/DevelopmentApprovalProcess/GeneticEngineering/GeneticallyEngineeredAnimals/ucm113605.htm.

Wald, George. 1979. "The Case against Genetic Engineering." In *The Recombinant DNA Debate,* edited by David A. Jackson and Stephen P. Stich, 127-28. Englewood Cliffs, NJ: Prentice Hall.

Wheeler, Matthew B. 2007. "Agricultural Applications for Transgenic Livestock." *Trends in Biotechnology* 25 (5): 204-10, doi:10.1016/j.tibtech.2007.03.006.

Whitney, Stuart L., Hendrika J. Maltby, and Jeanine M. Carr. 2004. "'This Food May Contain ...' What Nurses Should Know about Genetically Engineered Foods." *Nursing Outlook* 52 (5): 262-66, doi:10.1016/j.outlook.2004.03.003.

Wintermantel, William M., and James E. Schoelz. 1996. "Isolation of Recombinant Viruses between Cauliflower Mosaic Virus and a Viral Gene in Transgenic Plants under Conditions of Moderate Selection Pressure." *Virology* 223 (1): 156-64, doi: 10.1006/viro.1996.0464.

WIPO (World Intellectual Property Organization). 2006. "Bioethics and Patent Law: The Relaxin Case." *WIPO Magazine,* Issue 2/2006. http://www.wipo.int/wipo_magazine/en/2006/02/article_0009.html.

Cases Cited

AMP (Association for Molecular Pathology) v. Myriad Genetics, 11-725 (US 26 March 2012) (previously titled *Ass'n for Molecular Pathology v. US Patent and Trademark Office*), 653 F.3d 1329. http://www.law.stanford.edu/sites/default/files/event/266699/media/sls-public/AMP%20v.%20Myriad%20(CAFC%202011).pdf.

Association for Molecular Pathology et al. v. Myriad Genetics, Inc., et al., 569 U.S. (2013). http://www.supremecourt.gov/opinions/12pdf/12-398_1b7d.pdf.

Harvard College v. Canada (Commissioner of Patents), [2002] 4 S.C.R. 45, 2002 S.C.C. 76 *[Harvard].*

Hoffman v. Monsanto Canada Inc. 2007 SKCA 47 (Sask. C.A.) affirming (2005), 2005 SKQB 225 (Sask. Q.B.), leave to appeal to S.C.C. refused.

Monsanto Canada v. Schmeiser, [2004] 1 S.C.R. 902, 2004 S.C.C. 34.

Re Howard Florey Institute/Relaxin, EPO Opposition Division, [1995] E.P.O. R. 541 *[Howard Florey].*

Chapter 6 Consuming Carcinogens: Women and Alcohol

Nancy Ross, Jean Morrison,
Samantha Cukier, and Tasha Smith

Binge drinking for many women and girls occurs in a social environment in which the terms "rape culture" and "porn culture" fail to surprise us. Women and girls in Canada are in trouble. The links between alcohol abuse and women and girls who have experienced sexual victimization are staggering and profound (Brown 2008; Najavits 2002b; Poole and Greaves 2012). Many women who are susceptible to the lure of alcohol are plagued with a general feeling of inadequacy and a sense of disconnection (Alexander 2008; Maté 2008). This vulnerability is exploited ruthlessly by sophisticated advertising and marketing of alcohol (Jernigan 2011). Ann Dowsett Johnston (2013) points out that our fairy tale romance with alcohol has been conjured by experts in spin and has set the stage for skyrocketing problems with alcohol among women and girls. Alcohol corporations have been relentless in their development of marketing strategies and products that exclusively target women. Ironically, at the same time that women and girls are experiencing an epidemic rise in drinking, there is a corresponding taboo and stigma that often prevents them from acknowledging problems with alcohol. These unique hazards, frequently invisible, exist in a social environment that gives rise to problematic substance use among women.

Abuse of alcohol and other substances by women and girls has been called the "silent epidemic" (Najavits 2002a, 10) because it has been hidden and neglected for so long. Women are twice as likely as men to die from alcohol-related problems, and four times as many women die from illnesses related to substance dependence as those who die from breast cancer (Najavits 2002a). If these data included the many deaths from breast cancer that are attributable to alcohol use and abuse, the figures would be even higher. Alcohol is a dangerous and frequently deadly commodity for women. In Canada, alcohol consumption is the second-highest major risk factor for the burden of disease, surpassed only by tobacco (WHO 2009).

This chapter employs a population health approach in discussing the pathways to alcohol abuse among women and girls, the harms associated with this use, and the cultural and policy implications. Although it is recognized that many women and girls who abuse alcohol may also abuse tobacco products, prescribed medication, and illicit drugs, the chapter will focus on alcohol use alone. The research discussed here applies to both Canada generally and Nova Scotia specifically. The authors have been working in rural Nova Scotia, where the first provincial alcohol strategy in Nova Scotia, called "Changing the Culture of Alcohol Use in Nova Scotia," has been adopted (Nova Scotia Department of Health Promotion and Protection 2007).

The development of alcohol use disorders is often viewed as attempts at self-medicating, as many girls and woman experience depression, anxiety, and post-traumatic stress symptoms as a result of living in environments in which they are not safe (Brown 2008; Najavits 2002b). Substance abuse as a coping mechanism can create a vicious cycle that often leads to other problems (Brown 2008; Dube et al. 2002). The Fifty-eighth World Health Assembly declared that harmful drinking was among the foremost underlying causes of disease, domestic violence against women and children, disability, social problems, and premature death, and pointed out that alcohol was often a risk factor in the perpetuation of violence against women (WHO 2005). In addition, alcohol abuse can be conceived of as the ingestion of a toxic chemical. The World Health Organization (WHO) has described alcohol as a "carcinogenic beverage" (Baan et al. 2007).

Alcohol remains the substance most commonly used by women and girls, and recent international studies indicate that the gender gap in the prevalence of alcohol use is closing (Poole and Greaves 2007; Thomas 2012). Historically, boys and men have consumed more alcohol than girls and women; however, girls and many young women are now consuming alcohol at similar rates. Women and girls are drinking more and beginning to consume alcohol at younger ages (Adlaf, Begin, and Sawka 2005). This is cause for concern as they tend to experience a more rapid progression of harm from alcohol consumption than men, sometimes referred to as the "telescoping effect" (Mancinelli, Vitali, and Ceccanti 2009; Mann et al. 2005, 896; Poole and Dell 2005; Ceccanti and Vitali 2009). Recent studies confirm that alcohol use is linked to breast cancer and other cancers (e.g., Allen at al. 2009). Fetal Alcohol Spectrum Disorder (FASD) research also clearly demonstrates the developmental effects of alcohol on the fetus, as well as a link between alcohol use during pregnancy and a child's weakened immune system (Mancinelli, Binetti, and Ceccanti 2006). Recent research indicates a need for further study to refine our knowledge of how much alcohol in what stages of pregnancy is harmful to the developing fetus. Researchers have concluded that since it is not clear how much alcohol it takes to cause problems, the best advice is that women should avoid alcohol if they are pregnant or might become pregnant (LeWine 2013). All of the physical harms linked to alcohol consumption are compounded by environmental factors. In this chapter, a woman's environment includes the physical space in which she lives, as well as her "community" and the "cultural" factors to which she is exposed. For example, early experiences of childhood trauma, attachment difficulties, and other forms of abuse are linked to the development of substance abuse problems.

The Effect of the Environment

The pathway to alcohol abuse and dependence is inextricably linked to the environments in which women and girls live (Poole and Salmon n.d.). If we define environment in broad terms, as the Population Health Promotion Model (Hamilton and Bhatti 1996) suggests, many factors need to be considered. These factors operate together in a complex system that greatly influences people's lives and circumstances (see Parts 1, 3, and 4 of this volume for other discussions of social determinants of health). This chapter links the social determinants of health with the development of substance abuse, specifically alcohol abuse among women and young girls. This abuse is described as consumption of chemicals and the harm these chemicals produce in women's lives.

A woman's physical environment is an important factor in determining the degree to which her use or abuse of alcohol will harm her. Access to proper nutrition and adequate housing can, for example, decrease the impacts of a pregnant woman's alcohol consumption on fetal development (National Indian and Inuit Community Health

Representatives Organization n.d.). A literature review conducted by Mancinelli, Binetti, and Ceccanti (2006) concluded that the harm caused by environmental exposure to lead and other toxins is greater when the woman is abusing substances. It is ironic that even as more and more information about the harms of alcohol abuse becomes available, communities still seem to disregard it. A goal of this chapter is to make readers more aware of the impact of alcohol on women and girls.

Raising the issue of alcohol use is controversial, regardless of whether it is addressed with friends and family or as a matter of public policy. From an individual perspective, suggestions of curbing access to alcohol are often met with protest and the defence that drinking habits are the individual's responsibility. Those who explore these issues with a wide social lens suggest that reduced availability and higher prices are effective ways to lessen the debilitating effects of alcohol use in society; the less access people have to alcohol, the less they will consume and the fewer the harms that will be incurred (Babor et al. 2010). This is not to say that policy makers should advocate abstinence from alcohol. Rather, it is a request for improved information to help individuals and communities understand the addictive potential of alcohol's chemical properties and the fact that its abuse results in great harm, particularly for women. As explained in greater detail elsewhere in this chapter, women do not process alcohol as efficiently as men due to differences in biology. Consequently, they can become dependent on alcohol more quickly than men and suffer more severe health impacts sooner, a phenomenon referred to as the "telescoping effect." This knowledge is reflected in the Canadian Low Risk Drinking Guidelines, which state that binge drinking occurs at four standard drinks for women or five standard drinks for men in one sitting (Rehm, Patra, and Taylor 2007). An increase in current information from credible sources, disseminated within a supportive environment, will lead to more informed decision making and policies that incorporate a collaborative, women-centred, population health, and harm reduction approach.

We Live in an Alcohol World

In the province of Nova Scotia, the alcohol industry has successfully targeted women and girls with a new kind of advertising, aimed at increasing in-store sales and impulse purchases. For example, "shopper marketing" uses advertising at the point of sale to suggest that drinking alcohol will offer excitement, fun, and romance (Deloitte 2007).

Greater sales of alcohol to women have resulted in increased profits for the main supplier of alcohol in the province. After a conversation with Rick Perkins, Vice President of Communications and Corporate Responsibility for the Nova Scotia Liquor Corporation (NSLC), Lombardi (2008) noted that "relocating [NSLC] outlets next to supermarkets and redesigning stores to appeal more to women were major profit enhancing moves," and that "women shoppers can now roll right in with their grocery carts to pick up their selection [and] ... this tactic has increased their numbers by 45 percent and increased wine sales" (Lombardi 2008, 27). Stores selling alcohol have been transformed from a place to buy to a place to shop in order to attract more female customers.

Dr. David Jernigan (2009), director of the Center on Alcohol Marketing and Youth in Baltimore, Maryland, notes that the global alcohol industry is highly focused and innovative in its marketing activities. Global marketing strategies are especially effective among youth (Jernigan 2009). In the United States, for example:

> In the traditional media of television, magazines, radio and billboards, analysis of data from a longitudinal study of 1872 youth aged 15-26 years in 24 media markets found that for every additional advertisement young people reported seeing or hearing above an average of 23 per month, they drank 1 percent more, while every additional dollar spent per capita on alcohol advertising in their media market was associated with a 3 percent increase in young people's drinking. (Jernigan 2009, 4)

According to Jernigan, females are generally exposed to more advertising than males because of their greater exposure to magazines and other print material. In the United States, flavoured alcoholic beverages are most popular among young females. Jernigan (2009, 5) highlights the power of marketing to young women:

> A marked shift in beverage preference among 17 and 18 year old binge-drinking girls from beer to distilled spirits occurred between 2001 and 2005, during the same period that flavoured alcoholic beverages were introduced into the US market ... marketers spent more than $360 million on television advertisements for distilled spirits branded "flavored alcoholic beverages," such as Smirnoff Ice and Bacardi Silver.

There is more evidence of subtle pressure by the advertising industry in our daily lives. During a weekly meeting of a women's Life Issues Recovery Group in Bridgewater, Nova Scotia, one seventy-one-year-old expressed her feelings of living in an alcohol world: "Everywhere I go there is alcohol. I play golf and there is alcohol. I go to a wedding and there is alcohol. I go to any party and there is alcohol. The Christmas Season of parties is a very challenging time not to drink. I went to a fundraiser and they were serving wine" (Anonymous, personal communication, September 2008).

The availability of alcoholic beverages, its social acceptance, and its traditional use in the family result in a very low perception of the risk of alcohol abuse and make it a very difficult issue to address (Mancinelli, Binetti, and Ceccanti 2006).

Patterns of Use

Alcohol is the most common psychoactive substance consumed by women in Canada. Over the past fifteen years, there has been an increase in women's use of alcohol, especially in those who use it heavily. According to the 2004 Canadian Addiction Survey (Adlaf, Begin, and Sawka 2005), 76.8 percent of Canadian women and 82 percent of Canadian men over the age of fifteen reported that they had consumed alcohol. The percentage of women reporting frequent alcohol use increased with level of education. Those with the highest levels of income reported the highest rates of frequent but light drinking, whereas those with lower incomes showed the highest rates of frequent and heavy drinking.

The Alcohol Use Disorders Identification Test (AUDIT) is a tool used to identify hazardous alcohol consumption, harmful use patterns, and alcohol dependence. According to the 2005 *Nova Scotia Alcohol Indicators Report* (Graham 2005), 1 in 5 drinkers over the age of fifteen – or about 117,144 Nova Scotians – are high-risk drinkers, meaning that they attained a score of 8 or higher on the AUDIT. Depending on the data source used, high-risk or heavy drinking is usually defined as consumption of five or more drinks for either sex, or five or more drinks for men and four or more drinks for women, at one sitting. Using criteria tailored to women will aid in making the drinking patterns of women and girls visible so that more attention will be given to this issue.

The population of Nova Scotia in 2005 was reported to be 937,889, with approximately 152,062 under the age of fifteen (Nova Scotia

Finance 2005). The number of high-risk drinkers is therefore about 1 in 6.7. However, this is a low estimate because it does not take into account the rates of high-risk drinking for those under fifteen. For example, in British Columbia, female students aged fifteen or younger have a higher monthly rate of binge drinking (13.4 percent) than their male counterparts (11.8 percent) (Poole, Gonneau, and Urquhart 2010). This suggests a possible need for intervention and further assessment and referral to treatment in order to reduce the risks associated with individual alcohol use. The same report noted that an estimated 237,270 (1 in 3) Nova Scotians eighteen years and older have been harmed by another person's use of alcohol. The most-reported harms include being insulted or humiliated, verbally abused, and/or involved in a serious argument with a drinker (Graham 2005). If we speak of women's environments as including their social contexts, then the drinking of others in the family or community can be viewed as fostering a harmful environment. Sex-disaggregated data (data separated based on sex) that include a gender-based analysis would provide a more accurate picture of women's health burdens.

Canadian Low Risk Drinking Guidelines (Butt et al. 2011; CAMH 2009b) recommend no more than ten drinks per week for women and fifteen drinks per week for men (two and three drinks a day, respectively). Dr. Tim Stockwell and his team at the Centre for Addictions Research of British Columbia (CARBC) estimate that approximately 73 percent of alcohol consumption in Canada occurs above these guidelines (Stockwell, Sturge, and McDonald 2007).

While alcohol's steadfast acceptance is apparent in the province of Nova Scotia and in much of Canada, it is also a growing phenomenon worldwide, even in countries where there have traditionally been cultural taboos against alcohol consumption. For example, in India alone there are 65.2 million alcohol users, 50 percent of whom are reported to be hazardous drinkers, and their numbers are steadily increasing; in addition, the age of initiation has decreased from 19 years in 1986 to 13.5 years in 2006 (Dhar 2008). A study of trends in the prevalence of monthly alcohol use and lifetime drunkenness in twenty European countries, the Russian Federation, Israel, the United States, and Canada found that drunkenness rates rise strongly between the ages of 13 and 15 in all cases (Poole, Gonneau, and Urquhart 2010). Canada was one of five countries in which rates of drunkenness in this age range were, in many cases, more common among girls than among boys (Poole, Gonneau, and Urquhart 2010).

In several areas of the world, there are also trends in alcohol abuse associated with populations that have experienced oppression. For example, alcohol abuse is high among Canada's Indigenous populations (Alexander 2008) and among Burmese refugees in Thailand, where the substance dependence rate has reportedly reached 40-80 percent of the displaced "ethnic" population of Burma when they arrive as refugees in Thailand (DARE 2008). This is particularly true of refugees from the Karen State, where oppressive structures are particularly severe. It would appear that a loss of community and "ethnicity," combined with experiences of oppression, contributes to an environment that is ripe for the development of substance abuse (Alexander 2008; Campbell, Szumlinski, and Kippin 2009; DARE 2008; Ezard et al. 2011; Maté 2008).

This same dynamic and vulnerability to addictive substances is also observed with sex trade workers (Dabu 2007). Human trafficking is now described as the second-largest industry in the criminal world, second only to drug trafficking, with addiction highly linked to the industry (Polaris Project 2014; Vienna Forum to Fight Human Trafficking 2008). Several studies in Canada have demonstrated a strong correlation between substance abuse and the childhood sexual victimization of those working in the industry (Dabu 2007). In an interview, a former sex trade worker from Thailand estimated that half of all sex trade workers in her country were addicted to alcohol and other drugs (Erickson, King, and Young Women in Transit 2007; Nancy Ross, personal communication, April 2008).

There is also evidence that alcohol abuse increases during times of political instability. In Russia, a substantial decrease in life expectancy between 1989 and 1994 was followed by an increase in life expectancy during the years after the dissolution of the Soviet Union (1994-98). This shift was seen as resulting from a decrease in mean alcohol consumption by 19 percent (Shkolnikov, McKee, and Leon 2001). According to the 2010 edition of *Alcohol: No Ordinary Commodity,* a well-respected research and policy text: "Not only does alcohol account for most of the large fluctuations in Russian mortality, in recent years alcohol was a cause of more than half of all Russian deaths between ages 15 and 54 years (Babor et al. 2010, 67).

Alcohol as a Carcinogenic Substance

As of 2009, alcohol was considered the second leading risk factor contributing to the burden of disease and disability in high-income countries such as Canada (WHO 2009); this was a shift from WHO's 2002 report on global health risks, when it was ranked third. The International Agency for Research on Cancer (IARC) Monograph Working Group has classified alcoholic beverages as "carcinogenic to humans" (Baan et al. 2007, 293). Close to 4 percent of cancers worldwide are attributable to alcohol, with breast cancer representing 60 percent of such cancers in women (Boffetta et al. 2006).

It is important to note that although a number of studies have found cardio-protective effects in middle-aged men and women who participate in moderate alcohol use (e.g., Malinski et al. 2004; Mukamal et al. 2001; Rim and Moats 2007; Ronksley et al. 2011), the net impact of alcohol use in Canada is negative: based on the average volume of alcohol consumed in Canada, 866 net deaths (the number of deaths caused versus those prevented) were attributable to alcohol use among individuals seventy years and younger. Additionally, any cardio-protective effects are negated once a heavy pattern of consumption is introduced, such as binge drinking.

This information is not new. The findings follow many years of research on the relationship between alcohol and disease, especially the relationship between alcohol and cancer. A systematic review of studies on overall burden of disease attributable to alcohol indicated a causal link between consumption and injury, with chronic and acute conditions including multiple cancers, epilepsy, heart disease, depressive disorders, and others (Rehm et al. 2010). The WHO Working Group, among others, has noted a causal connection between alcohol consumption and breast cancer (Baan et al. 2007; Beral 2002; Boffetta et al. 2006; Ellison et al. 2001; Singletary and Gapstur 2001; Smith-Warner et al. 1998). Research on potential mechanisms that would explain the increased risk is well under way (Seitz et al. 2012; Vogel and Taioli 2006). The Nurses' Health Study, which followed 105,986 women over twenty-eight years, looked at the associations of cancer risk with changing patterns of alcohol consumption through adult life using multiple alcohol intake assessment updates over time. The study found significant risk of breast cancer for women who consumed as few as three to six drinks per week; the relative risk of breast cancer increased with an increase in the amount of alcohol consumed (Allen et al. 2009; Chen et al. 2011).

Another review of studies on the epidemiology of alcohol and breast cancer found increased risk of breast cancer among light alcohol drinkers as well as heavy alcohol drinkers (Seitz et al. 2012), with a 4 percent increased risk of breast cancer among those who consumed as little as one drink per day, and a 40–50 percent increased risk in women who drank three or more drinks per day. Based on the evidence, the relative risk of breast cancer increases between 7.1 and 10 percent for each additional 10 grams per day of alcohol consumed (less than one standard drink) (Seitz et al. 2012).

Other Health Implications

Girls and women suffer greater physical harm than men from alcohol consumption (Poole and Dell 2005). Part of this difference is due to the impact of female hormones, different processing of alcohol in women's stomachs, less water, and more fat in women's bodies, resulting in slower rates of alcohol metabolism (Carroll et al. 2004). Alcohol is also more concentrated in the blood of women, causing more harm to bodily organs, and it takes women less time than men to become addicted to alcohol (Najavits 2002a).

Historically, research into women's health and substance use has received little attention, with most studies generalizing both sexes (Fattore, Altea, and Fratta 2008). However, in a twelve-year study of 13,285 men and women aged thirty to seventy-nine years, a group of researchers from Denmark identified an increase in women's risk of liver disease after 7 to 13 standard drinks per week, compared with 14 to 27 drinks for men. Consuming 28 to 41 standard drinks per week resulted in 17 times the risk of liver disease in women and only 7 times the risk in men (Becker et al. 2003). An early study conducted in Ottawa (Nanji and French 1987) investigated alcohol-related disorders and the male-to-female ratio of cirrhosis of the liver as well as harm to other organs. The ratio was significantly correlated with cerebrovascular disease (disease of blood vessels supplying the brain) as well as cancers of the esophagus, liver, and lung. The study concluded that "other organs and/or disease processes in women may be affected more by alcohol than in men" (Nanji and French 1987, 89).

Several studies also indicate that the risk of heart damage from substance abuse is greater in women than in men (Urbano-Márquez et al. 1995), as is the risk of damage to the brain (Hommer et al. 2001; Mann et al. 1992). Many studies have also indicated health issues related to reproduction. Heavy drinking during the reproductive years has been shown to disrupt normal menstrual cycling and reproductive function, sometimes leading to infertility and increased risk of spontaneous abortion (Mancinelli, Binetti, and Ceccanti 2006). Mancinelli, Binetti, and Ceccanti (2006) argue that alcohol abuse also amplifies the harmful effects of exposure to other toxins in our environment, such as lead, and that "female alcoholics are a population heavily at risk for lead exposure" (34). They also point out that thiamine (vitamin B) is dramatically reduced in alcoholics, with particularly severe thiamine deficiency in alcoholic women. Women's increased vulnerability to lead exposure is important because, as Weuve and colleagues (2009) found, cumulative exposure to lead, even at low levels found in community settings, may have adverse consequences for women's cognition when they get older.

When Mancinelli, Vitali, and Ceccanti (2009) updated their research on women, alcohol, and the environment, they continued to stress that all the available gender studies on alcohol show greater severity of alcohol-related damage, including brain damage, in females than in males. The differences are due to physiological peculiarities that make women more vulnerable to the effects of alcohol. They note that the current tendency to begin consuming alcohol at younger ages, together with the growing number of women drinking excessively, is increasing the alcohol-related risks to women's health and justifying the need for better, gender-based studies of alcohol use and abuse.

Despite the increased health risks, a thirty-year review of the literature relating to alcohol abuse concluded that women are less likely than men to seek and enter treatment programs (Greenfield et al. 2007). Information garnered from Women's Services Coordinators at Addiction Services in Nova Scotia revealed that the differences in health implications of alcohol consumption for women often come as a surprise, even to other health professionals. Educating others about the impact of alcohol on women's health, especially chronic disease, may help persuade people that alcohol use is a chronic disease vector, not a moral issue. Building awareness may lead to more women being screened, and more women seeking and receiving help for substance abuse issues.

Is It All in the Genes?

The debate about whether addiction arises from a genetic predisposition has a lengthy history (Maté 2008). Research studies on twins are often cited as evidence that there is a genetic predisposition to substance abuse. Dr. Gabor Maté, in his book *In the Realm of Hungry Ghosts* (2008), argues that the experiences of a mother during her pregnancy impact the developing fetus. He writes that women pregnant at the time of the 9/11 World Trade Center collapse, and those who suffered post-traumatic stress disorder (PTSD), passed on their stress effects to their newborns. At one year of age, these infants had abnormal levels of the stress hormone cortisol. The greatest change in infants occurred in those whose mothers were in the last three months of their pregnancy when the event took place. He states: "The fact that the stage of pregnancy a woman was at when the tragedy occurred was correlated with the degree of cortisol abnormality, suggests that we are looking at an in-utero effect" (Maté 2008, 206).

The results of this and other studies offer a compelling argument that while individuals develop predispositions to struggles with addictive substances, these predispositions result from adverse effects in fetal development and early childhood, not from a gene. Maté believes three factors need to coincide for substance abuse to develop: a susceptible organism, a drug with addictive potential, and stress. Through stories of his patients in Vancouver's Downtown Eastside related to sexual abuse, physical abuse, neglect, and damaged attachments, he demonstrates how pre-and postnatal environments can be recreated from one generation to the next in a way that would impair a child's healthy development without any genetic contribution. Maté suggests that the greater the pain and loss experienced in childhood, the greater the predisposition to issues with substance abuse. In fact, "parenting styles are often inherited epigenetically – that is, passed on biologically, but not through DNA transmission from parent to child" (Maté 2008, 207). He believes the reason that genetic assumptions have such broad acceptance is society's tendency to look for an easy answer, thereby preventing us from feeling a sense of responsibility or guilt about the poor quality of life of those who have a predisposition to dependence on an addictive substance. He states that "we are spared from having to look at how our social environment supports, or does not support the parents of young children; at how social attitudes, prejudices and policies burden, stress and exclude certain segments of the population and thereby increase their propensity for addiction" (Maté 2008, 208).

The Adverse Childhood Effects study, carried out by Felitti and associates over many years and resulting in numerous publications, looked at the impact of childhood trauma on health status and chronic diseases later in life (Felitti 2002, 2004; Felitti et al. 1998), including dependence on addictive substances. The study was based on a sample of 17,421 participants and was a collaborative effort between Kaiser Permanente and the US Centers for Disease Control and Prevention. It explored the relationship between eight childhood events and subsequent adult health. The adverse events were: recurrent and severe physical abuse; recurrent and severe emotional abuse or sexual abuse; growing up with someone who abused substances or was incarcerated, depressed, mentally ill, suicidal, or institutionalized; having a mother who was treated violently; or lacking the presence of both biological parents (Felitti 2004, 4). Results indicated that these adverse events were common in the population, and "they had a powerful correlation to adult health a half century later" (Felitti 2002, 45). A positive correlation between adverse childhood events and later alcohol abuse was observed for both men and women (Felitti 2004, 7).

The results "document the conversion of traumatic emotional experiences in childhood into organic disease later in life" (Felitti 2002, 2). The positive association of Adverse Childhood Events (ACE) with substance use disorders and related health concerns is almost unheard of in epidemiological studies (Felitti 2002). Dr. Felitti reflects that "our attention is typically focused on tertiary consequences, far downstream, while the primary issues are well protected by social convention and taboo." He points out that health care is often limited to the smallest part of the problem as medication is prescribed (7). Felitti (2002) challenges health care providers worldwide to reconsider how they see the origins of organic disease. His research points to the influence of our early environment on our health, and the roles of substance abuse and trauma on the pathway to illness. Those who experience more childhood adversity are challenged in their access to opportunities for health and well-being. This evidence is consistent with the findings that women who engage in substance abuse or binge drinking are more likely to have experienced childhood sexual abuse. Those who have been victims of adolescent sexual or physical assault are at increased risk of binge drinking (Timko et al. 2008).

Trauma and Alcohol Abuse

Concurrent disorders, traumatic experiences, and alcohol abuse can become a vicious cycle. In one Canadian study of women who abused alcohol, 90 percent of participants reported experiencing traumatic events. One of the key reasons the participants used alcohol was to self-medicate the painful feelings related to these incidents (Brown 2008). The incidence of alcohol dependency has been found to be up to fifteen times greater for women impacted by violence than for the general public (Logan et al. 2002). Women are more likely than men to drink alcohol in response to negative and stressful situations (Greenfield et al. 2010). Najavits (2002b) looked at gender differences in the experience of PTSD. She found that women in treatment for substance abuse were two to three times more likely to have a diagnosis of PTSD than men in the same circumstances (Najavits 2002a).

It is estimated that at least half of Canadian women have experienced domestic violence at some time in their lives (Johnson 2005). Romito, Turan, and De Marchi (2005) found that women who experience violence at the hands of their partners are six times more likely to be depressed and four times more likely to use psychoactive drugs than women who do not have this experience. Recognizing and addressing the differences between men and women who abuse alcohol is a critical part of any strategy that aims to change our culture of alcohol use.

Young Women

Research indicates that women and girls are drinking more alcohol than at any other time in history (Mancinelli, Binetti, and Ceccanti 2006). Between 2003 and 2008, the number of Canadian women aged twenty to thirty-four who reported monthly risky drinking increased by almost 20 percent. In this same time period, monthly risky drinking declined for men but increased for women by 9.1 percent (Poole, Gonneau, and Urquhart 2010). American and European epidemiological studies indicate that by the end of the 1990s, the ratio of at-risk drinking for male and female adolescents (teenagers aged twelve to seventeen years) was nearly 1:1 (European Institute of Women's Health, 2013; Mancinelli, Binetti, and Ceccanti 2006). Anecdotally, many young women now attempt to "drink men under the table" or at least stay in the running. Additionally, some young college women see their heavy drinking as a form of feminism, to demonstrate that women can do what men can do (Ehrenreich 2002). This rise in binge drinking has unique hazards. The harms are well documented, and include placing oneself at risk for unwanted sexual activity (Testa et al. 2006) and a higher risk of exposure to violence (Richardson and Budd 2003). The 2004 Canadian Campus Survey (Adlaf, Demers, and Gliksman 2005) reported that 12.8 percent of women attending college and universities had unplanned sexual activity due to alcohol.

In 2008, in the small town of Bridgewater, Nova Scotia, a mother had her sixteen-year-old daughter taken by ambulance to the emergency room for alcohol poisoning. She learned from the emergency room nurses that her daughter had been the fifth young person treated for alcohol poisoning that weekend (Anonymous, personal communication, August 2008). No brief interventions were available for these youths, and no information was given about the services offered by the local Addiction Services department. Unfortunately, many young girls and women do not access Addiction Services and many do not even know that such support exists (Schrans, Schellinck, and McDonald 2008). Heavy drinking has become a societal norm, and its long-term health and social effects are often not revealed until later on in life (Nova Scotia Department of Health Promotion and Protection 2007).

The prevalence of female student binge drinking in the United Kingdom may be as high as 63 percent (Guise and Gill 2007), yet many girls and young women do not see this form of drinking as problematic. This form of denial was revealed in a study on binge drinking among undergraduate students in Scotland, in which respondents "crucially positioned themselves outside the categories of 'serious' or 'anti-social' drinkers" (Guise and Gill 2007, 895). Binge drinking in the United Kingdom is a huge concern and often appears to result in violence (Richardson and Budd 2003). In 2006 and 2007, 87,200 females were arrested for violence, twice as many as in 2002. Many argue that this has been fuelled by the government's lax approach towards binge drinking (Slack 2008). Robert Whelan of the Civitas think tank blamed such increases in violence from substance abuse on a breakdown in society's values: "These problems have been ignored or mishandled for so long we have become a more savage society. It is very difficult to see a way out of it now" (Slack 2008, 1).

Women of Child-Bearing Age

There is a rapid growth in research demonstrating the adverse effects of alcohol on fetal development (Mancinelli, Binetti, and Ceccanti 2006). "Fetal Alcohol Spectrum Disorder" (FASD) is an umbrella term used to denote a range of harms to the fetus caused by maternal alcohol consumption during pregnancy, including permanent and irreversible central nervous system damage. Such damage can affect an individual's physical makeup as well as cognitive and behavioural functioning. The degree to which prenatal alcohol exposure harms a fetus depends on numerous factors, including genetics, maternal characteristics, nutrition, environment, developmental timing, reactions to other drugs, and duration and extent of alcohol exposure (Fraser 2008). If a woman has access to good nutrition, safe housing, and medical care, these protective factors may mitigate some of the harm caused by substance abuse (National Indian and Inuit Community Health Representatives Organization n.d.). Research presented at the Fetal Alcohol Canadian Expertise (FACE) Research Roundtable in 2008 indicated that heavy drinking metabolizes methanol (which is present in most alcoholic beverages) to produce formic acid, a substance that can kill brain cells. These findings led to the conclusion that drinking during pregnancy exposes an unborn child to potential brain damage. The same researchers also found that folic acid may help prevent cell death caused by formic acid, and may be a future option in prevention of FASD. FASD is considered the leading cause of developmental and mental disabilities worldwide (Stade, Clark, and D'Agostino 2004).

Pregnant women who consume alcohol may do so for a variety of reasons. Some studies suggest that as many as 50 percent of pregnancies are unplanned (Finer and Zoina 2014). Thus, many women may consume alcohol without realizing that they are pregnant. Given the increase in binge drinking among younger women, the possibility of harm to the fetus is of great concern. The 2004 Canadian Campus Survey (Adlaf, Demers, and Gliksman 2005) reported that 33.4 percent of women drank more than once a week, and that 42.4 percent demonstrated at least one indicator of harmful drinking according to the AUDIT screening tool. With women drinking this much at this frequency, it is likely that many may drink without knowing they are pregnant. In a 2001 study, Ira Chasnoff revealed that, despite concern over the consequences of prenatal alcohol and drug exposure, obstetricians continued to place low priority on advising patients about substance use during pregnancy. Although 97 percent of obstetricians reported asking their patients about alcohol use, 80 percent said that they told patients that limited alcohol use posed an insignificant threat to the pregnancy and developing fetus. In fact, 4 percent of the obstetricians surveyed stated that eight drinks or more per week was a "safe" level of alcohol consumption for a pregnant woman (Chasnoff 2001, 3).

This information was misleading: recent evidence indicates that even one drink per week can place a child at increased risk for delinquent behaviour and other problems later in life (Chasnoff 2001). The Centre for Addiction and Mental Health (CAMH 2009a), a WHO collaborating centre, categorically states that no amount of alcohol is safe during pregnancy. The 2004 Canadian Campus Survey found, however, that although 94 percent of women believed that drinking during pregnancy could be harmful to the developing fetus, 69 percent stated that either a small or moderate amount of alcohol consumed during pregnancy would be considered safe (Adlaf, Demers, and Gliksman 2005). Further efforts are needed to educate the medical community and the general public on this issue (Mancinelli, Binetti, and Ceccanti 2006, 33).

Older Women

Seniors are one of the fastest-growing populations in Canada. Alcohol is the most commonly used psychoactive substance among seniors. Findings from the 2004 Canadian Addiction Survey (Adlaf, Begin, and Sawka 2005) suggest that over 13 percent of women over sixty-five years of age drink alcohol four or more times a week. Health professionals frequently qualify alcohol abuse problems among seniors as either "early onset" or "late onset." The latter suggests that beginning to abuse alcohol at a later age signifies that something may have happened in an individual's life to trigger this heavier use. This abuse is regarded as a form of self-medication for problems such as grief, retirement, and other losses associated with aging. The socially isolated environments in which older women live can also contribute to their substance abuse (Blow and Barry 2005).

Many seniors experience higher rates of depression and anxiety and are frequently given medication, such as sedatives, to help them cope with these concerns. Older women are prescribed sedatives at twice the rate of men (Currie 2003). Studies indicate that 20-50 percent of all women over sixty may be prescribed sedatives or sleeping pills, and that long-term use increases with age. The combination of this type of medication with alcohol often produces negative impacts, including intensified effects of alcohol, higher rates of injury, and hazardous side effects such as drowsiness, dizziness, problems with coordination, unusual behaviour, and memory problems. It may also slow heart rate and breathing (Monsom and Schoenstadt 2007). Mancinelli, Binetti, and Ceccanti (2006) argue that heavy drinking among seniors may increase the risk for Alzheimer's disease, particularly in women, who are often more vulnerable to alcohol-induced brain damage.

As people age, the body's ability to process alcohol diminishes (Health Canada 2001). This reality adds to a growing concern over the quantity of alcohol that elderly people choose to drink. Health care professionals working in hospitals of the South Shore District Health Authority in Bridgewater, Nova Scotia, comment on the growing numbers of seniors entering emergency rooms under the influence of alcohol. These seniors often suffer from falls and fractures resulting from substance use. It is these consequences that are documented in medical charts, making it difficult to retrieve or research substance abuse as an underlying health issue among seniors, and allowing the problem to go undetected and under-acknowledged. As seniors constitute a larger portion of the population, the social determinants that impact their quality of life should be given more attention.

Screening for Alcohol Abuse

There is a growing body of literature tracking women's unique experiences with substance use, including entry into use, progression of severity, paths of treatment and relapse, and maintenance of changes (Brady, Back, and Greenfield 2009). High-quality systematic reviews of women and alcohol abuse are emerging but still scarce. To render visible the different experiences of women and girls, we need screening and instruments that are sensitive to gender and sex differences. To understand the impact of alcohol, we need to capture the profound differences in the life experiences of women and men.

Alcohol screening for women is rife with complications that relate to stigma and the fear that children will be apprehended by Child Welfare. When women begin to abuse alcohol, few questions tend to be asked (Schorling and Buchsbaum 1997) and there is a missed opportunity to distribute information about risk factors and health implications. Downplaying or sidestepping questions about alcohol use diminishes discomfort in the short run, but in the long run it endangers women's health and slows the collection of substantive evidence that would encourage both policy makers and health care professionals to take action. As a result, information about women, particularly those who are marginalized, is scarce, minimized, or missing. Meanwhile, the alcohol industry continues to promote its products through various media, reinforcing the view that drinking is routine (O'Malley and Valverde

2004). This is of particular concern for women, considering that even before alcohol was declared carcinogenic (Baan et al. 2007), Najavits (2002a) reviewed the literature and noted that women were two times more likely than men to die of alcohol-related problems (McCrady and Raytek 1993) and four times more likely to die of an addiction-related illness than to die of breast cancer (Blumenthal 1998).

In 2010, the *Journal of Obstetrics and Gynaecology Canada* published the Alcohol Use and Pregnancy Consensus Clinical Guidelines to improve the screening and recording of alcohol use (Carson et al. 2010). Prenatal exposure to alcohol is the leading cause of preventable neurodevelopmental disabilities in Canada (Murthy et al. 2009; Stade et al. 2006), and evidence on maternal drinking can assist in diagnosing Fetal Alcohol Spectrum Disorders (Chudley et al. 2005). The guidelines recommend, therefore, that screening for alcohol use be done on *all* women of child-bearing age. Because many pregnancies are unplanned, universal screening can promote conversations for all women of child-bearing age about risky drinking and its impacts on pregnancy. This screening initiative may contribute to improved data on prevalence of abuse, which in turn could inform Canadian alcohol policies, improve the health and well-being of women and girls, and diminish the occurrence of FASD.

Unfortunately, stigma about substance abuse can be magnified during the screening process when a woman is pregnant. Discussions in these circumstances require great sensitivity and a forthright ability to ask direct questions in a non-judgmental fashion: even seasoned health professionals can struggle to be compassionate towards women who abuse substances while pregnant. Screening for substance use is improved when health professionals confront their own personal biases and prejudices (Westermeyer, Mellman, and Alarcon 2006) and consider the reality of women's lives, including past trauma and earlier stresses (Brown 2008; Najavits 2002a, 2002b). The Alcohol Use and Pregnancy Consensus Clinical Guidelines (Carson et al. 2010) provide information for health professionals on dealing with substance abuse in a respectful, supportive, and non-judgmental fashion.

The screening of women for high-risk alcohol has sometimes been complicated by the use of questionnaires originally designed to detect harmful alcohol abuse in men (Bradley et al. 1998). Men have different patterns of drinking, different thresholds for harm from alcohol consumption, and different consequences, including less stigma associated with alcohol abuse. Some questionnaires have now been adapted for women by using lower cut-off points for assessment of hazardous alcohol use (Chang 2001). Considering the difference that brief screenings can make in the lives of women and girls, it is valuable for health professionals to invest time in creating a safe environment and a gender-sensitive process for identifying risky substance use.

Stigma and Social Exclusion in the Rural Context

In examining society's obligation to provide the opportunity for health and well-being, Norman Daniels (2008) identifies social exclusion as a barrier to accessing health, and argues that in many ways social exclusion is a matter of distributive justice, raising questions about who is under-represented, who is invisible, and who lacks access to health services for social reasons. Social exclusion often works in subtle ways that are difficult to identify.

The 2006 Kirby and Keon report *Out of the Shadows at Last: Transforming Mental Health, Mental Illness and Addiction Services in Canada,* for the Standing Senate Committee on Social Affairs, Science and Technology, observed that people affected by substance abuse did not come forward to comment on their experiences when given the opportunity. Although the report was meant to be a blueprint for Canada's mental health and addiction services, only 20 of its over 500 pages discussed substance use specifically, and less than one page focused on women. People were willing to come forward to address the topic of mental illness, but the format did not capture many responses about substance abuse. It is imperative that information collection methodologies in Canadian health care consider and compensate for those who may be socially excluded.

Exclusion from services can occur when data used to inform policies do not consider who is missing and why, or when individuals are unable to access services due to personal or social barriers. Women tend to have lower incomes (Statistics Canada 2012) and are therefore less able to obtain the support they require, such as child care and affordable transportation, in order to use services. People who live in rural areas often find that public transit can be sporadic or unavailable. Those who live outside of the main towns may not have access to reliable vehicles to reach health services.

Stigma is the most difficult barrier to remove, and it is exacerbated by assumptions that a woman who abuses substances may be a "bad mother" (Finkelstein 1994). This stigma may leave women isolated and hiding their alcohol use. It is daunting to communicate with health care and Addiction Services if there is the possibility that Child Welfare will be contacted for suspected neglect or abuse of a child.

Stigma is particularly threatening in rural communities, where anonymity is a challenge. It is important to maintain a good reputation for the sake of employment, renting accommodations, and even the community's expectations of other family members. When a woman seeks help in a rural area, she will probably be speaking to a health professional who is familiar to someone in her circle of family and friends. Health care providers and patients encounter one another at the market, the grocery store, school, sporting events, local amenities, and formal and informal gatherings. Additionally, those working in the rural health care system who suffer from substance abuse may have to travel for hours to receive treatment, or risk their professional reputations.

The Canadian Rural Partnership Analysis (1991-2001) concluded that those living in rural and small-town Nova Scotia are unable to access the social determinants of health required to achieve the same level of health as their urban counterparts (De Peuter and Sorensen 2005). According to the group Rural Nova Scotia Impacting Policy (RCIP 2006), between 60 and 70 percent of Nova Scotians live in rural settings, compared with an average of 20 percent throughout the rest of Canada (De Peuter and Sorensen 2005). In rural Nova Scotia, it is challenging to access a primary health care provider (Lombard 2005). This is particularly significant for those with alcohol use issues, as primary health care providers are the first ones Nova Scotians approach for help and treatment (Schrans, Schellinck, and McDonald 2008).

In addition to these barriers, there can also be a lack of representation of diverse groups among rural women. For example, African Nova Scotians, who have lived in the province for many generations, continue to be under-represented in health care delivery, research, and the design and implementation of health policy (Bernard 2005; Sharif, Atul, and Amaratunga 2000). Rural women from marginalized groups seeking services must contend with a combination of geographic, racial and ethnic, and gendered barriers. Bernard (2005) reports that racism has a cumulative health effect on black women in Nova Scotia, resulting in a higher risk of chronic diseases, including depression, substance abuse, and a host of other stress-related diseases.

Compared with non-Aboriginal women, Aboriginal women have greater experience with marginalization, violence, and intergenerational trauma (Halseth 2013). Such experiences increase their vulnerablity to employing harmful coping mechanisms such as alcohol abuse. Knowing that there will be an understanding of the impact of race, culture, and trauma with an attentiveness to the differences among women will make it easier to enter and remain in treatment. When there is increased vulnerability, we need to be attentive and responsive to changes that will increase opportunities for health and well-being.

There are many reasons that diverse groups have difficulty obtaining treatment for alcohol use disorders. For lesbian, gay, and transgendered women, there is a fear of being "outed" if they seek treatment and disclose their sexual orientation. Health Canada identified several barriers to women with issues of substance use in rural communities:

- lack of services for people who struggle with addiction
- lack of women-centred services
- lack of treatment programs for women who were sexually and physically abused as children
- limited education of professionals on the challenges that rural women with addictions face.

Besides these factors, clinicians may not have an understanding of the cultural issues that may impact these groups:

> A Health Canada study of immigrant women and substance use ... identified a number of issues related to services access and barriers for an immigrant woman experiencing a substance use problem. These included lack of expertise and resources on the part of both mainstream health organizations and immigrant aid or ethno-cultural organizations to address women's substance use problems. (Health Canada 2007, Section 6.5.4)

Stigma, trauma, fear of losing children, decreased access to health care providers, and lower income are barriers to services when alcohol use becomes a problem for women. Attention needs to be paid to our diverse cultures, races, histories, and the trauma in our lived experiences. The lens of social exclusion and gender analysis will help render visible the changes necessary to improve access to the opportunities for health and well-being of women and girls involved with alcohol.

Access to the Social Determinants of Health

Canada has been a leader in recognizing that influences on health are broader than genetics and lifestyle choices. The primary determinants of health are most often social (Raphael 2003). Although these social factors are beginning to be addressed at the policy level, health care workers and the media continue to stress healthy lifestyles and biomedical processes as the most important approaches to improving one's health (Raphael 2003). Social determinants of health include but are not limited to the following: income and social status, social support networks, education, employment and working conditions, physical environment, social environment, biology and genetic endowment, coping skills and personal health practices, healthy child development, health services, gender, and culture (PHAC 2010). (See Parts 1, 3, and 4 in this volume for other discussions of social determinants of health.)

Leaders of the World Health Organization (WHO 2008) remind us that social disparities, those gaps or inequalities in social and economic status between groups in a given population, greatly affect population health status as a whole. Women and men have different roles and opportunities. These differences have led to the recognition of gender as one of the social determinants of health. We see the influence of gender with chronic disease. Although women live longer than men, they live with more chronic disease from a younger age (Pan American Health Organization 2011). The consequent increase in medications, challenging finances, and increased physical limitations contribute to stressful living. Combining some of the medications for chronic disease with alcohol can be dangerous. Recognizing, understanding, and addressing gender differences will lead to better programs, policies, and health care outcomes for the entire population (PHAC 1998).

The primary social determinant of health in Canada is income (PHAC 2010; Chapters 1, 3, 10, and 11). Low income is associated with having fewer choices and resources to influence life situations, particularly stressful situations. Income is connected to safe housing, adequate child care, opportunities for good nutrition, transportation, and educational opportunities. A population that experiences increased levels of stress will also have increased levels of chronic disease, including problems with mental health and substance abuse. It is well documented that stress and negative emotions induce cravings in alcohol-dependent individuals (Fox and Sinha 2009).

As discussed in Chapters 1, 3, 10, and 11, among others, women often experience poverty disproportionately to men (Morris and Gonsalves 2005; Statistics Canada 2012). The average total income of a Canadian woman is 64 percent of that earned by a man (Statistics Canada 2012). Statistics Canada (2012) underlines this disparity:

> The difference in women's and men's earnings is found in both the professional and non-professional occupational groupings. For example, women in medicine and health-related occupations earned about 57 percent as much as men in those occupations; women in business and finance occupations earned about 59 percent as much as their male counterparts. In the non-professional occupations such as sales and service, women earned about 57 percent of men working in these occupations.

As an example, the Nova Scotia Child Poverty Report Card (2007) states that 90 percent of single-parent families in Nova Scotia are led by women, and 25 percent of these women receive social assistance. Single mothers in Atlantic Canada were also reported to have had 2.7 hours per day less free time in 2005 than in 1999 (Pannozzo et al. 2009). With lower incomes, single women working outside the home may encounter difficulties finding affordable and appropriate child care. When problems develop in families where resources are tight and stress is already high, the impact is strong.

WHO (2008) emphasizes the need to address inequities in early child development opportunities. They note that inequities in early life lead to inequities in health in adulthood. Single parents suffer the stress of wanting to give their children a good start while frequently dealing with less access to the social determinants of health. The targeting of women of child-bearing age for increased sales of alcohol may place an extra burden on families.

Stress is a robust predictor of relapse for alcohol-dependent individuals (Fox and Sinha 2009). Stress comes from many sources, including income. The elevated stress resulting from social barriers to health and well-being diminishes an individual's sense of self-efficacy and hope for change. An attitude of "why bother" is a major trigger for initiating or returning to alcohol abuse. A cycle of stress, discouragement, alcohol use, and diminished coping skills can be difficult to break. Stress is exacerbated in those individuals who do not have a financial buffer if they encounter job loss, illness, disability, or other unforeseen circumstances. It is difficult to know the number of women who are forced to choose between staying in abusive relationships and poverty. Similarly, if a woman is financially dependent on friends and family who are also drinking partners, her choice to pursue sobriety may be unsupported. The lack of adequate income to respond to life's demands has a cumulative deleterious effect on women's health and can contribute to a higher frequency of chronic disease. When alcohol is combined with the medications for chronic disease, further complications can arise.

A population health approach encourages us to see that each time we work towards equality and a sense of self-efficacy for women, we are working towards improving the health of all Canadians (Hayward and Colman 2003). The more we approach the improvement of health by addressing social injustice, the less blame and stigma are assigned to individual health problems such as alcohol abuse. It also helps us understand that when problems develop among those who have less access to the social determinants of health, there is less to buffer the harms. If we value childhood development highly in our country, we need to ask ourselves why we tolerate the proliferation of alcohol marketing directed towards women of child-bearing age.

Financial Cost of Alcohol Use

As Canada's second leading risk factor for the global burden of injury and disease, alcohol use comes with a significant social and economic price tag. The social cost is difficult to estimate and is often paid by family, friends, and communities. Alcohol use leads to higher economic costs for health care and policing. It diminishes productivity in the workplace and consequent economic progress, and ultimately places a greater burden on taxpayers. In a health care system consistently challenged by funding constraints, it is important to consider the revenue that governments receive from the sale of alcohol vis-à-vis the social, health, and policing costs associated with alcohol abuse.

Although there is the potential for improving governments' income from alcohol and for reducing alcohol-related harms, public opinion needs to change before stricter alcohol policies will be accepted. The public needs more information about sex and gender differences with regard to alcohol. This need can be met by funding future research agendas about women and alcohol and using the power of media.

Increased attention to how to protect those most at risk will help reduce the social and economic costs of alcohol. Unfortunately, public opinion is strongly influenced by the deep pockets of the alcohol industry and by alcohol marketing. The alcohol industry has been paying attention to the potential for increased sales to women, and has developed and successfully marketed more products. Women's rate of alcohol consumption is rising faster than that of men (Thomas 2012).

Make the Healthier Choice the Easier Choice

Our relationship with alcohol is complex. Alcohol is highly politicized and is used on a regular basis for a number of reasons. It acts as a social lubricant or a coping instrument, and is usually a welcome addition to dinner. Alcohol is part of our everyday lives. In a time of plenty and in a society and culture of excess, bigger is almost always perceived as better. Our environment tempts us and encourages us to follow the path of least resistance; the healthier choice is rarely the easier choice (Michael Vallis, personal communication, October 2008).

Consider how easy it is for us to pick up hamburgers, french fries, and pizza, and how expensive and time-consuming it is to prepare homemade food. Think of how many more candies you will eat because you bought the jumbo pack, and how easy it is to continue eating. Think of the path of least resistance, and how our society is helping us along. Now consider a few examples related to alcohol.

In Nova Scotia, for example, alcohol became more accessible to the public when in July 2007 it became available for purchase at alcohol outlets seven days a week instead of six. International research has shown that increasing the availability of alcohol has led to an increase in the amount of alcohol purchased and consumed in a population (Babor et al. 2010). This is exactly what has happened in Nova Scotia: according to its CEO, the Nova Scotia Liquor Corporation (NSLC) "exceeded its financial target for the third year in a row. Driving these results was the impressive growth in our wine and beer categories, combined with increased access through the expansion of agency stores and Sunday shopping ... It is reasonable to conclude that the NSLC's targeted Sunday openings has produced the desired lift in sales" (NSLC 2008, 1, 4). Through increased availability and accessibility, Nova Scotia has seen an increase in sales of alcohol and in government revenue. Is the government of Nova Scotia aware that increased sales also lead to increased harm (Babor et al. 2010; Tombourou et al. 2007; WHO 2011)? In making alcohol more accessible, companies and governments are condoning and promoting our culture of excess and making the healthier choice the more difficult one.

The province of British Columbia provides an example of a government trying to put health first. Increasing the minimum price at which alcohol can be sold has been shown to reduce levels of consumption, thereby reducing levels of harm (Babor et al. 2010). A 2012 study looked at how levels of alcohol consumption were impacted by increases in the minimum price of alcoholic beverages over a twenty-year period (1989-2010) in British Columbia (Stockwell et al. 2012). Results showed that as the price of alcohol increased, levels of consumption decreased. A 10% increase in the price of spirits, wine, and beer reduced the consumption of each of these types of alcohol at significant levels. The consumption of spirits decreased by 6.8 percent ($p = .004$), wine by 8.9 percent ($p = .033$), and beer by 1.5 percent ($p = .043$).

Another example is Ontario, the only province in Canada that adjusts prices for all alcoholic beverages based on alcohol content: those with higher alcohol content have higher price points than those with lower alcohol content. Other provinces are moving in this direction, however. For example, Saskatchewan uses a number of strength categories of alcoholic beverages when setting prices, and Quebec uses strength of alcohol content to set prices for beer (Giesbrecht et al. 2013).

Conclusion and Recommendations

Research is increasingly revealing connections between alcohol and women's health. Alcohol is implicated in over sixty disease categories (Rehm et al. 2006a), including causal relationships with breast cancer and colorectal cancer (Baan et al. 2007). These links to cancer and disease are confounded by the particular vulnerabilities that many women with abuse issues have in other areas of their lives. Alcohol abuse is heightened in cases of domestic violence, adverse childhood events, instances of injury to oneself and others, and risky sexual behaviour (Poulin and Graham 2001). Discussing prevention efforts, Maté (2008, 419) suggests that our efforts should begin during childhood development or earlier, "in the social recognition that nothing is more important for the future of our culture than the way children develop. There has to be much more support for pregnant women. Early prenatal visits should be an opportunity not only for blood tests, physical exams and nutritional advice, but also for a stress inventory of a woman's life."

Regulation of access and availability of alcohol and taxation of alcohol have been identified by international research as best practices in reducing alcohol-related harms (Babor et al. 2010). The greatest increase in alcohol sales in Canada has occurred in provinces in which regulation of alcohol has been eroded (Giesbrecht et al. 2011). The types of policies described above can help governments transform the health of their populations (Rehm et al. 2006b). As we use policy to help change our environment, healthier choices will become the easier choices and public behaviour will be transformed.

Throughout this chapter, we have presented the unique experiences of women related to health concerns, trends, social determinants of health, trauma, social exclusion, and stigma. We see from both Gabor Maté's work and the Adverse Childhood Events study that such circumstances can influence our vulnerability to alcohol abuse long before we reach the legal drinking age. Canadian women continue to lack the same access to health opportunities as men, and if they turn to alcohol to cope with life stresses, they experience greater stigma. This is a problem of both distributive and social justice that all of us share responsibility for solving.

Knowing that alcohol consumption leads to a large proportion of *preventable* morbidity and mortality, Canada must take a stand. Canadian regulations and policies need to reflect the mounting evidence of the health consequences of harmful patterns of alcohol consumption, and to emphasize the protection of vulnerable populations. The following are our recommendations for bringing these about.

Revise Regulations that Govern the Advertising of Alcohol

Considering the recent identification of alcohol as a carcinogenic chemical, with profound and far-reaching effects on the health of Canadians, Health Canada's Drug Strategy and Controlled Substances Programme, under the legislated mandate of the Canadian Centre on Substance Abuse Act,[1] should oversee the revision of regulations governing the advertising of alcohol.

At present, advertising is regulated by the Canadian Radio-television Telecommunications Commission (CRTC) through Advertising Standards

Canada (ASC). All alcohol advertisements to be placed on Canadian television or radio must be cleared by the ASC, whose team follows the CRTC's regulations, last amended in 1996 (CRTC 1996). Considering the risks of moderate alcohol consumption for women, and the links found between increased exposure to alcohol advertising and increased levels of consumption, especially among young people (Anderson et al. 2009; British Medical Association Board of Science 2009; Gordon, Hastings, and Moodie 2010; Smith and Foxcroft 2009), we recommend firmer guidelines for advertising and marketing techniques that target young girls and women as both purchasers and consumers of alcohol. Public health experts with knowledge of the damaging effects of alcohol advertising should be included in regulatory design and decision-making processes to ensure proper guideline development, monitoring, and enforcement.

The Drug Strategy and Controlled Substances Programme should also oversee the development of alcohol advertising and marketing guidelines for the Internet and cell phones. There are currently no guidelines governing advertising on Internet social media sites such as Facebook or on cell phone applications, often promoted on beer bottles, that engage consumers in trivia, polling questions, and more. While online age verification requirements exist, they are not enforced, and any child can enter a birth date of legal drinking age in order to access advertisements, alcohol-brand websites, or phone applications.

Considering the enormity of the task of regulating alcohol advertising over the Internet, monitoring and regulating could initially be limited to Canadian-content websites, such as those of alcohol companies and social media. Guidelines should include restrictions on where online advertising takes place; for example, advertising should be limited on any sport, school, or online gaming websites frequented by or catering to young people. There should also be stricter regulations concerning the content of Canadian alcohol organization sites, including prohibition of coupons for deeply discounted alcohol or of content glamourizing alcohol use. These regulations should also prohibit ads that specifically target women or young people.

The Public Health Agency of Canada (PHAC) should take responsibility for the alcohol-education messages that are required as part of the current CRTC guidelines for organizations releasing alcohol advertisements. It is irresponsible to have the alcohol industry promote its product while simultaneously educating about its harms: this is a clear conflict of interest. Researchers have found that strategies used by industry to fulfill their social responsibility mandates, such as campaigns against drinking and driving and school-based campaigns, are ineffective against harms related to alcohol use (Babor et al. 2010; Bond, Daube, and Chikritzhs 2009, 2010). The most effective age- and gender-appropriate messages about alcohol abuse will come from PHAC employees who have no stake in the sale of alcohol.

The World Health Organization released its first Global Strategy to Reduce the Harmful Use of Alcohol in 2011. Among its ten recommended areas of intervention is a section on limiting exposure to alcohol advertising through increased regulation of such advertising in new media, alcohol sponsorships of events, and other protective measures regarding alcohol marketing. By creating a public health mandate to regulate alcohol advertising, the Public Health Agency of Canada and Health Canada would be following the lead of the WHO.

Implement Gender-Specific Policy, Prevention, and Treatment Programming

McGourty and Chasnoff (2003) argue that gender-specific treatment is key to a woman's success in treatment and her ongoing recovery. A Canadian women's health policy brief by the BC Centre of Excellence for Women's Health (2005) notes the great international interest in women's substance use treatment, prevention, and policy. The document points to a recent United Nations report advocating gender-specific prevention, outreach, treatment, and harm reduction approaches. More attention needs to be focused on women as a group at particular risk for alcohol purchase and consumption and on how this influences their opportunities for health and well-being.

Implement Universal Screening

McGourty and Chasnoff (2003) note that issues of substance abuse have been largely neglected in policy. They suggest that substance abuse issues be regarded as health care concerns, not social problems, and recommend screening of the entire population instead of target populations suspected of substance abuse. The Alcohol Use and Pregnancy Consensus Clinical Guidelines (2010) provide a fine example of both universal screening *and* recording of alcohol use.

Research Nutritional Supplementation as a Harm Reduction Approach

More research is necessary to adequately inform women about nutritional supplementation as a part of harm reduction, such as vitamin B as a thiamine replacement for women abusing alcohol, and folic acid for pregnant women in particular.

Implement Gender-Based Analysis of Harms

Health Canada's commitment (2011) to "gender-based analysis" has been helpful in tracking differences in alcohol use patterns between men and women based on the availability of particular social determinants of health. To date, most of the analysis has focused on biological differences through sex-segregated data. We recommend further exploration into a gender- or sociologically based analysis and the inclusion of both sex- and gender-based analysis in future alcohol policies, especially with regard to the current epidemic of binge drinking among young women and girls.

Include All Groups in Designing Research and Policies

Since our health care systems require evidence in order to fund services, it is important to design methodologies that consider and compensate for those who may be socially excluded. Future research agendas and policies should consider those who have not been counted. Mothers fly under the radar when it comes to alcohol; do we really want to ignore this group? Similarly, those who are numbing traumatic experiences with alcohol will not readily step up to participate in studies of this topic.

Make the Healthier Choice the Easier Choice

Creating environments that are most conducive to healthy behaviours will have the greatest impact on public health, as opposed to focusing only on individual prevention and treatment programming (Frieden 2010). The best way to create healthy environments is to enact policies that make the healthier behavioural choice the easier choice. Over thirty years of research by eminent alcohol policy experts indicates that policies to reduce access to alcohol will limit alcohol purchases and consumption and the resulting harms (Babor et al. 2010). Such policies include limiting the days and hours that alcohol retail outlets are open, increasing the price of alcohol or the taxes on alcohol, and enforcing a minimum legal drinking age.

These strategies are cost-effective ways of restricting access and enable consumers to make the healthier choice. When alcohol outlets close earlier in the day or are not open every day of the week, or when alcohol costs more, people will consume less. When people consume less, fewer harms are incurred. Limiting access to alcohol through these measures will lead to substantial reductions in alcohol-related harms.

Note

1 *Canadian Centre on Substance Abuse Act,* R.S.C., 1985, c. 49 (4th Supp.) (http://laws-lois.justice.gc.ca/eng/acts/C-13.4/index.html).

References

Adlaf, Ed M., Patricia Begin, and Ed Sawka, eds. 2005. *Canadian Addiction Survey (CAS): A National Survey of Canadians' Use of Alcohol and Other Drugs, Prevalence of Use and Related Harms – Detailed Report.* Ottawa: Canadian Centre on Substance Abuse. http://www.ccsa.ca/Resource%20Library/ccsa-004028-2005.pdf.

Adlaf, Edward M., Andrée Demers, and Louis Gliksman. 2005. *Canadian Campus Survey 2004.* Toronto: Centre for Addiction and Mental Health. http://www.camh.ca/en/research/research_areas/community_and_population_health/Documents/CCS_2004_report.pdf.

Alexander, Bruce. 2008. *The Globalization of Addiction: A Study in Poverty of the Spirit.* New York: Oxford University Press.

Allen, Naomi E., Valerie Beral, Delphine Casabonne, et al. 2009. "Moderate Alcohol Intake and Cancer Incidence in Women." *Journal of the National Cancer Institute* 101 (5): 296-305, doi:10.1093/jnci/djn514.

Anderson, Peter, Avalon de Bruijn, Kathryn Angus, et al. 2009. "Impact of Alcohol Advertising and Media Exposure on Adolescent Alcohol Use: A Systematic Review of Longitudinal Studies." *Alcohol and Alcoholism* 44 (3): 229-43, doi:10.1093/alcalc/agn115.

Baan, Robert, Kurt Straif, Yann Grosse, et al. 2007. "Carcinogenicity of Alcoholic Beverages." *Lancet Oncology* 8 (4): 292-93, doi:10.1016/S1470-2045(07)70099-2.

Babor, Thomas, Raul Caetano, Sally Casswell, et al. 2010. *Alcohol: No Ordinary Commodity: Research and Public Policy.* 2nd ed. Oxford: Oxford University Press.

BC Centre of Excellence for Women's Health. 2005. *Girls, Women, Substance Use and Addiction: Women's Health Policy Brief.* Vancouver: BC Centre of Excellence for Women's Health. http://www.bccewh.bc.ca.

Becker, Ulrik, Allan Deis, Thorkild I.A. Sørensen, et al. 2003. "Prediction of Risk of Liver Disease by Alcohol Intake, Sex and Age: A Prospective Population Study." *Journal of Hepatology* 23 (5): 1025-29, doi:10.1053/jhep.1996.v23.pm0008621128.

Beral, Valerie. 2002. "Alcohol, Tobacco and Breast Cancer: Collaborative Reanalysis of Individual Data from 53 Epidemiological Studies, Including 58,515 Women with Breast Cancer and 95,067 Women without the Disease." *British Journal of Cancer* 87: 1234-45, doi:10.1038/sj.bjc.6600596.

Bernard, Wanda Thomas. 2005. "Black Women's Health in Nova Scotia: One Woman's Story." In *Surviving in the Hour Of Darkness: The Health and Wellness of Women of Colour and Indigenous Women,* edited by G. Sophie Harding, 47-70. Calgary: University of Calgary Press.

Blow, Frederic C., and Kristen Lawton Barry. 2005. *Use and Misuse of Alcohol among Older Women.* Washington, DC: National Institute on Alcohol Abuse and Alcoholism. http://pubs.niaaa.nih.gov/publications/arh26-4/308-315.htm.

Blumenthal, Susan J. 1998. "Welcome from the US Public Health Service." In *Drug Addiction Research and the Health of Women: Executive Summary,* edited by Cora Lee Wetherington and Adele B. Roman, 4-8. Rockville, MD: US Department of Health and Human Services – National Institute on Drug Abuse.

Boffetta, Paolo, Mia Hashibe, Carlo La Vecchia, et al. 2006. "The Burden of Cancer Attributable to Alcohol Drinking." *International Journal of Cancer* 119 (4): 884-87, doi:10.1002/ijc.21903.

Bond, Laura, Mike Daube, and Tanya Chikritzhs. 2009. "Selling Addictions: Similarities in Approaches between Big Tobacco and Big Booze." *Australasian Medical Journal* 3 (6): 325-32, doi:10.4066/AMJ.2010.363.

–. 2010. "Access to Confidential Alcohol Industry Documents: From 'Big Tobacco' to 'Big Booze.'" *Australasian Medical Journal* 1 (3): 1-26, doi:10.4066/AMJ.2009.43.

Bradley, Katharine A., Jodie Boyd-Wickizer, Suzanne H. Powell, et al. 1998. "Alcohol Screening Questionnaires in Women: A Critical Review." *Journal of the American Medical Association* 280 (2): 166-71, doi:10.1001/jama.280.2.166.

Brady, Kathleen T., Sudie E. Back, and Shelly F. Greenfield. 2009. *Women and Addiction: A Comprehensive Handbook.* New York: Guilford Press.

British Medical Association Board of Science. 2009. *Under the Influence: The Damaging Effect of Alcohol Marketing on Young People.* London: British Medical Association.

Brown, Catrina. 2008. *It's Not Cut and Dry: Women's Experiences of Alcohol Use, Depression, Anxiety and Trauma.* Halifax: Dalhousie University.

Butt, Peter, Doug Beirness, Frank Cesa, et al. 2011. *Alcohol and Health in Canada: A Summary of Evidence and Guidelines for Low-Risk Drinking.* Ottawa: Canadian Centre on Substance Abuse.

CAMH (Centre for Addiction and Mental Health). 2009a. "Alcohol, Pregnancy and Breast-Feeding." http://www.camh.net/About_Addiction_Mental_Health/Drug_and_Addiction_Information/Women_and_Alcohol/pregnancy_breastfeed.html.

–. 2009b. "Canada's Low-Risk Alcohol Drinking Guidelines." http://www.camh.net/About_Addiction_Mental_Health/Drug_and_Addiction_Information/low_risk_drinking_guidelines.html.

Campbell, J., K. Szumlinski, and T. Kippin. 2009. "Contribution of Early Environmental Stress to Alcoholism Vulnerability." *Alcohol* [online] 43 (7): 547-54. http://www.sciencedirect.com.

Carroll, Marilyn E., Wendy J. Lynch, Megan E. Roth, et al. 2004. "Sex and Estrogen Influence Drug Abuse." *Trends in Pharmacological Sciences* 25 (5): 273-79. doi:10.1016/j.tips.2004.03.011.

Carson, George, Lori Vitale Cox, Joan Crane, et al. 2010. "Alcohol Use and Pregnancy Consensus Clinical Guidelines." *Journal of Obstetrics and Gynaecology Canada* 32 (8): S1-S31.

Cecanti, Mauro, and Mario Vitali. 2009. "Alcoholics with a History of Heroin Consumption: Clinical Features and Chronology of Substance Abuse." *Heroin Addiction and Related Clinical Problems* 11 (3): 35-38.

Chang, Grace. 2001. "Alcohol-Screening Instruments for Pregnant Women." *Alcohol Research and Health* 25 (3): 204-9.

Chasnoff, Ira J. 2001. *The Nature of Nurture: Biology, Environment, and the Drug Exposed Child*. Chicago: National Training Institute Publishing.

Chen, Wendy Y., Bernard Rosner, Susan E. Hankinson, et al. 2011. "Moderate Alcohol Consumption during Adult Life, Drinking Patterns, and Breast Cancer Risk." *Journal of the American Medical Association* 306 (17): 1884-90, doi:10.1001/jama.2011.1590.

Chudley, Albert E., Julianne Conry, Jocelynn L. Cook, et al. 2005. "Fetal Alcohol Spectrum Disorder: Canadian Guidelines for Diagnosis." *Canadian Medical Association Journal* 172 (5S): S1-S21, doi:10.1503/cmaj.1040302.

CRTC (Canadian Radio-television Telecommunications Commission). 1996. "Code for Broadcast Advertising of Alcoholic Beverages." http://www.crtc.gc.ca/eng/general/codes/alcohol.htm.

Currie, Janet C. 2003. *Manufacturing Addiction: The Over-Prescription of Benzodiazepines and Sleeping Pills to Women in Canada*. Vancouver: BC Centre of Excellence for Women's Health. http://www.benzo.org.uk/amisc/benzobrief.pdf.

Dabu, Sheila. 2007. "Stress Plagues Sex-Trade Workers." *Toronto Star,* 20 August. http://www.thestar.com/living/article/247824.

Daniels, Norman. 2008. *Just Health*. Cambridge: Cambridge University Press.

DARE (Drug and Alcohol Recovery and Education Network). 2008. *DARE Network Annual Report 2007*. http://www.darenetwork.com/media/publications/2007DARE_Annual_Report.pdf.

De Peuter, Jennifer, and Marianne Sorensen. 2005. *Rural Nova Scotia Profile: A Ten-Year Census Analysis (1991-2001)*. Ottawa: Government of Canada – Canadian Rural Partnerships.

Deloitte. 2007. *Shopper Marketing: Capturing a Shopper's Mind, Heart and Wallet*. Association of Food, Beverage and Consumer Products Companies. http://www.deloitte.com/assets/Dcom-Slovenia/Local%20Assets/Documents/Shopper_Marketing_survey2007(6).pdf.

Dhar, Aarti. 2008. "Framework Convention on Alcohol Control Mooted." *The Hindu,* 30 July. http://www.hindu.com/2008/07/30/stories/2008073061241600.htm.

Dowsett Johnston, Ann. 2013. *Drink: The Intimate Relationship between Women and Alcohol*. New York: HarperCollins.

Dube, Shanta R., Robert F. Anda, Vincent J. Felitti, et al. 2002. "Adverse Childhood Experiences and Personal Alcohol Abuse as an Adult." *Addictive Behaviors* 27 (5): 713-25, doi:10.1016/S0306-4603(01)00204-0.

Edwards, Peggy. 2004. *The Social Determinants of Health: An Overview of the Implications for Policy and the Role of the Health Sector*. Ottawa: Health Canada. http://www.hc-sc.gc.ca/hppb/phdd/pdf/overview_implications/01_overview_e.pdf.

Ehrenreich, Barbara. 2002. "Viewpoint: Libation as Liberation?" *Time,* 1 April. http://www.time.com/time/magazine/article/0,9171,1002113,00.html.

Ellison, R. Curtis, Yuqing Zhang, Christine E. McLennan, et al. 2001. "Exploring the Relation of Alcohol Consumption to Risk of Breast Cancer." *American Journal of Epidemiology* 154 (8): 740-47, doi:10.1093/aje/154.8.740.

Erickson, Patricia G., Katharine King, and Young Women in Transit (Ywit). 2007. "On the Street: Influences of Homelessness in Young Women." In *Highs and Lows: Canadian Perspectives on Women and Substance Use,* edited by Nancy Poole and Lorraine Greaves, 51-58. Toronto: Centre for Addiction and Mental Health.

European Institute of Women's Health. 2013. "Women and Alcohol in the European Union. Gender and Chronic Disease Policy Briefing." http://www.eurocare.org/resources/factsheets/alcohol_and_women/women_and_alcohol_in_the_eu.

Ezard, N., E. Oppenheimer, A. Burton, et al. 2011. "Six Rapid Assessments of Alcohol and Other Substances Displaced by Conflict." *Conflict and Health* 5 (1): 1-12.

Fattore, Liana, Silvia Altea, and Walter Fratta. 2008. "Sex Differences in Drug Addiction: A Review of Animal and Human Studies." *Women's Health* 4 (1): 51-65, doi: 10.2217/17455057.4.1.51.

Felitti, Vincent J. 2002. "The Relationship between Adverse Childhood Experiences and Adult Health: Turning Gold into Lead." *Permanente Journal* 6 (1): 44-47.

–. 2004. *The Origins of Addiction: Evidence from the Adverse Childhood Experiences Study*. San Diego: Department of Preventive Medicine, Kaiser Permanente Medical Care Program. http://www.acestudy.org/files/OriginsofAddiction.pdf.

Felitti, Vincent J., Robert F. Anda, Dale Nordenberg, et al. 1998. "Relationship of Childhood Abuse and Household Dysfunction to Many of the Leading Causes of Death in Adults: The Adverse Childhood Experiences (ACE) Study." *American Journal of Preventive Medicine* 14 (4): 245-58.

Finer, Lawrence B., and Mia R. Zoina. 2014. "Shifts in Intended and Unintended Pregnancies in the United States, 2001-2008." *American Journal of Public Health* 104 (S1): 43-48.

Finkelstein, Norma. 1994. "Treatment Issues for Alcohol- and Drug-Dependent Pregnant and Parenting Women." *Health and Social Work* 19 (1): 7-15, doi:10.1093/hsw/19.1.7.

Fox, Helen C., and Rajita Sinha. 2009. "Sex Differences in Drug-Related Stress-System Changes: Implications for Treatment in Substance-Abusing Women." *Harvard Review of Psychiatry* 17 (2): 103-19, doi:10.1080/10673220902899680.

Fraser, Charlotte. 2008. "Victims and Fetal Alcohol Spectrum Disorder (FASD): A Review of the Issues." *Victims of Crime Research Digest* 1: 24-28.

Frieden, Thomas R. 2010. "A Framework for Public Health Action: The Health Impact Pyramid." *American Journal of Public Health* 100 (4): 590-95, doi:10.2105/AJPH.2009.185652.

Giesbrecht, Norman, Timothy Stockwell, Perry Kendall, et al. 2011. "Alcohol in Canada: Reducing the Toll through Focused Interventions and Public Health Policies." *Canadian Medical Association Journal* 183 (4): 450-55, doi:10.1503/cmaj.100825.

Giesbrecht, Norman, Ashley Wettlaufer, Nicole April, et al. 2013. *Strategies to Reduce Alcohol-Related Harms and Costs in Canada: A Comparison of Provincial Policies*. Toronto: Centre for Addiction and Mental Health.

Gordon, Ross, Gerard Hastings, and Crawford Moodie. 2010. "Alcohol Marketing and Young People's Drinking: What the Evidence Base Suggests for Policy." *Journal of Public Affairs* 10 (1-2): 88-101, doi:10.1002/pa.338.

Graham, Linda. 2005. *Alcohol Indicators Report: A Framework of Alcohol Indicators Describing the Consumption of Use, Patterns of Use, and Alcohol-Related Harms in Nova Scotia*. Halifax: Nova Scotia Department of Health Promotion. *http://www.gov.ns.ca/ohp/publications/AlcoholFullFINAL.pdf*.

Greenfield, Shelly F., Sudie E. Back, Katie Lawson, et al. 2010. "Substance Abuse in Women." *Psychiatric Clinics of North America* 33 (2): 339-55, doi:10.1016/j.psc.2010.01.004.

Greenfield, Shelly F., Audrey J. Brooks, Susan M. Gordon, et al. 2007. "Substance Abuse Treatment Entry, Retention, and Outcome in Women: A Review of the Literature." *Drug Alcohol Dependence* 86 (1): 1-21, doi:10.1016/j.drugalcdep.2006.05.012.

Guise, Jennifer M.F., and Jan S. Gill. 2007. "Binge Drinking? 'It's Good, It's Harmless Fun': A Discourse Analysis of Accounts of Female Undergraduate Drinking in Scotland." *Health Education Research* 22 (6): 895-906, doi:10.1093/her/cym034.

Halseth, R. 2013. *Aboriginal Women in Canada: Gender, Socio-Economic Determinants of Health and Initiatives to Close the Wellness Gap*. Prince George, BC: National Collaborating Centre for Aboriginal Health.

Hamilton, Nancy, and Tariq Bhatti. 1996. "Population Health Promotion: An Integrated Model of Population Health and Health Promotion." Public Health Agency of Canada, http://www.phac-aspc.gc.ca/ph-sp/php-psp/index-eng.php.

Hayward, Karen, and Ronald Colman. 2003. *The Tides of Change: Addressing Inequity and Chronic Disease in Atlantic Canada*. Halifax: Population and Public Health Branch, Atlantic Regional Office, Health Canada. http://gpiatlantic.org/pdf/health/inequity.pdf.

Health Canada. 2001. *Best Practices: Treatment and Rehabilitation for Women with Substance Use Problems*. Ottawa: Health Canada. http://www.hc-sc.gc.ca/hc-ps/alt_formats/hecs-sesc/pdf/pubs/adp-apd/bp_women-mp_femmes/women-e.pdf.

–. 2007. "Best Practices Treatment for Women with Substance Abuse Problems." http://www.hc-sc.gc.ca/hl-vs/pubs/adp-apd/bp_women-mp_femmes/litreview-examendoc-eng.php.

–. 2011. "Sex and Gender Based Analysis. Health Canada's Gender Based Analysis Policy." Ottawa: Health Canada. http://www.hc-sc.gc.ca/hl-vs/gender-genre/analys/index-eng.php.

Hommer, Daniel W., Reza Momenan, Erica Kaiser, et al. 2001. "Evidence for a Gender-Related Effect of Alcoholism on Brain Volumes." *American Journal of Psychiatry* 158 (2): 198-204, doi:10.1176/appi.ajp.158.2.198.

Jernigan, David H. 2009. "The Global Alcohol Industry: An Overview." *Addiction* 104 (Supp. 1): 6-12, doi:10.1111/j.1360-0443.2008.02430.x.

–. 2011. "Framing a Public Health Debate over Alcohol Advertising: The Center on Alcohol Marketing and Youth 2002-2008." *Journal of Public Health Policy* 32 (2): 165-79.

Johnson, Holly. 2005. *Assessing the Prevalence of Violence against Women in Canada*. Geneva: UN Division for the Advancement of Women in Collaboration with Economic Commission for Europe and World Health Organization. http://www.un.org/womenwatch/daw/egm/vaw-stat-2005/docs/expert-papers/johnson.pdf.

Kirby, Michael J.L., and Wilbert Joseph Keon. 2006. *Out of the Shadows at Last: Transforming Mental Health, Mental Illness and Addiction Services in Canada*. The Standing Senate Committee on Social Affairs, Science and Technology. Ottawa: Parliament of Canada.

LeWine, Howard. 2013. "Study: No Connection between Drinking Alcohol Early in Pregnancy and Birth Problems." Harvard Health Publications, http://www.health.harvard.edu/blog/study-no-connection-between-drinking-alcohol-early-in-pregnancy-and-birth-problems-201309106667.

Logan, T.K., Robert Walker, Jennifer Cole, et al. 2002. "Victimization and Substance Use among Women: Contributing Factors, Interventions, and Implications." *Review of General Psychology* 6 (4): 325-97.

Lombard, Amélie Catherine. 2005. *Retention of Health Professionals in Rural Nova Scotia*. Halifax: Rural Communities Impacting Policy (RCIP) Project. http://www.ruralnovascotia.ca/documents/rural%20health/retention05%20full%20report.pdf.

Lombardi, Rosie. 2008. "Converting Shoppers into Buyers: Retailers and Manufacturers Are Paying More Attention to Shopper Marketing for One Simple Reason – It Works." *Canadian Retailer Magazine* 27. http://www.onestopmedia.com/about-us/press/converting-shoppers-buyers.

Malinski, Maciej K., Howard D. Sesso, Francisco Lopez-Jimenez, et al. 2004. "Alcohol Consumption and Cardiovascular Disease Mortality in Hypertensive Men." *Archives of Internal Medicine* 164 (6): 623-28.

Mancinelli, Rosanna, Roberto Binetti, and Mauro Ceccanti. 2006. "Female Drinking, Environment and Biological Markers." *Annali Dell'Istituto Superiore di Sanita* 42 (1): 31-38.

Mancinelli, Rosanna, Mario Vitali, and Mauro Ceccanti. 2009. "Women, Alcohol and the Environment: An Update and Perspectives in Neuroscience." *Functional Neurology* 24 (2): 77-81.

Mann, Karl, K. Ackermann, B. Croissant, et al. 2005. "Neuroimaging of Gender Differences in Alcohol Dependence: Are Women More Vulnerable?" *Alcoholism: Clinical and Experimental Research* 29 (5): 896-901, doi:10.1097/01.ALC.0000164376.69978.6B.

Mann, Karl, A. Batra, A Günther, et al. 1992. "Do Women Develop Alcoholic Brain Damage More Readily Than Men?" *Alcoholism: Clinical and Experimental Research* 16 (6): 1052-56, doi:10.1111/j.1530-0277.1992.tb00698.x.

Maté, Gabor. 2008. *In the Realm of Hungry Ghosts: Close Encounters with Addiction*. Toronto: Alfred A. Knopf Canada.

McCrady, Barbara S., and Helen Raytek. 1993. "Women and Substance Abuse: Treatment Modalities and Outcomes." In *Women and Substance Abuse,* edited by Edith S. Lisansky Gomberg and Ted D. Nirenberg. Norwood, NJ: Ablex Publishing.

McGourty, Richard F., and Ira J. Chasnoff. 2003. *The Power beyond Measure*. Chicago: National Training Institute Publishing.

Monsom, K., and A. Schoenstadt. 2007. "Alcohol and Ativan." eMedTV, http://anxiety.emedtv.com/ativan/alcohol-and-ativan.html.

Morris, Marika, and Tahira Gonsalves. 2005. *Women and Poverty Fact Sheet*. 3rd ed. Ottawa: Canadian Research Institute for the Advancement of Women.

Mukamal, Kenneth J., Malcolm Maclure, James E. Muller, et al. 2001. "Prior Alcohol Consumption and Mortality Following Acute Myocardial Infarction." *Journal of the American Medical Association* 285 (15): 1965-70.

Murthy, Poornima, Swamy Kudlur, Sanju George, et al. 2009. "A Clinical Overview of Fetal Alcohol Syndrome." *Addictive Disorders and Their Treatment* 8 (1): 1-12, doi:10.1097/ADT.0b013e318163b062.

Najavits, Lisa M. 2002a. *A Women's Addiction Workbook: Your Guide to In-Depth Healing*. Oakland, CA: New Harbinger Publications.

–. 2002b. *Seeking Safety: A Treatment Manual for PTSD and Substance Abuse*. New York: Guilford Press.

Nanji, Amin A., and Samuel W. French. 1987. "Female to Male Mortality Ratios for Alcohol-Related Disorders: Possible Indicator of Susceptibility in Different Sexes." *Advances in Alcohol and Substance Abuse* 6 (3): 89-95, doi:10.1300/J251v06n03_08.

National Indian and Inuit Community Health Representatives Organization. n.d. "Sheway: An Oasis for Women in Vancouver's Downtown Eastside." http://www.niichro.com/fas/fas_9.html.

Nova Scotia Department of Health Promotion and Protection. 2007. *Changing the Culture of Alcohol Use in Nova Scotia: An Alcohol Strategy to Prevent and Reduce the Burden of Alcohol-Related Harm in Nova Scotia*. Halifax: Department of Health Promotion and Protection, Addiction Services. http://www.gov.ns.ca/ohp/publications/Alcohol_Strategy.pdf.

Nova Scotia Finance. 2005. *Nova Scotia at a Glance, 2005*. Economics and Statistics Division. http://www.novascotia.ca/finance/publish/facts/2005/NS-At-A-Glance.pdf.

NSLC (Nova Scotia Liquor Corporation). 2008. "NSLC Announces Year End Results." News release. http://www.mynslc.com/Documents/News%20Releases/2008/July%208,%202008%20-%20NSLC%20announces%20year-end%20results.pdf.

O'Malley, Pat, and Mariana Valverde. 2004. "Pleasure, Freedom and Drugs: The Uses of 'Pleasure' in Liberal Governance of Drug and Alcohol Consumption." *Sociology* 38 (1): 25-42, doi:10.1177/0038038504039359.

Pan American Health Organization. 2011. "Women and Men Face Different Chronic Disease Risks." http://www.paho.org/hq/index.php?option=com_content&view=article

&id=7731%3Awomen-and-men-face-different-chronic-disease-risks&catid=4243%3Ahsd0107x-cd-media-center&lang=en.

Pannozzo, Linda, Ronald Colman, Nathan Ayer, et al. 2009. *Measuring Sustainable Development: Application of the Genuine Progress Index to Nova Scotia. The 2008 Nova Scotia GPI Accounts – Indicators of Genuine Progress.* Glen Haven, NS: GPI Atlantic. http://www.gpiatlantic.org/pdf/integrated/gpi2008.pdf.

PHAC (Public Health Agency of Canada). 2010. "What Determines Health?" http://www.phac-aspc.gc.ca/ph-sp/determinants/index-eng.php.

Polaris Project. 2014. "Polaris Project: For a World without Slavery." http://www.polarisproject.org/human-trafficking/overview.

Poole, Nancy, and Colleen Anne Dell. 2005. *Girls, Women and Substance Use.* Ottawa: Canadian Centre on Substance Abuse and BC Centre of Excellence for Women's Health. http://www.ccsa.ca/Resource%20Library/ccsa-011142-2005.pdf.

Poole, Nancy, Ginny Gonneau, and Cristine Urquhart. 2010. "Preventing Heavy Alcohol Use on the Part of Girls and Young Women." Slide presentation. BC Women's and BC Centre of Excellence for Women's Health. http://www.bccewh.bc.ca/news-events/documents/GirlsandAlcohol_Presentation_April2010.pdf.

Poole, Nancy, and Lorraine Greaves. 2007. "Introduction." In *Highs and Lows: Canadian Perspectives on Women and Substance Use,* edited by Nancy Poole and L. Greaves, xix-xxv. Toronto: Centre for Addiction and Mental Health. http://www.camh.net/Publications/Resources_for_Professionals/Highs_Lows/highs_lows_introduction.pdf.

–, eds. 2012. *Becoming Trauma Informed.* Toronto: Centre for Addiction and Mental Health.

Poulin, Christiane, and Linda Graham. 2001. "The Association between Substance Use, Unplanned Sexual Intercourse and Other Sexual Behaviours among Adolescent Students." *Addiction* 96 (4): 607-21, doi:10.1046/j.1360-0443.2001.9646079.x.

Provincial Health Officer of British Columbia and CARBC (Centre for Addictions Research of British Columbia). 2008. *A Proposal for Changes to BC Liquor Prices in Order to Reduce Harm from Alcohol Consumption.* Victoria: University of Victoria.

Raphael, Dennis. 2003. "Addressing the Social Determinants of Health in Canada: Bridging the Gap between Research Findings and Public Policy." *Policy Options* 24 (3): 35-40. http://www.irpp.org/po/archive/mar03/raphael.pdf.

RCIP (Rural Communities Impacting Policy). 2006. "Demographics." In *Painting the Landscape of Rural Nova Scotia.* http://www.ruralnovascotia.ca/RCIP/PDF/demographics.pdf.

Rehm, Jürgen, Dolly Baliunas, Guillerme L.G. Borges, et al. 2010. "The Relation between Different Dimensions of Alcohol Consumption and Burden of Disease: An Overview." *Addiction* 105 (5): 817-43, doi:10.1111/j.1360-0443.2010.02899.x.

Rehm, Jürgen, Norman Giesbrecht, Jayadeep Patra, et al. 2006a. "Estimating Chronic Disease Deaths and Hospitalizations Due to Alcohol Use in Canada in 2002: Implications for Policy and Prevention Strategies." *Preventing Chronic Disease* 3 (4): 1-19. http://www.cdc.gov/pcd/issues/2006/oct/05_0009.htm.

Rehm, Jürgen, Norman Giesbrecht, Svetlana Popova, et al. 2006b. *Overview of Positive and Negative Effects of Alcohol Consumption – Implications for Preventive Polices in Canada.* Toronto: Centre of Addiction and Mental Health Research Document Series 168.

Rehm, Jürgen, Jayadeep Patra, and Ben Taylor. 2007. "Harm, Benefits, and Net Effects on Mortality of Moderate Drinking of Alcohol among Adults in Canada in 2002." *Annals of Epidemiology* 17 (5): S81-S86, doi:10.1016/j.annepidem.2007.01.018.

Richardson, Anna, and Tracey Budd. 2003. "Young Adults, Alcohol, Crime and Disorder." *Criminal Behaviour and Mental Health* 13 (1): 5-16, doi:10.1002/cbm.527.

Rim, Eric, and Caroline Moats. 2007. "Alcohol and Coronary Heart Disease: Drinking Patterns and Mediators of Effect." *Annals of Epidemiology* 17 (5): S3-S7, doi:10.1016/j.annepidem.2007.01.002.

Romito, Patrizia, Janet Molzan Turan, and Margherita De Marchi. 2005. "The Impact of Current and Past Interpersonal Violence on Women's Mental Health." *Social Sciences and Medicine* 60 (8): 1717-27, doi:10.1016/j.socscimed.2004.08.026.

Ronksley, Paul E., Susan E. Brien, Barbara J. Turner, et al. 2011. "Association of Alcohol Consumption with Selected Cardiovascular Disease Outcomes: A Systematic Review and Meta-Analysis." *BMJ* 342: d671, doi:10.1136/bmj.d671.

Rutman, Deborah. 2011. *Substance Using Women with FASD and FASD Prevention: Service Providers' Perspectives on Promising Approaches in Substance Use Treatment and Care for Women with FASD.* Victoria, BC: University of Victoria.

Schorling, John B., and David G. Buchsbaum. 1997. "Screening for Alcohol and Drug Abuse." *Medical Clinics of North America* 81 (4): 845-65.

Schrans, Tracy, Tony Schellinck, and K. McDonald. 2008. *Culture of Alcohol Use in Nova Scotia.* Halifax: Nova Scotia Department of Health Promotion and Protection. http://www.gov.ns.ca/hpp/publications/2008-Alcohol-Benchmark-Report.pdf.

Seitz, Helmut, Claudio Pelucchi, Vincenzo Bagnardi, et al. 2012. "Epidemiology and Pathophysiology of Alcohol and Breast Cancer: Update." *Alcohol and Alcoholism* 47 (3): 204-12, doi: 10.1093/alcalc/ags011.

Sharif, Najma R., Atul A. Dar, and Carol Amaratunga. 2000. *Ethnicity, Income and Access to Health Care in the Atlantic Region: A Synthesis of the Literature.* Halifax: Maritime Centre for Excellence in Women's Health. http://www.dal.ca/content/dam/dalhousie/pdf/ace-women-health/ACEWH_ethnicity_income_access_to_health_care_lit_review.pdf.

Shkolnikov, Vladimir, Martin McKee, and David A. Leon. 2001. "Changes in Life Expectancy in Russia in the Mid-1990s." *Lancet* 357 (9260): 917-21, doi:10.1016/S0140-6736(00)04212-4.

Singletary, Keith W., and Susan M. Gapstur. 2001. "Alcohol and Breast Cancer: Review of Epidemiologic and Experimental Evidence and Potential Mechanisms." *Journal of the American Medical Association* 286 (17): 2143-51.

Slack, James. 2008. "Menace of the Violent Girls: Binge-Drinking Culture Fuels Surge in Attacks by Women." *Daily Mail,* 14 August. http://www.dailymail.co.uk/news/article-1039963/Menace-violent-girls-Binge-drinking-culture-fuels-surge-attacks-women.html.

Smith, Lesley A., and David R. Foxcroft. 2009. "The Effect of Alcohol Advertising, Marketing and Portrayal on Drinking Behaviour in Young People: Systematic Review of Prospective Cohort Studies." *BMC Public Health* 9: 51-62, doi:10.1186/1471-2458-9-51.

Smith-Warner, Stephanie A., Donna Spiegelman, Shiaw-Shyuan Yaun, et al. 1998. "Alcohol and Breast Cancer in Women: A Pooled Analysis of Cohort Studies." *Journal of the American Medical Association* 279 (7): 535-40. http://jama.jamanetwork.com/article.aspx?articleid=187252.

Stade, Brenda, Karen Clark, and Danielle D'Agostino. 2004. "Fetal Alcohol Spectrum Disorder and Homelessness: Training Manual." *Journal of FAS International* 2: e10. http://www.motherisk.org/JFAS_documents/FAS_Street_Level.pdf.

Stade, Brenda, Bonnie Stevens, Wendy Unger, et al. 2006. "Health-Related Quality of Life of Canadian Children and Youth Prenatally Exposed to Alcohol." *Health and Quality of Life Outcomes* 4: 81, doi:10.1186/1477-7525-4-81.

Statistics Canada. 2012. *Women in Canada: A Gender Based Statistical Report. Economic Well-Being.* Ottawa: Statistics Canada. http://www.statcan.gc.ca.

Stockwell, Tim, Jodi Sturge, and Scott McDonald. 2007. *Patterns of Risky Alcohol Use in British Columbia: Results of the 2004 Canadian Addiction Survey – Bulletin 1*. Victoria: Centre for Addictions Research of BC. http://carbc.ca/Portals/0/PropertyAgent/558/Files/20/CARBCBulletin1.pdf.
Testa, Maria, Carole VanZile-Tamsen, Jennifer A. Livingston, et al. 2006. "The Role of Women's Alcohol Consumption in Managing Sexual Intimacy and Sexual Safety Motives." *Journal of Studies on Alcohol* 67 (5): 665-74.
Thomas, G. 2012. *Levels and Patterns of Alcohol Use in Canada*. Alcohol Price Policy Series: Report 1. Ottawa: Canadian Centre on Substance Abuse.
Timko, Christine, Anne Sutkowi, Joanne Pavao, et al. 2008. "Women's Childhood and Adult Adverse Experiences, Mental Health, and Binge Drinking: The California Women's Health Survey." *Substance Abuse Treatment Prevention and Policy* 3 (1): 15-24. http://www.biomedcentral.com/content/pdf/1747-597X-3-15.pdf.
Toumbourou, J.W., T. Stockwell, C. Neighbors, et al. 2007. "Interventions to Reduce Harm Associated with Adolescent Substance Use." *Lancet,* Adolescent Health Series, 369 (9570): 1391-1401.
Urbano-Márquez, Alvaro, Ramon Estruch, Joaquin Fernández-Solá, et al. 1995. "The Greater Risk of Alcoholic Cardiomyopathy and Myopathy in Women Compared with Men." *Journal of the American Medical Association* 274 (2): 149-54.
Vienna Forum to Fight Human Trafficking. 2008. "Health and Human Trafficking: Working Together for Responses to the Survivor's Health Care Needs." http://www.ungift.org/docs/ungift/pdf/vf/aidmemoirs/health%20and%20human%20trafficking1.pdf.
Vogel, Victor G., and Emanuela Taioli. 2006. "Have We Found the Ultimate Risk Factor for Breast Cancer?" *Journal of Clinical Oncology* 24 (12): 1791-94, doi:10.1200/JCO.2005.05.4122.
Westermeyer, Joseph, Lisa Mellman, and Renato Alarcon. 2006. "Cultural Competence in Addiction Psychiatry." *Addictive Disorders and Their Treatment* 5 (3): 107-19, doi:10.1097/01.adt.0000210719.10693.6c.
Weuve, Jennifer, Susan A. Korrick, Marc A. Weisskopf, et al. 2009. "Cumulative Exposure to Lead in Relation to Cognitive Function in Older Women." *Environmental Health Perspectives* 117 (4): 574-80.
WHO (World Health Organization). 2002. *The World Health Report, 2002: Reducing Risks, Promoting Healthy Life*. Geneva: WHO Press. http://www.who.int/whr/2002/en/.
–. 2005. *Fifty-eighth World Health Assembly*. Geneva: WHO Press. http://apps.who.int/gb/ebwha/pdf_files/WHA58-REC1/english/A58_2005_REC1-en.pdf.
–. 2008. "Improve Daily Living Conditions." In *Social Determinants of Health: Closing the Gap in a Generation – How?* Geneva: WHO Press. http://www.who.int/social_determinants/thecommission/finalreport/closethegap_how/en/index1.html.
–. 2009. *Global Health Risks: Mortality and Burden of Disease Attributable to Selected Major Risks*. Geneva: WHO Press. http://www.who.int/healthinfo/global_burden_disease/GlobalHealthRisks_report_full.pdf.
–. 2011. *Evidence for Gender Responsive Actions to Prevent and Manage Injuries and Substance Abuse: Young People's Health as a Whole-of-Society Response*. Copenhagen: WHO Press. http://www.sante.public.lu/publications/sante-fil-vie/enfance-adolescence/evidence-gender-responsive-actions-prevent-injuries-substance-abuse/evidence-gender-responsive-actions-prevent-injuries-substance-abuse.pdf.

Part 3: Hormones as the "Messengers of Gender"?

Chapter 7 The Impact of Phthalates on Women's Reproductive Health

Maria P. Velez, Patricia Monnier,
Warren G. Foster, and William D. Fraser

The process of industrialization that characterized the twentieth century, along with increased participation of women in the workforce, has resulted in a dramatic rise in the exposure of women to environmental toxicants, both in Canada and worldwide. Some of these chemicals, commonly referred to as "endocrine-disrupting chemicals" (EDCs), can alter hormone functions essential to development and reproduction, and lead to adverse health effects in an organism or its offspring (Damstra et al. 2002). Animal models have shown that exposures might also impact subsequent generations (Anway and Skinner 2006). These chemicals are persistent in the environment and include natural and synthetic hormones, plant constituents, pesticides, compounds used in the plastics industry and in consumer products, and by-products of other industrial processes (Damstra et al. 2002).

Phthalates are a group of synthetic EDCs used as plasticizers in industrial production. They have recently received special attention because of their high-volume production, ubiquitous environmental presence, and possible association with adverse reproductive health outcomes (CDC 2009). Human exposure to phthalates may result in changes to tissues in the testes and reduced sperm counts; lower levels of testosterone; increased prenatal mortality; decreases in fetal growth, birth weight, and anogenital distance (the distance between the anus and the base of the penis); and a number of biological malformations (NTP-CERHR 2006).

Most phthalate studies to date have focused on male exposures and reproductive effects. There continues to be limited research exploring the effects of phthalates on women's health, despite the fact that female reproductive function and development are equally, if not uniquely, susceptible to exposure to hormonally active chemicals. Evidence is now emerging showing that females are more highly exposed to phthalates than males. For example, the Fourth US National Report on Human Exposure to Environmental Chemicals examined data from the National Health and Nutrition Examination Survey (NHANES, 1999-2000, 2001-2, 2003-4) on urinary levels of phthalates in a subsample of the US population aged six years and older (CDC 2009). The analysis identified gender differences in the concentration of the phthalates under assessment, with higher urinary phthalate levels observed in women than in men. This pattern was also observed more recently in the Canadian Health Measures Survey (2007-9), where biomonitoring of phthalate metabolites in urine samples of 3,236 individuals ranging from six to forty-nine years of age reported significantly higher concentrations in females than in males (Saravanabhavan et al. 2013).

In this chapter, we review the available scientific literature on phthalate exposure and women's health. We argue that there is now enough evidence to demonstrate a potentially important connection between these exposures and adverse reproductive health effects in women during specific times of development that requires further investigation. We also examine social factors that can influence phthalate exposure and biological and psychological health outcomes. We conclude with a look at what is currently being done to address phthalates in the environment, and provide recommendations for future research and policy initiatives that can lead to a greater understanding of the gender differences in phthalate exposures, in order to ensure the health and safety of men and women in Canada.

Phthalates: Routes of Exposure

Phthalates are a group of industrial, synthetic chemicals used to impart flexibility and resilience to polyvinyl chloride (PVC) plastics (CDC 2009). Many consumer products contain phthalates, including vinyl flooring, adhesives, detergents, lubricating oils, solvents, automotive plastics, some medical pharmaceuticals, plastic bags, blood-storage bags, intravenous medical tubing, children's toys, and personal care products such as soap, shampoo, deodorants, fragrances, hair spray, and nail polish (CDC 2009). Di(2-ethylhexyl) phthalate (DEHP) is the most abundant phthalate in the environment, and mono-(2-ethylhexyl) phthalate (MEHP) is its primary metabolite, created by the breakdown of DEHP in the body (NTP-CERHR 2006).

Because phthalates are not chemically bound to PVC plastics, they gradually separate from consumer products over time. As a result, humans can experience phthalate exposure during their entire life cycle, whether it be through digestion, inhalation, dermal contact, the indirect leaching of phthalates into other products, or general environmental contamination (Heudorf, Mersch-Sundermann, and Angerer 2007). Although phthalates tend to rapidly metabolize in the body, their widespread use and ubiquitous presence in dust, soil, and indoor and outdoor air results in continuous human exposure to these chemicals and possible health effects as a result. Additionally, as phthalates metabolize in the body they are better able to interfere with biological function (Frederiksen, Skakkebaek, and Andersson 2007).

Experimental studies have indicated that the effect of phthalates on human health is very much related to the time of exposure. As mentioned in Chapters 1-4, 9, and 11, exposures during *critical windows of vulnerability* may have adverse effects in the short and long term. Phthalate exposure during periods regulated by steroid hormone levels, such as ovarian and testicular development in the womb and during puberty, pregnancy, and menopause, can lead to reproductive effects later in life. Evidence is also building regarding the cumulative health effects of simultaneous exposure to multiple phthalates and other EDCs (Borch et al. 2004; Hotchkiss et al. 2004). For example, studies have demonstrated associations between male reproductive malformations in rats and the presence of a combination of phthalates and EDCs in the body (Hotchkiss et al. 2004).

Possible Developmental and Reproductive Effects

A body of evidence addressing the effect of phthalate exposure on human health has been emerging. While men have been the focus of many phthalate studies up to this point, new research is beginning to reveal the distinct health effects that girls and women experience from exposure, due to female biology and critical windows of vulnerability that are gendered in nature. Below is a compilation of current scientific research on the unique associations between phthalate exposure and biological health. We explore the possible developmental and reproductive effects of exposure during key periods of development, including the perinatal period (before and after birth), puberty, and adulthood. Although this review includes studies on both men and women, a concerted effort has been made to focus on women's health outcomes in particular.

Perinatal Period

A number of studies have examined the relationship between perinatal phthalate exposure and reproductive development. This stage of development is thought to be particularly vulnerable to chemical exposure because of the greater surface area-to-volume ratio and higher calorie intake per kilogram of weight in infants, leading to increased dermal and oral exposure compared with adults (Lottrup Poulsen et al. 2006) (see also Chapter 1 of this volume). Studies performed on rodents have demonstrated that in utero phthalate exposure (exposure in the womb) can cause adverse effects in the male reproductive system (Barlow and Foster 2003; Borch et al. 2006; Gray et al. 2000). Some of these effects are probably caused by the alteration of cell development and function in the testes, leading to a reduction in the production of testosterone (Barlow et al. 2003; Borch et al. 2004, 2006; Parks et al. 2000). Swan (2008) and Swan and colleagues (2005) examined anogenital distance, fetal sex development, and anogenital index (anogenital distance relative to birth weight) in boys who experienced prenatal phthalate exposure. These studies suggested that phthalates might interfere in the reproductive development of male infants (Swan 2008; Swan et al. 2005). Huang and colleagues (2009) reported shorter anogenital distances in a group of female infants whose mothers had higher concentrations of phthalates in their amniotic fluid compared with those whose mothers had low concentrations (13.9 mm versus 17.6 mm). In another study, Main and colleagues (2006) evaluated the influence of phthalates found in human breast milk on newborn boys. Results pointed to abnormal changes in hormone levels and activity, also leading to alterations in male reproductive development. In addition, the NHANES surveys indicated that children aged six to eleven years excreted higher concentrations of most phthalates than did older age groups (CDC 2009). These findings parallel those of Koch, Drexler, and Angerer (2004), which demonstrated higher levels of MEHP, mono-(2-ethly-5-oxohexyl) phthalate (MEOHP), and mono-(2-ethlyl-5-hydroxyhexyl) phthalate (MEHHP) in nursery school children (aged two to six years) compared with their teachers or parents.

There has been little research on the impact of phthalates on female anogenital distance or the development of the female pelvic floor. The integrity of both the female pelvis and anogenital distance is essential for proper reproductive function (Rizk and Thomas 2000), so any effects on these areas as a result of phthalate exposure is a cause for concern (Borch et al. 2006). Evidence shows a modest link between exposure and adverse effect that requires further study. While one study reported a significant increase in the anogenital distance of female offspring of pregnant rats exposed to diisobutyl phthalate (DiBP) (Borch et al. 2006), Boberg and colleagues (2011) and Guerra and colleagues (2010) did not replicate this finding using DiBP and DBP, respectively.

Puberty

Most processes associated with the absorption, distribution, metabolism, and elimination of substances in the body reach maturity in adolescents (twelve to eighteen years old) (Makri et al. 2004). At puberty, adolescents go through important biological changes that make them especially susceptible to phthalate exposure, including growth spurts and hormonal fluctuations. Animal studies have indicated that exposure to phthalates during puberty may trigger unique and lasting adverse health effects unlike those seen during adulthood (Makri et al. 2004). A report by the Environmental Working Group (2008), which looked at studies done on pubescent rats, observed that rats at this stage of development appear to be more sensitive to testicular toxicity induced by exposure to phthalates compared with adults, and that this may be the result of differences in the absorption and metabolism of phthalate compounds (Environmental Working Group 2008).

Animal studies on phthalates and female reproductive function have reported toxic effects of exposure on the ovaries during puberty. One study demonstrated that high oral doses of DEHP administrated to adult female rats twice a week at 1,400 mg/kg/day led to a decrease in the duration of hormone cycles in offspring (Hirosawa et al. 2006). Grande and colleagues (2006) reported that adult female rats exposed to doses of DEHP as low as 15 mg/kg/day throughout pregnancy and lactation had female offspring with significant delays in the onset of puberty. They also evaluated the possible effects of in utero and lactational DEHP exposure on female reproductive function later in life (Grande et al. 2007). The highest dose tested (405 mg/kg/day) led to an increase in the number of eggs developed that do not undergo ovulation during a menstrual cycle. These findings suggest that phthalate exposure can inhibit estrogen production and activity, as well as increase male reproductive hormones in females.

On the other hand, the epidemiological evidence of the effect of phthalates on the onset of puberty is controversial. A study conducted on young girls in Puerto Rico found higher concentrations of phthalate compounds in the blood of those who experienced premature breast development (thelarche) compared with controls (Colon et al. 2000). Likewise, higher urinary levels of monomethyl phthalate (MMP) were reported in Taiwanese girls with premature thelarche and central precocious puberty compared with normal controls (Chou et al. 2009). These two studies must be interpreted with caution, however, because of their small sample sizes and the possibility of environmental phthalate contamination (Kay, Chambers, and Foster 2013). In addition, premature sexual development is associated with increased female reproductive hormones, a finding that contradicts the results described in the abovementioned animal studies. Moreover, three studies with larger sample sizes and a rigorous assessment of participants reported no association between various phthalate metabolites and accelerated breast development (Frederiksen et al. 2012; Mouritsen et al. 2013; Wolff et al. 2010). Of note, two of these studies found an association between phthalate exposure and delayed pubic hair growth in girls (a marker of anti-androgenic activity) (Frederiksen et al. 2012; Wolff et al. 2010). Although this finding was not reproduced recently by Mouritsen and colleagues (2013), the authors reported lower adrenal androgen levels in the serum of girls with excretion of DBP metabolites above the mean. We therefore conclude that the evidence is too weak to support an association between phthalate exposure and precocious puberty.

Adulthood

Thyroid Function

Thyroid hormones are produced by the thyroid gland and are responsible for the regulation of growth and metabolism. They are also essential for fetal development, reproduction, and brain and organ development. Experimental studies have indicated possible adverse effects of phthalate exposure on the thyroid gland in living organisms (in vivo) and in thyroid tissue in laboratory experiments (in vitro) (Pereira, Mapuskar, and Vaman Rao 2007; Sugiyama et al. 2005). A study of pregnant Taiwanese women found that increases in DBP exposure during their second trimester was associated with a decline in maternal thyroid hormone levels (Huang et al. 2007). Significant correlations were found between elevated urinary phthalate metabolites and reductions in the levels of thyroid hormones. Another study indicated an association between MEHP exposure and decreased thyroid hormone in adult men (Meeker, Calafat, and Hauser 2007). Further research is required to understand the ways in which phthalates interact with thyroid hormones and affect thyroid function.

Endometriosis

Endometriosis is a health condition that occurs in adult women, where cells from the lining of the uterus (endometrium) grow in other parts of the body. The reported prevalence of endometriosis in women of reproductive age (twenty to forty years old) varies wildly, ranging from 2 to 22 percent, depending on diagnostic criteria and/or population characteristics. Endometriosis has been found in 38 percent of infertile women (Guo and Wang 2006).

Endometriosis results from a number of genetic and environmental factors. Given that endometriosis is an estrogen-dependent disease, it is thought that estrogen and estrogen-mimicking hormones in the body could contribute to its development and progression (Huhtinen et al. 2012). Previous work in non-human primates has shown that exposure to EDCs, such as the chemical compound 2,3,7,8-tetrachlorodibenzo-*p*-dioxin (TCDD), is associated with an increased prevalence of endometriosis (Rier 2002). Other animal experiments have also found connections between dioxin and dioxin-like compounds and this disease (Rier and Foster 2002).

Two hospital-based case-control studies on humans have reported elevated concentrations of DEHP in the blood (plasma) of patients with endometriosis. In one study, fifty-five women diagnosed with endometriosis were found to have higher concentrations of plasma DEHP compared with twenty-four controls (0.57 µg/mL versus 0.18 µg/mL, respectively) (Cobellis et al. 2003). Specifically, 92.6 percent of the women had detectable DEHP and/or MEHP in the peritoneal fluid. Another study reported that middle-aged, infertile women suffering from endometriosis had elevated levels of a number of phthalate compounds in their bodies, including di-*n*-butyl phthalate (DnBP), butyl benzyl phthalate (BBP), di-*n*-octyl phthalate (DnOP), and DEHP, compared with thirty-eight women without endometriosis (Reddy et al. 2006).

Such evidence suggests that phthalates may disrupt hormone activity or act as immune toxins, thereby affecting the response of the endometrium to steroids and resulting in endometriosis. These results should be interpreted with caution, however, given the small sample sizes, lack of true control populations, and failure to adequately adjust for potential confounders (for example, elevated levels of DEHP may reflect increased contact with products found in the health profession). A lack of overall experimental data to support the association between phthalate exposure and endometriosis illustrates a major gap in the literature that should be addressed.

Time to Pregnancy (TTP) and Infertility

A cohort study involving 6,302 women reported increased odds of prolonged TTP (>6 months) after occupational exposure to phthalates as measured by self-reported exposure to a list of chemicals (OR = 2.16 [1.02-4.57]) (Burdorf et al. 2011). Recently, Tranfo and colleagues (2012) measured the urinary concentrations of a number of phthalates in fifty-six infertile couples recruited from an assisted reproduction centre in Italy, and compared them with a control group of fertile couples. The analysis revealed significantly higher phthalate concentrations in cases versus controls. The differences remained statistically significant after stratification by gender. This evidence suggests a negative impact of phthalate exposure on fecundity, but further studies with larger sample sizes and biomonitoring assessment of exposure are needed to determine the effect of phthalates on fecundity.

Pregnancy

Between 1992 and 1994, Toft and colleagues (2012) assessed the associations between phthalate metabolites and pregnancy loss. This did this by examining urine samples from 128 women who were planning their first pregnancy. The study found a significant association between elevated MEHP exposure and pregnancy loss (Toft et al. 2012). Moreover, Latini and colleagues (2003) measured DEHP and MEHP concentrations in the umbilical cord blood of 84 infants. Newborns exposed to

MEHP showed a lower gestational age (the age between conception and birth) than infants who were not exposed (38.16 weeks versus 39.35 weeks) (Latini et al. 2003). From these findings, the authors hypothesized that MEHP can influence the timing of the onset of labour. Conversely, a study by Adibi and colleagues (2009) examining DEHP exposure in 283 pregnant women found that women with DEHP urinary concentrations of 75 percent had pregnancies approximately two days longer than women with concentrations of 25 percent. Increases in MEHP urinary concentrations correlated with delivery at 41 weeks or later and less chance of preterm delivery (Adibi et al. 2009). The findings suggest that DEHP may interfere with hormone signals influencing the timing of childbirth. Further research is needed to clarify the divergent results reported by these studies (Adibi et al. 2009; Latini et al. 2003).

It has been proposed that the survival of a population is predominantly controlled by a female's average reproductive activity over a lifetime, and the number of female offspring resulting from this activity (Gurney 2006). The study proposes that in many circumstances, toxicants such as phthalates are likely to produce direct adverse effects on female fertility, even at low levels of exposure. These effects occur at levels far below those that might impact male fertility and any male influences on pregnancy outcomes (Gurney 2006). As a result, even minor reproductive health effects in females as a result of phthalate exposure can have a much larger impact on reproduction at the population level. This theory should be addressed in further studies.

Phthalate Exposure and Women's Social Determinants of Health

While studies of environmental exposure are often conducted in isolation, a focus on biological factors as well as social, economic, and cultural practices can significantly influence exposure and any subsequent health effects (Weiss and Myers 2001). As also discussed in Chapters 1, 2, 3, 8, 9, and 11, women of reproductive age may be exposed to phthalates in unique ways due to behaviour and lifestyle. Occupation, diet and physical activity, use of personal care, cosmetic, or home cleaning products, hobbies, stress, wearing of jewellery, and use of prescription and over-the-counter medications can all add to or affect the concentration and duration of a woman's exposure to phthalates (Arbuckle 2006). These multiple routes of phthalate exposure can cause the toxicants to accumulate in the body and intensify effects on biological tissues and processes, thereby increasing the potential risk of toxicity or development of health problems.

Further, many women work in occupations through which they experience elevated phthalate exposures. For example, cosmetology, a profession disproportionately held by women, routinely exposes workers to a wide range of environmental chemicals through inhalation or skin absorption, which can lead to a number of adverse health effects (Peters et al. 2010). Even in circumstances where men and women share the same job title, women may have differing workplace exposures resulting from the segregation of tasks by gender (Arbuckle 2006; Chapter 10). Additionally, protective workplace clothing, gear, and equipment are often designed using men's measurements and sizes, and could account for further differences in exposure between the sexes (Arbuckle 2006).

Outside of the work environment, women of reproductive age can experience elevated phthalate exposures through contact with a variety of commonly used cosmetics and household products. A woman's use of personal care products, and her patterns of exposure from these products, depends on her age, ethnic or socio-economic status, and other social variables. Many cosmetic products are directly applied to the skin or mucous membranes, such as the nose, mouth, eyes, or ears. Although the skin provides a protective barrier to some extent, certain ingredients, such as phthalates and other EDCs, can penetrate the skin and be distributed systemically. For example, one study has demonstrated detectable increases in levels of phthalates in the urine of adult males after the application of body cream (Janjua et al. 2007, 2008). Some cosmetic products are applied via spray, presenting the possibility of exposure through inhalation. In fact, a study on pregnant women found positive correlations between concentrations of phthalates in the air and concentrations found in the urine, suggesting inhalation as a route of exposure in this subpopulation (Adibi et al. 2003).

Packaging and storage devices present another possible source of phthalate exposure for women (NTP-CERHR 2006). Dietary DEHP exposure results from the bioaccumulation of phthalate compounds in different foods, as well as from the leaching of phthalate from packaging during various stages of food processing and storage (Latini, De Felice, and Verrotti 2004). Phthalates have also been detected in mineral water stored in plastic water bottles, although the levels observed may not represent a significant exposure pathway compared with reference doses

(Montuori et al. 2008). Some studies have suggested that the use of certain types of medical equipment can also result in high levels of phthalate exposure. Research on pregnant women in hospital has demonstrated a link between DEHP exposure and the leaching of DEHP into solutions stored in PVC medical devices (NTP-CERHR 2006).

A number of studies have examined the association between possible sources of phthalate exposure and risks to newborn babies (Weuve et al. 2006). This not only has significance for gendered development and reproductive function due to the particular vulnerabilities linked with infantile exposure but can also impact maternal health, as women are predominantly responsible for overseeing the health and well-being of children (see Chapters 2 and 3). Consequently, any risk to the health of a child is a burden that women must carry, and can lead to impacts on their own health as well. Women frequently make decisions based on what will result in the least amount of exposure for their children (Chapter 2). Before the implementation of new techniques for quantifying phthalate levels in the body, positive associations were found between phthalates in newborns and exposure to different medical procedures and devices (Pak and McCauley 2007). With the advent of new analytical techniques, some studies show even stronger associations between phthalates in newborns and possible medical sources of exposure. Calafat and colleagues (2004) assessed DEHP exposure in six premature newborns who had undergone medical intervention procedures. The study not only found positive urinary phthalate concentrations but also measured levels that were twenty-six times higher than in the general population in the US. Another study involving fifty-four infants hospitalized in intensive care units found that exposure to PVC plastic medical devices containing DEHP was associated with urinary concentrations of its metabolite MEHP (Green et al. 2005). Further analyses of the same population found associations between DEHP-containing medical devices and two other metabolites, MEHHP and mono-(2-ethyl-5-oxohexyl) phthalate (MEOP) (Weuve et al. 2006). Loff and colleagues (2002) found a steep temperature gradient in relation to the leaching of DEHP from equipment in newborn intensive care units. This correlation with temperature is significant considering that the temperature in most intensive care facilities is around 30°C and the temperature within an incubator is around 37°C.

Another source of phthalate exposure for newborns and infants is through the application of infant care products. One study explored the association between phthalate concentrations in an infant's urine and a mother's application of care products to the infant's skin over a twenty-four-hour time span (Sathyanarayana et al. 2008). Nine phthalate compounds were measured in the urine of 163 infants during times when specific lotions, powders, and shampoos were used. In contrast, no associations were found during periods when infants were in contact with plastic toys or pacifiers. Further studies are needed to confirm these findings. Finally, high levels of phthalates have been detected in indoor household dust (Hwang et al. 2008). Associations have been made between exposure to dust containing phthalates and respiratory conditions in infants and adolescents, such as rhinitis, eczema, and asthma (Bornehag et al. 2004).

Social Disparities: Unequal Distribution of Exposures, and Susceptibility among Women

As discussed elsewhere in this volume, disadvantaged populations experience greater susceptibility to environmental hazards because of compounding health hazards (poor nutrition, poor housing, poor health care, and higher levels of environmental contaminants) (Gupta and Ross 2007; Haluza-DeLay et al. 2009). Exposure to environmental toxicants varies according to race, income, gender, age, and other demographic characteristics (Sexton, Olden, and Johnson 1993), in addition to susceptibilities related to genetics or physiologic differences related to sex. For example, data from the NHANES surveys revealed that non-Hispanic blacks and Mexican Americans have higher levels of mono-ethyl phthalate (MEP), mono-isobutyl phthalate (MiBP), and MEHP in their urine than non-Hispanic whites (CDC 2009). These data were not stratified by sex. Marginalized groups discriminated by demographic characteristics are at greater risk of experiencing more toxic exposures, including phthalate exposures, and are more vulnerable to toxic effects resulting from a limited access to power and the living conditions and services needed to address contaminants in their communities (Haluza-DeLay et al. 2009).

Socio-economic status continues to be the leading determinant of health in Canada, and poverty has become a growing reality for many individuals. Although all social classes use consumer products containing phthalates, those with lower socio-economic status are more likely to be exposed to products fabricated offshore that are less controlled by regulatory authorities, such as cosmetics, children's toys, and food containers, among others. As discussed in Chapters 1, 2, 3, 10, and 11, the majority of people living in poverty are women, particularly single mothers and their children. The combined effect of socio-economic status on product use, along with other nutritional and metabolic characteristics, can therefore make women especially vulnerable to phthalate exposure.

Social and Emotional Consequences of Reproductive Health Effects Resulting from Exposure

Adverse health effects stemming from phthalate exposure can have significant social and psychological consequences for women. For example, the capacity to conceive children is important for many women and can contribute to a woman's social and psychological well-being. In utero exposure to phthalates can lead to infertility, as a result of impaired ovarian function or the development of disorders such as endometriosis. The inability to conceive can cause stress and anxiety for women and their partners (Cousineau and Domar 2007). Moreover, women who are able to conceive may experience anxiety about the potential adverse health effects of phthalates on their pregnancies. Finally, research on endometriosis indicates that chronic pelvic discomfort and menstrual pain associated with the disorder can have negative psychological and social impacts on women (Jones, Jenkinson, and Kennedy 2004). A qualitative study found that physical, sexual, and social performance, energy levels, emotional well-being, and employment success and status were all factors in a woman's life that might be affected by endometriosis (Jones, Jenkinson, and Kennedy 2004). For example, the physical pain of endometriosis can cause women to withdraw socially, discourage them from pursuing intimate relationships, or prevent them from being at work on a regular basis.

Current Policies and Programs to Address Phthalate Production and Exposure

Several governments worldwide have begun to address the manufacture and use of phthalates in products, and their adverse reproductive health effects, through a number of policy initiatives. The European Union has banned a number of phthalates from cosmetics (Council of the European Communities 2007), as well as phthalate compounds used in the manufacture of children's toys or articles that might be handled by children (European Parliament 2005). Phthalates have been identified by the US National Toxicology Program (NTP) as reproductive toxicants that require further study (Moorman et al. 2000; NTP-CERHR 2006). Since August 2008, the Consumer Product Safety Improvement Act[1] has banned phthalates from children's products across the United States.

In Canada, the Canadian Environmental Protection Act, 1999 (CEPA)[2] requires that every new chemical substance made in Canada or imported from other countries be assessed against specific criteria (see Chapter 3). In 1998, Health Canada conducted a risk assessment of children's teething products containing diisononyl phthalate (DiNP). The assessment concluded that DiNP posed a health risk to children under a year old who may suck or chew on soft vinyl products for prolonged periods of time. The guidelines following from the assessment recommended that manufacturers, importers, distributors, and retailers eliminate DiNP from all soft vinyl teethers, rattles, and other products that might come into contact with the mouths of young children (Health Canada 2006). Similarly, in 2007, Health Canada's Consumer Product Safety Bureau proposed a consultation to prohibit the use of DEHP in plastic toys and products likely to make contact with the mouths of children under three years of age (Health Canada 2009). In 2010, Health Canada announced new phthalate regulations under the Hazardous Products Act (2009)[3] that restrict the allowable concentrations of DEHP, BBP, and dibutyl phthalate (DBP) to no more than 1,000 mg/kg (0.1 percent) in the soft vinyl of all children's toys and child care articles, as well as the concentrations of DiNP, diisodecyl phthalate (DiIDP), and di-*n*-octyl phthalate (DNOP) for toys and articles that might be placed in the mouth of a child under four years of age.[3]

On the monitoring side, two human biomonitoring studies being conducted under the Government of Canada's Chemicals Management Plan (see Chapter 3) may shine further light on phthalate exposure and concentrations within the body. Begun in 2007, the Canadian Health Measures Survey (CHMS) is measuring the levels of environmental chemicals, including phthalates, in 5,000 Canadians between the ages of three and seventy-nine (Health Canada 2013). Results from the first two cycles of the CHMS (2007-09 and 2009-11) have been released (Health Canada 2013), including phthalates biomonitoring by male/female categories in each age group, as previously mentioned (Saravanabhavan et al. 2013). The Maternal-Infant Research on Environmental Chemicals (MIREC) study, an initiative launched in 2006 and of particular interest to women's health, is developing a national profile of exposures to environmental contaminants in pregnant women and their offspring (Arbuckle et al. 2013). The study follows 2,000 pregnant women and their offspring from the first trimester of pregnancy through delivery, and up to five years after birth on a subset of the cohort. Each woman makes regularly scheduled clinical visits in order to complete questionnaires, provide medical history, and provide samples of maternal blood, urine, hair, and breast milk; cord blood; and infant blood and stools. The biomonitoring data collected may be used in future risk assessment and management of hazardous substances, and will provide new knowledge on the specific toxic effects of phthalate exposure on mothers and fetuses. The MIREC study may also provide an important step in building support and funding for further evidence-based research and regulatory policies that reduce phthalate exposure and improve women's reproductive health. Long-term studies beyond MIREC that consider impacts on current and future offspring could lead to a greater understanding of phthalate exposure and its effect on the reproductive health of multiple generations.

Future Directions

Despite these important steps in policy and research, government bodies and scientists still have a long way to go in addressing phthalate production and exposure and protecting the health of all Canadians, particularly women. The potential for adverse effects of phthalate exposure on environmental and human health, together with socio-economic implications and costs to industry, make it clear that scientists, government, and industry need to conduct the necessary research to ensure that regulatory decisions are evidence-based and made in a timely manner (Foster 2008).

Although much of the data presented in this chapter have originated from experimental animal studies, such experiments provide a window into the kinds of human health effects that can occur. Limitations in the translation of results from animal studies to humans suggest that a number of factors may not be accounted for in assessing the harm of phthalate exposure to human health, such as a human's immersion in complex social and cultural settings. Gender differences are also important for assessing the adverse health effects of environmental chemicals in general, and particularly for phthalates, which are essential compounds in many products with which women come in contact (cosmetics, household products, and so on). Future research efforts should include biomonitoring in both animal and human studies, and should seek to measure comparable indicators of biological toxicity in order to facilitate appropriate human health risk assessments and explore human health outcomes (Ritter and Arbuckle 2007). Population studies that evaluate human reproductive impairment are also needed. Since human health studies are expensive and logistically difficult, resources must be prioritized in conducting this kind of research. Given the importance of fetal, childhood, and gendered windows of vulnerability, experimental animal and human studies should also be designed to adequately assess time-specific exposures and detect both early and delayed adverse health outcomes.

So far, much of the research on phthalate exposures and reproductive health has focused on males. The evidence discussed in this chapter points to the existence of sex and gender differences in phthalate exposure, and the significant if not harmful effects of exposure on women's reproductive health. There is a need for studies that consider exposure from a gendered perspective and examine women's unique susceptibilities to exposure based on social and environmental factors. A women-centred research agenda would take into account distinctive exposure patterns among women; improve methods to test chemicals for developmental toxicity in women; and lead to more clinical, epidemiological, and toxicological studies that identify women's health outcomes at important periods of development and transformation, including menstruation, pregnancy, and menopause (Wigle et al. 2007). Continued research in this area will enable government and policy makers to make more effective decisions on the management, regulation, and elimination of phthalates in Canada, ensuring the health and safety of current and future generations.

Notes

1 *Consumer Product Safety Improvement Act of 2008,* Public Law 110-314, 122 Stat. 3016 (http://www.cpsc.gov/cpsia.pdf).

2 *Canadian Environmental Protection Act, 1999,* S.C. 1999, c. 33 (http://laws-lois.justice.gc.ca/eng/acts/C-15.31/index.html).

3 *Phthalates Regulations,* SOR/2010-298 (http://laws-lois.justice.gc.ca/eng/regulations/SOR-2010-298/index.html).

References

Adibi, Jennifer J., Russ Hauser, Paige L. Williams, et al. 2009. "Maternal Urinary Metabolites of Di-(2-Ethylhexyl) Phthalate in Relation to the Timing of Labor in a

US Multicenter Pregnancy Cohort Study." *American Journal of Epidemiology* 169 (8): 1015-24, doi:10.1093/aje/kwp001.

Adibi, Jennifer J., Frederica P. Perera, Wieslaw Jedrychowski, et al. 2003. "Prenatal Exposures to Phthalates among Women in New York City and Krakow, Poland." *Environmental Health Perspectives* 111 (14): 1719-22.

Anway, Matthew D., and Michael K. Skinner. 2006. "Epigenetic Transgenerational Actions of Endocrine Disruptors." *Endocrinology* 147 (6 Suppl): S43-S49, doi:10.1210/en.2005-1058.

Arbuckle, T.E., W.D. Fraser, M. Fisher, et al. 2013. "Cohort Profile: The Maternal-Infant Research on Environmental Chemicals Research Platform." *Paediatric and Perinatal Epidemiology* 27 (4): 415-25, doi:10.1111/ppe.12061.

Arbuckle, Tye E. 2006. "Are There Sex and Gender Differences in Acute Exposure to Chemicals in the Same Setting?" *Environmental Research* 101 (2): 195-204, doi:10.1016/j.envres.2005.08.015.

Barlow, Norman J., and Paul M.D. Foster. 2003. "Pathogenesis of Male Reproductive Tract Lesions from Gestation through Adulthood Following In Utero Exposure to Di(n-Butyl) Phthalate." *Toxicologic Pathology* 31 (4): 397-410, doi:10.1080/01926230390202335.

Barlow, Norman J., Suzanne L. Phillips, Duncan G. Wallace, et al. 2003. "Quantitative Changes in Gene Expression in Fetal Rat Testes Following Exposure to Di(n-Butyl) Phthalate." *Toxicological Sciences* 73 (2): 431-41, doi:10.1093/toxsci/kfg087.

Boberg, J., S. Christiansen, M. Axelstad, et al. 2011. "Reproductive and Behavioral Effects of Diisononyl Phthalate (DINP) in Perinatally Exposed Rats." *Reproductive Toxicology* 31 (2): 200-9, doi:10.1016/j.reprotox.2010.11.001.

Borch, Julie, Marta Axelstad, Anne Marie Vinggaard, et al. 2006. "Diisobutyl Phthalate Has Comparable Anti-Androgenic Effects to Di-n-Butyl Phthalate in Fetal Rat Testis." *Toxicology Letters* 163 (3): 183, doi:10.1016/j.toxlet.2005.10.020.

Borch, Julie, Ole Ladefoged, Ulla Hass, et al. 2004. "Steroidogenesis in Fetal Male Rats Is Reduced by DEHP and DINP, but Endocrine Effects of DEHP Are Not Modulated by DEHA in Fetal, Prepubertal and Adult Male Rats." *Reproductive Toxicology* 18 (1): 53-61, doi:10.1016/j.reprotox.2003.10.011.

Bornehag, Carl-Gustaf, Jan Sundell, Charles J. Weschler, et al. 2004. "The Association between Asthma and Allergic Symptoms in Children and Phthalates in House Dust: A Nested Case-Control Study." *Environmental Health Perspectives* 112 (14): 1393-97, doi:10.1289/ehp.7187.

Burdorf, A., T. Brand, V.W. Jaddoe, et al. 2011. "The Effects of Work-Related Maternal Risk Factors on Time to Pregnancy, Preterm Birth and Birth Weight: The Generation R Study." *Occupational and Environmental Medicine* 68 (3): 197-204, doi:10.1136/oem.2009.046516.

Calafat, Antonia M., Larry L. Needham, Manori J. Silva, et al. 2004. "Exposure to Di-(2-Ethylhexyl) Phthalate among Premature Neonates in a Neonatal Intensive Care Unit." *Pediatrics* 113 (5): e429-e34, doi:10.1289/ehp.0800265.

CDC (Centers for Disease Control and Prevention, US Department of Health and Human Services). 2009. *Fourth National Report on Human Exposure to Environmental Chemicals*. Atlanta: National Center for Environmental Health. http://www.cdc.gov/exposurereport/pdf/fourthreport.pdf.

Chou, Y.Y., P.C. Huang, C.C. Lee, et al. 2009. "Phthalate Exposure in Girls during Early Puberty." *Journal of Pediatric Endocrinology and Metabolism* 22 (1): 69-77.

Cobellis, L., Giuseppe Latini, Claudio De Felice, et al. 2003. "High Plasma Concentrations of Di-(2-Ethylhexyl) Phthalate in Women with Endometriosis." *Human Reproduction* 18 (7): 1512-15, doi:10.1093/humrep/deg254.

Colon, Ivelisse, Doris Caro, Carlos J. Bourdony, et al. 2000. "Identification of Phthalate Esters in the Serum of Young Puerto Rican Girls with Premature Breast Development." *Environmental Health Perspectives* 108 (9): 895-900, doi:10.1289/ehp.00108895.

Council of the European Communities. 2007. Council Directive 76/768 of 27 July 1976 on the Approximation of the Laws of the Member States Relating to Cosmetic Products (76/768/EEC; "Cosmetics Directive"). http://eur-lex.europa.eu/LexUriServ/LexUriServ.do?uri=CONSLEG:1976L0768:20070919:en:PDF.

Cousineau, Tara M., and Alice D. Domar. 2007. "Psychological Impact of Infertility." *Best Practice and Research Clinical Obstetrics and Gynaecology* 21 (2): 293, doi:10.1016/j.bpobgyn.2006.12.003.

Crowne, E.C., S.M. Shalet, W.H. Wallace, et al. 1991. "Final Height in Girls with Untreated Constitutional Delay in Growth and Puberty." *European Journal of Pediatrics* 150 (10): 708-12.

Damstra, Terri, Sue Barlow, Aake Bergman, et al., eds. 2002. *Global Assessment of the State-of-the-Science of Endocrine Disruptors.* Geneva: World Health Organization, International Programme on Chemical Safety. http://www.who.int/ipcs/publications/new_issues/endocrine_disruptors/en/.

Environmental Working Group. 2008. "Teen Girls' Body Burden of Hormone-Altering Cosmetics Chemicals." http://www.ewg.org/research/teen-girls-body-burden-hormone-altering-cosmetics-chemicals.

European Parliament. 2005. "P6_TA(2005)0266: Phthalates in Toys and Childcare Articles ***II" (Recommendation for second reading for adopting a directive relating to restrictions on the marketing and use of certain dangerous substances and preparations). http://www.europarl.europa.eu/sides/getDoc.do?pubRef=-//EP//NONSGML+TA+P6-TA-2005-0266+0+DOC+PDF+V0//EN.

Foster, Warren G. 2008. "Environmental Estrogens and Endocrine Disruption: Importance of Comparative Endocrinology." *Endocrinology* 149 (9): 4267-68, doi:10.1210/en.2008-0736.

Frederiksen, H., K. Sorensen, A. Mouritsen, et al. 2012. "High Urinary Phthalate Concentration Associated with Delayed Pubarche in Girls." *International Journal of Andrology* 35 (3): 216-26, doi:10.1111/j.1365-2605.2012.01260.x.

Frederiksen, Hanne, Niels Erik Skakkebaek, and Anna-Maria Andersson. 2007. "Metabolism of Phthalates in Humans." *Molecular Nutrition and Food Research* 51 (7): 899-911, doi:10.1002/mnfr.200600243.

Gluckman, Peter D., and Mark A. Hanson. 2006. "Evolution, Development and Timing of Puberty." *Trends in Endocrinology and Metabolism* 17 (1): 7-12, doi:10.1016/j.tem.2005.11.006.

Grande, Simon Wichert, Anderson J.M. Andrade, Chris E. Talsness, et al. 2006. "A Dose-Response Study Following In Utero and Lactational Exposure to Di(2-Ethylhexyl) Phthalate: Effects on Female Rat Reproductive Development." *Toxicological Sciences* 91 (1): 247-54, doi:10.1093/toxsci/kfj128.

–. 2007. "A Dose-Response Study Following In Utero and Lactational Exposure to Di-(2-Ethylhexyl) Phthalate (DEHP): Reproductive Effects on Adult Female Offspring Rats." *Toxicology* 229 (1-2): 114-22, doi:10.1016/j.tox.2006.10.005.

Gray, L. Earl Jr., Joseph Ostby, Johnathan Furr, et al. 2000. "Perinatal Exposure to the Phthalates DEHP, BBP, and DINP, but Not DEP, DMP, or DOTP, Alters Sexual Differentiation of the Male Rat." *Toxicological Sciences* 58 (2): 350-65, doi:10.1093/toxsci/58.2.350.

Green, Ronald, Russ Hauser, Antonia M. Calafat, et al. 2005. "Use of Di(2-Ethylhexyl) Phthalate-Containing Medical Products and Urinary Levels of Mono(2-Ethylhexyl)

Phthalate in Neonatal Intensive Care Unit Infants." *Environmental Health Perspectives* 113 (9): 1222-25, doi:10.1289/ehp.7932.

Guerra, M.T., W.R. Scarano, F.C. de Toledo, et al. 2010. "Reproductive Development and Function of Female Rats Exposed to Di-Eta-Butyl-Phthalate (DBP) In Utero and during Lactation." *Reproductive Toxicology* 29 (1): 99-105, doi:10.1016/j.reprotox.2009.10.005.

Guo, Sun-Wei, and Yuedong Wang. 2006. "Sources of Heterogeneities in Estimating the Prevalence of Endometriosis in Infertile and Previously Fertile Women." *Fertility and Sterility* 86 (6): 1584-95, doi:10.1016/j.fertnstert.2006.04.040.

Gupta, Shamali, and Nancy A. Ross. 2007. "Under the Microscope: Health Disparities within Canadian Cities." *Health Policy Research Bulletin* (14): 23-28. Health Canada, http://www.hc-sc.gc.ca/sr-sr/pubs/hpr-rpms/bull/2007-people-place-gens-lieux/index-eng.php.

Gurney, William S.C. 2006. "Modeling the Demographic Effects of Endocrine Disruptors." *Environmental Health Perspectives* 114 (Suppl. 1): 122-26, doi:10.1289/ehp.8064.

Haluza-DeLay, Randolph, Pat O'Riley, Peter Cole, et al. 2009. "Introduction. Speaking for Ourselves, Speaking Together: Environmental Justice in Canada." In *Speaking for Ourselves: Environmental Justice in Canada,* edited by Julian Agyeman, Peter Cole, Randolph Haluza-DeLay, and Pat O'Riley, 1-26. Vancouver: UBC Press.

Health Canada. 2006. *Industry Guide to Canadian Safety Requirements for Children's Toys and Related Products.* Ottawa: Health Canada. http://publications.gc.ca/collections/collection_2007/hc-sc/H128-1-06-469E.pdf.

–. 2009. Proposal for legislative action on di(2-ethylhexyl) phthalate under the Hazardous Products Act. *Canada Gazette* 143 (25): 1867. http://gazette.gc.ca/rp-pr/p1/2009/2009-06-20/pdf/g1-14325.pdf.

–. 2013. "The Canadian Health Measures Survey: Cycle 3 (2012-2013)." http://www.hc-sc.gc.ca/ewh-semt/contaminants/human-humaine/chms-ecms-eng.php.

Heudorf, Ursel, Volker Mersch-Sundermann, and Jürgen Angerer. 2007. "Phthalates: Toxicology and Exposure." *International Journal of Hygiene and Environmental Health* 210 (5): 623, doi:10.1016/j.ijheh.2007.07.011.

Hirosawa, Narumi, Kazuyuki Yano, Yuko Suzuki, et al. 2006. "Endocrine Disrupting Effect of Di-(2-Ethylhexyl) Phthalate on Female Rats and Proteome Analyses of Their Pituitaries." *Proteomics* 6 (3): 958-71, doi:10.1002/pmic.200401344.

Hotchkiss, A.K., L.G. Parks-Saldutti, J.S. Ostby, et al. 2004. "A Mixture of the 'Antiandrogens' Linuron and Butyl Benzyl Phthalate Alters Sexual Differentiation of the Male Rat in a Cumulative Fashion." *Biology of Reproduction* 71 (6): 1852-61, doi:10.1095/biolreprod.104.031674.

Huang, Po-Chin, Pao-Lin Kuo, Yen-Yin Chou, et al. 2009. "Association between Prenatal Exposure to Phthalates and the Health of Newborns." *Environment International* 35 (1): 14-20, doi:10.1016/j.envint.2008.05.012.

Huang, Po-Chin, Pao-Lin Kuo, Yue-Liang Guo, et al. 2007. "Associations between Urinary Phthalate Monoesters and Thyroid Hormones in Pregnant Women." *Human Reproduction* 22 (10): 2715-22, doi:10.1093/humrep/dem205.

Huhtinen, K., M. Ståhle, A. Perheentupa, et al. 2012. Estrogen Biosynthesis and Signaling in Endometriosis. *Molecular and Cellular Endocrinology* 358 (2): 146-54, doi:10.1016/j.mce.2011.08.022.

Hwang, Hyun-Min, Eun-Kee Park, Thomas M. Young, et al. 2008. "Occurrence of Endocrine-Disrupting Chemicals in Indoor Dust." *Science of the Total Environment* 404 (1): 26-35, doi:10.1016/j.scitotenv.2008.05.031.

Janjua, Nadeem Rezaq, Hanne Frederiksen, Niels E. Skakkebaek, et al. 2008. "Urinary Excretion of Phthalates and Paraben after Repeated Whole-Body Topical Application in Humans." *International Journal of Andrology* 31 (2): 118-30, doi:10.1111/j.1365-2605.2007.00841.x.

Janjua, Nadeem Rezaq, Gerda Krogh Mortensen, Anna-Maria Andersson, et al. 2007. "Systemic Uptake of Diethyl Phthalate, Dibutyl Phthalate, and Butyl Paraben Following Whole-Body Topical Application and Reproductive and Thyroid Hormone Levels in Humans." *Environmental Science and Technology* 41 (15): 5564-70, doi:10.1021/es0628755.

Jones, Georgina, Crispin Jenkinson, and Stephen Kennedy. 2004. "The Impact of Endometriosis upon Quality of Life: A Qualitative Analysis." *Journal of Psychosomatic Obstetrics and Gynecology* 25 (2): 123-33, doi:10.1080/01674820400002279.

Kaltiala-Heino, Riittakerttu, Elise Kosunen, and Matti Rimpela. 2003. "Pubertal Timing, Sexual Behaviour and Self-Reported Depression in Middle Adolescence." *Journal of Adolescence* 26 (5): 531-45, doi:10.1016/S0140-1971(03)00053-8.

Kaltiala-Heino, Riittakerttu, Mauri Marttunen, Paivi Rantanen, et al. 2003. "Early Puberty Is Associated with Mental Health Problems in Middle Adolescence." *Social Science and Medicine* 57 (6): 1055-64, doi:10.1016/S0277-9536(02)00480-X.

Karapanou, Olga, and Anastasios Papadimitriou. 2010. "Determinants of Menarche." *Reproductive Biology and Endocrinology* 8: 115, doi:10.1186/1477-7827-8-115.

Kay, V.R., C. Chambers, and W.G. Foster. 2013. "Reproductive and Developmental Effects of Phthalate Diesters in Females." *Critical Reviews in Toxicology* 43 (3): 200-19, doi:10.3109/10408444.2013.766149.

Koch, Holger M., Hans Drexler, and Jurgen Angerer. 2004. "Internal Exposure of Nursery-School Children and Their Parents and Teachers to Di(2-Ethylhexyl) Phthalate (DEHP)." *International Journal of Hygiene and Environmental Health* 207 (1): 15-22, doi:10.1078/1438-4639-00270.

Latini, Giuseppe, Claudio De Felice, Giuseppe Presta, et al. 2003. "In Utero Exposure to Di-(2-Ethylhexyl) Phthalate and Duration of Human Pregnancy." *Environmental Health Perspectives* 111 (14): 1783-85, doi:10.1289/ehp.6202.

Latini, Giuseppe, Claudio De Felice, and Alberto Verrotti. 2004. "Plasticizers, Infant Nutrition and Reproductive Health." *Reproductive Toxicology* 19 (1): 27, doi:10.1016/j.reprotox.2004.05.011.

Loff, S., F. Kabs, U. Subotic, et al. 2002. "Kinetics of Diethylhexyl-Phthalate Extraction from Polyvinylchloride-Infusion Lines." *Journal of Parenteral and Enteral Nutrition* 26 (5): 305-9.

Lottrup Poulsen, Grete, Anna-Maria Andersson, H. Leffers, et al. 2006. "Possible Impact of Phthalates on Infant Reproductive Health." *International Journal of Andrology* 29 (1): 172-80; discussion 181-85, doi:10.1111/j.1365-2605.2005.00642.x.

Main, K.M., G.K. Mortensen, M.M. Kaleva, et al. 2006. "Human Breast Milk Contamination with Phthalates and Alterations of Endogenous Reproductive Hormones in Infants Three Months of Age." *Environmental Health Perspectives* 114 (2): 270-76, doi:10.1289/ehp.8075.

Makri, A., M. Goveia, J. Balbus, et al. 2004. "Children's Susceptibility to Chemicals: A Review by Developmental Stage." *Journal of Toxicology and Environmental Health, Part B: Critical Reviews* 7 (6): 417-35, doi:10.1080/10937400490512465.

Meeker, John D., Antonia M Calafat, and Russ Hauser. 2007. "Di(2-Ethylhexyl) Phthalate Metabolites May Alter Thyroid Hormone Levels in Men." *Environmental Health Perspectives* 115 (7): 1029-34, doi:10.1289/eph.9852.

Montuori, Paolo, E. Jover, M. Morgantini, et al. 2008. "Assessing Human Exposure to Phthalic Acid and Phthalate Esters from Mineral Water Stored in Polyethylene Terephthalate and Glass Bottles." *Food Additives and Contaminants, Part A: Chemistry,*

Analysis, Control, Exposure and Risk Assessment 25 (4): 511-18, doi:10.1080/02652030701551800.

Moorman, William J., Heinz W. Ahlers, Robert E. Chapin, et al. 2000. "Prioritization of NTP Reproductive Toxicants for Field Studies." *Reproductive Toxicology* 14 (4): 293-301, doi:10.1016/S0890-6238(00)00089-7.

Mouritsen, A., H. Frederiksen, K. Sorensen, et al. 2013. "Urinary Phthalates from 168 Girls and Boys Measured Twice a Year during a 5-Year Period: Associations with Adrenal Androgen Levels and Puberty." *Journal of Clinical Endocrinology and Metabolism* 98 (9): 3755-64, doi:10.1210/jc.2013-1284.

NTP-CERHR (National Toxicology Program, Center for the Evaluation of Risks to Human Reproduction). 2006. *NTP-CERHR Monograph on the Potential Human Reproductive and Developmental Effects of Di(2-Ethylhexyl) Phthalate (DEHP).* NIH Publication 06-4476. Research Triangle Park, NC: CERHR. http://ntp.niehs.nih.gov/ntp/ohat/phthalates/dehp/DEHP-Monograph.pdf.

Pak, Victoria M., and Linda A. McCauley. 2007. "Risks of Phthalate Exposure among the General Population: Implications for Occupational Health Nurses." *AAOHN (American Association of Occupational Health Nurses) Journal* 55 (1): 12-17.

Parks, Louise G., Joe S. Ostby, Christy R. Lambright, et al. 2000. "The Plasticizer Diethylhexyl Phthalate Induces Malformations by Decreasing Fetal Testosterone Synthesis during Sexual Differentiation in the Male Rat." *Toxicological Sciences* 58 (2): 339-49, doi:10.1093/toxsci/58.2.339.

Pereira, Contzen, Kranti Mapuskar, and C. Vaman Rao. 2007. "A Two-Generation Chronic Mixture Toxicity Study of Clophen A60 and Diethyl Phthalate on Histology of Adrenal Cortex and Thyroid of Rats." *Acta Histochemica* 109 (1): 29-36, doi:10.1016/j.acthis.2006.09.008.

Peters, Claudia, Melanie Harling, Madeleine Dulon, et al. 2010. "Fertility Disorders and Pregnancy Complications in Hairdressers: A Systematic Review." *Journal of Occupational Medicine and Toxicology* 5: 24, doi:10.1186/1745-6673-5-24.

Reddy, B. Satyanarayana, Roya Rozati, B.V. Reddy, et al. 2006. "Association of Phthalate Esters with Endometriosis in Indian Women." *BJOG (An International Journal of Obstetrics and Gynecology)* 113 (5): 515-20, doi:10.1111/j.1471-0528.2006.00925.x.

Rier, S.E. 2002. "The Potential Role of Exposure to Environmental Toxicants in the Pathophysiology of Endometriosis." *Annals of the New York Academy of Sciences* 955: 201-12; discussion 230-32, 396-406, doi:10.1111/j.1749-6632.2002.tb02781.x.

Rier, Sherry, and Warren G. Foster. 2002. "Environmental Dioxins and Endometriosis." *Toxicological Sciences* 70 (2): 161-70, doi:10.1093/toxsci/70.2.161.

Ritter, Leonard, and Tye E. Arbuckle. 2007. "Can Exposure Characterization Explain Concurrence or Discordance between Toxicology and Epidemiology?" *Toxicological Sciences* 97 (2): 241-52, doi:10.1093/toxsci/kfm005.

Rizk, Diaa E.E., and Letha Thomas. 2000. "Relationship between the Length of the Perineum and Position of the Anus and Vaginal Delivery in Primigravidae." *International Urogynecology Journal* 11 (2): 79-83, doi:10.1007/s001920050074.

Saravanabhavan, G., M. Guay, E. Langlois, et al. 2013. "Biomonitoring of Phthalate Metabolites in the Canadian Population through the Canadian Health Measures Survey (2007-9)." *International Journal of Hygiene and Environmental Health* 216 (6): 652-61, doi:10.1016/j.ijheh.2012.12.009.

Sathyanarayana, Sheela, Catherine J. Karr, Paula Lozano, et al. 2008. "Baby Care Products: Possible Sources of Infant Phthalate Exposure." *Pediatrics* 121 (2): e260-e268, doi:10.1542/peds.2006-3766.

Sexton, Ken, Kenneth Olden, and B.L. Johnson. 1993. "'Environmental Justice': The Central Role of Research in Establishing a Credible Scientific Foundation for Informed Decision Making." *Toxicology and Industrial Health* 9 (5): 685-727.

Sugiyama, Shin-Ichiro, Naoyuki Shimada, Hiroyuki Miyoshi, et al. 2005. "Detection of Thyroid System-Disrupting Chemicals Using In Vitro and In Vivo Screening Assays in *Xenopus laevis*." *Toxicological Sciences* 88 (2): 367-74, doi:10.1093/toxsci/kfi330.

Swan, Shanna H. 2008. "Environmental Phthalate Exposure in Relation to Reproductive Outcomes and Other Health Endpoints in Humans." *Environmental Research* 108: 177-84, doi:10.1016/j.envres.2008.08.007.

Swan, Shanna H., Katharina M. Main, Fan Liu, et al. 2005. "Decrease in Anogenital Distance among Male Infants with Prenatal Phthalate Exposure." *Environmental Health Perspectives* 113 (8): 1056-61, doi:10.1289/eph.8100.

Toft, G., B.A. Jonsson, C.H. Lindh, et al. 2012. "Association between Pregnancy Loss and Urinary Phthalate Levels around the Time of Conception." *Environmental Health Perspectives* 120 (3): 458-63, doi:10.1289/ehp.1103552.

Tranfo, Giovanna, Lidia Caporossi, Enrico Paci, et al. 2012. "Urinary Phthalate Monoesters Concentration in Couples with Infertility Problems." *Toxicology Letters* 213 (1): 15-20, doi:10.1016/j.toxlet.2011.11.033.

Weiss, Bernard, and John Peterson Myers. 2001. "Social, Economic and Cultural Context Influence the Expression of Exposure to Neurotoxicants: Session IV Summary and Research Needs." *Neurotoxicology* 22 (5): 559-61, doi:10.1016/S0161-813X(01)00070-5.

Weuve, Jennifer, Brisa N. Sanchez, Antonia M. Calafat, et al. 2006. "Exposure to Phthalates in Neonatal Intensive Care Unit Infants: Urinary Concentrations of Monoesters and Oxidative Metabolites." *Environmental Health Perspectives* 114 (9): 1424-31, doi:10.1289/ehp.8926.

Wigle, Donald T., Tye E. Arbuckle, Mark Walker, et al. 2007. "Environmental Hazards: Evidence for Effects on Child Health." *Journal of Toxicology and Environmental Health, Part B: Critical Reviews* 10 (1-2): 3-39, doi:10.1080/10937400601034563.

Wolff, M.S., S.L. Teitelbaum, S.M. Pinney, et al. 2010. "Investigation of Relationships between Urinary Biomarkers of Phytoestrogens, Phthalates, and Phenols and Pubertal Stages in Girls." *Environmental Health Perspectives* 118 (7): 1039-46, doi:10.1289/ehp.0901690.

Chapter 8 Plastics Recycling and Women's Reproductive Health

Aimée L. Ward and Annie Sasco

Endocrine-disrupting chemicals (EDCs) are a group of chemical substances found in a range of products and in a number of settings, including the home, the workplace, and the outdoors, where they may linger for an undetermined amount of time. They are particularly prevalent in a number of plastic consumer goods, many of which are recyclable or disposable. EDCs from these products can be released into the environment, either leaching from plastic waste through landfill liners or being burned in incinerators.

As described in the Introduction and in Chapters 3, 9, 10, and 11, EDCs in the body can disrupt hormones and bind to hormone receptor sites at key stages of development (TEDX 2011). Whereas natural hormones are carried by specialized proteins until they are required, EDCs remain unattached to other molecules in the blood stream and can more readily react with receptors (Brisken 2008; Edginton and Ritter 2009). As a result, EDCs impede the ability of cells, tissues, and organs to communicate hormonally, and exposure to these chemicals may result in a broad variety of health problems (McKinlay et al. 2008). Attention to the health risks of EDCs has increased over time. Since the start of the millennium, over a thousand studies involving EDCs have been published, indicating increasing concern among researchers worldwide. In 1998, the US Toxic Substances Control Act of 1976[1] established the Endocrine Disruptor Screening and Testing Advisory Committee to prioritize, screen, and test 70,000 such chemicals suspected of causing harm to human health (Daston, Cook, and Kavlock

2003; US EPA 1998). There are also some indications that endocrine disruption is being seen as a pressing concern by risk assessors in Canada (Chapter 3). A number of studies are now showing associations between EDCs and many illnesses, particularly disorders related to the female reproductive tract and cancers. Nevertheless, there continues to be limited scientific evidence to support causal relationships, and the diversity of potential targets in the body, along with the complexity of feedback mechanisms, further complicates the evaluation of the consequences of EDC exposures (Zacharewski 1998).

Given the prevalence and persistence of EDCs in the environment (over 100,000 EDCs are registered with the European Union alone) and their links to adverse health effects, EDCs deserve serious attention for their potential threats to public health (Brisken 2008). This is of particular significance considering the recycling and disposal of EDC-containing plastics. There are few regulations that address the production, management, or life cycles (the life of a product from manufacture or "birth" to product disposal or "death") of EDC-containing plastics, and those policies currently in place are inadequate in addressing toxic exposure from these substances. In this chapter, we explore EDCs in plastics and examine the risks associated with their recycling and disposal. Although EDC exposure can affect the health of all people, our focus is on impacts to women's health. We highlight the need for "upstream" solutions in plastics recycling: policies that target the systems that produce chemical waste (particularly in the form of plastics), and the societal and political structures that allow production and consumption to continue in its current unsustainable form. We argue that Canada's leadership in this arena could have a positive effect on health at a global scale.

EDCs and Human Health

The adverse health effects of EDCs have been discussed for almost fifty years. In 1962, Rachel Carson's book *Silent Spring* drew attention to the health impacts of environmental contaminants by using graphic depictions of birth defects, effects on fertility, and the threat to the food chain. Little did Carson know that this would be the beginning of a great debate (Carson 1962). Colborn, Dumanoski, and Myers's *Our Stolen Future* (1996), and Krimsky's *Hormonal Chaos* (2002) followed suit, broadening research on the effects of EDC exposure on humans and animals. The use of EDCs in high-volume consumer products and

TABLE 8.1 Common endocrine-disrupting chemicals and where they can be found

Chemical	Typically found In ...
Diethylstilbestrol (DES)	Synthetic estrogen
Methoxychlor	Insecticide
DDT	Insecticide
Vinclozolin	Fungicide
Polychlorinated biphenyls (PCBs)	Electrical transformers, capacitors, toxic waste sites, and the food chain (although PCBs are no longer manufactured, they still exist)
Atrazine	Herbicide
Dioxin	By-product of industrial processes
Bisphenol A (BPA)	Baby bottles, as protective coatings on food containers, and for composites and sealants in dentistry
4-nonylphenol (NP)	Domestic consumer products

Sources: Calafat et al. 2005; Ikezuki et al. 2002; Schönfelder et al. 2002; Solomon and Schettler 2000.

subsequent releases of EDCs into the environment mean that humans are exposed to EDCs on a daily basis through a number of routes. A study of urine samples from 394 subjects reported that the EDCs bisphenol A (BPA) and 4-nonylphenol (NP) were found in 95 percent of samples tested (Calafat et al. 2005), and low-levels of these EDCs have been detected in nearly 100 percent of human blood samples (Calafat et al. 2005; Solomon and Schettler 2000). EDCs go by many different names. Table 8.1 lists some common EDCs and where they are generally found (Calafat et al. 2005; Ikezuki et al. 2002; Schönfelder et al. 2002; Solomon and Schettler 2000).

With an increasing number of studies linking EDCs to human health risks, government-based reviews have attempted to understand and address EDC exposures. In 1997 and 1998, the United States Environmental Protection Agency (US EPA) mandated increased attention to EDC research and took the issue to the US Congress for an independent evaluation of scientific research on the harmful effects of EDCs on human health and of possible policy outcomes (US EPA 1997, 1998). The National Research Council followed in 1999 with a review that examined the risks and uncertainties related to EDC exposure, and helped to develop a science-based model for assessment (National Research Council 1999). In 2002, a report by the World Health Organization (WHO) concluded that although there existed only weak evidence linking EDC exposure to adverse health effects in humans, further investigation was required and the precautionary principle needed to be an important consideration in any future policy decisions on the issue (Damstra et al. 2002; Chapters 1, 3, and 4). The report recommended that more large-scale epidemiological studies be carried out in low- and middle-income countries in order to bring about a more global understanding of exposure, as levels of EDCs found in these regions have been significant, especially in children, a particularly vulnerable segment of the population (Damstra et al. 2002; Phillips et al. 2008; Rogan and Chen 2005). Although there is no evidence that large-scale epidemiological studies have taken place in low- and middle-income countries, the past ten years have seen a shift in the WHO recommendations. Since the 2002 report, intensive scientific work has improved understanding of the impact of EDCs. A new WHO report, released in 2013, has established that EDCs can work together, producing additive effects, even in very low doses (WHO/UNEP 2013). This report states that "it is reasonable to suspect that EDCs are adversely affecting human female reproductive health" (22). EDC-related sex ratio imbalances that result in fewer male offspring are also noted in the report (WHO/UNEP 2013), which concludes that "EDCs have the capacity to interfere with tissue and organ development and function ... this is a global threat that needs to be resolved" (227). Again, more studies are called for in countries all over the world (WHO/UNEP 2013).

The Question of Dose

The issue of dose-response, or how much EDC exposure (dose) is necessary to produce an effect (response), is critical to understandings of public health epidemiology and endocrine disruption. Because development and growth rely on a very delicate balance of hormone production and activity, the presence of synthetic hormones in the body, even at low doses, can have measurable effects on biological function (Damstra et al. 2002; Hirabayashi and Inoue 2011; Sekizawa 2008). As a consequence, many adverse health outcomes can arise from low-level exposure to EDC compounds (see Chapters 4, 7, and 10; Vandenberg et al. 2012). For example, hundreds of in vitro and in vivo animal studies regarding the mechanisms and effects of low doses of BPA have been published over the last decade. Further, studies on humans indicate that a typical person has measurable blood, tissue, and urine levels of BPA that far exceed levels in low-dose animal trials (vom

Saal et al. 2007; Weltje, vom Saal, and Oehlmann 2005). Having measurable levels of BPA in the body is increasingly identified as a health concern of unknown magnitude, as many studies now indicate the potential for the chemical to have no "safe" level of exposure, meaning that any exposure could trigger a harmful outcome. For this reason, many scientists and researchers have begun to call for a stricter adherence to the precautionary principle in the production, management, and reduction of EDCs in the environment. The precautionary principle, simply put, means that precaution should be taken regardless of the level of exposure or uncertainty of harm.

EDCs and Women's Reproductive Health

In conjunction with these various inquiries and activities, a growing body of evidence demonstrates the sex-specific impacts and gender differences of EDC exposure, with current research indicating predominant effects on the thyroid and on the production and function of sex steroids (Arbuckle 2006; Mantovani 2006; Sugiyama et al. 2005). A study in Quebec found that two EDCs, polychlorinated biphenyl (PCB) and organochlorine pesticides, negatively affected thyroid homeostasis in women to a greater degree than in men (Abdelouahab et al. 2008). As discussed in Chapters 1, 3, and 9, EDCs are stored in body fat, and since women tend to have higher fat-to-muscle ratios than men, they may experience a greater buildup of EDCs in their bodies over time (Cantarero and Aguirre 2010).

Studies have indicated a number of potential female reproductive health problems from EDC exposure. These include infection or inflammation of the reproductive organs; reproductive abnormalities and cancers, such as vaginal clear cell carcinoma and uterine fibroid tumours; a higher probability of developing endometriosis and various other ovarian and uterine pathologies; decreased fertility and fecundability (the probability that a couple will conceive during a specific time period); shorter pregnancy duration, suppressed lactation, and recurrent miscarriage; increased incidences of breast cancer; polycystic ovary syndrome (a condition in which women have an imbalance of female sex hormones); thyroid disorders and obesity; as well as mood-related disorders and negative effects on sexual desire and function (Latini et al. 2003; McLachlan, Simpson, and Martin 2005; Newbold et al. 2007; O'Leary et al. 2004; Sasco 2003; Sasco, Kaaks, and Little

2003; Solomon and Schettler 2000; Sugiura-Ogasawara et al. 2005; Yang, Park, and Lee 2006).

Recent studies have also made associations between maternal EDC exposure and effects on the health of their unborn children, leading to greater understanding of the link between female reproductive health and the health of future generations, and the particular vulnerability of the fetus and infant to EDC exposure (Doucet et al. 2009; Edginton and Ritter 2009; Huang et al. 2009; Introduction and Chapters 3 and 7). Potential effects on offspring include a declining male-to-female sex ratio, birth defects caused by disruptions in early fetal development, early puberty and breast development, and increased incidences of cancer and reproductive disorders later in life (Caserta et al. 2008; Colborn 2004; McLachlan, Simpson, and Martin 2005; Salehi et al. 2008a; van Larebeke et al. 2008; Wise et al. 2005; Yang, Park, and Lee 2006). For example, studies have confirmed that the widely used flame retardant polybrominated diphenyl ether (PBDE) – typically found in meat, fish, and dairy products, as well as in upholstered furniture and bedding, foam padding often found in baby products, and computers, televisions, and other electronics – can cross the placental barrier and accumulate in the liver of the fetus (Edginton and Ritter 2009). The key liver enzyme required to eliminate PBDE and other EDCs from the body is much lower in babies, with enzyme activity estimated at only 5 percent of that of a mature liver. These findings suggest that babies could be carrying concentrations of PBDE ten times higher than in adults (Edginton and Ritter 2009). The US EPA has issued an official statement that exposures to PBDE and other EDCs can pose serious health risks to infants (Doucet et al. 2009; US EPA 2008b, 2008c).

Breast cancer and early puberty are of particular interest when considering women's and children's health, as they have been most strongly linked to EDC exposure in the literature. They are the focus of further discussion below. Figure 8.1 provides detailed information on how BPA affects women's health via the breasts and the reproductive tract, illustrating where BPA enters the body and the effects that BPA may have on those organs and their functioning (Maffini et al. 2006).

Breast Cancer

The incidence of breast cancer has increased considerably over the last few decades. The US National Cancer Institute reported a rise in the

FIGURE 8.1 Effects of perinatal exposure to BPA and their underlying mechanisms

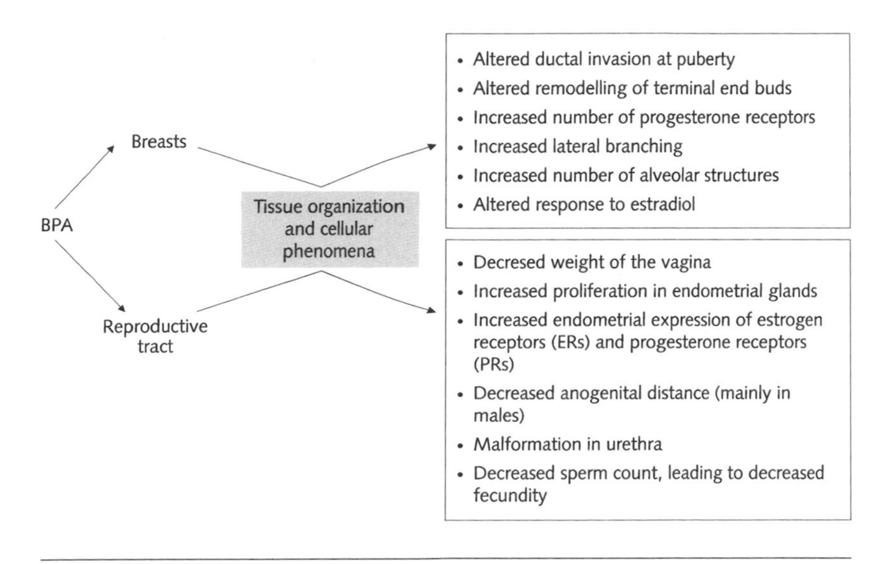

Source: Maffini et al. 2006.

prevalence of breast cancer in all age and ethnic groups in the United States between 1973 and 1999 (Ries et al. 2002). In 1940, a woman's risk of being diagnosed with breast cancer over the course of her entire lifetime was 1 in 22; that risk is now 1 in 9 (CCS 2012; Ries et al. 2002). As discussed in Chapters 9 and 10, breast cancer remains the most common cancer among women, regardless of socio-economic status or geographical location, and is the primary cause of cancer deaths among women globally (Bray, McCarron, and Parkin 2004). In the year 2000, breast cancer killed 370,000 women worldwide, and 1,050,000 new cases were reported, over half of which occurred in low- and middle-income countries (Sasco, Kaaks, and Little 2003).

As of 2012, breast cancer continues to be the most frequently diagnosed cancer among women in Canada, accounting for about 30 percent of all newly diagnosed cancer cases each year (CCS 2012; Salehi et al. 2008a, 2008b). It also ranks second in mortality at 14 percent (CCS 2012). Approximately 22,700 new cases of breast cancer are expected within the year, and current estimates suggest that 1 in 29 women will die from it (CCS 2012).

These elevated occurrences parallel the significant increase in the presence of man-made, persistent chemicals in the environment (Brisken 2008). While known risk factors for breast cancer account for about 40 percent of cases (including genetic or familial, reproductive or hormonal, and lifestyle-based), unknown risk factors related to the environment continue to carry more weight (Madigan et al. 1995; Salehi et al. 2008a). Steadily increasing diagnoses of breast cancer therefore point to the considerable influence of environmental factors, such as chemical and EDC exposure, on women's health (Sasco, Kaaks, and Little 2003; Salehi et al. 2008b). Indeed, evidence linking breast cancer and exposure to hormone-mimicking chemicals has caused mounting concern, leading to further epidemiological studies examining the ways in which these chemicals disrupt hormonal pathways in breast development (Maffini et al. 2006; Sasco 2003).

Take the case of polychlorinated biphenyls (PCBs), chlorinated organic compounds that were used in a variety of applications, including flame retardants, paints, plasticizers, and in the electrical industry, until they were banned in 1977. Despite the ban, PCBs persist and are widespread in the environment (Shields 2006). Women can be exposed to PCBs through their diet, particularly through the ingestion of fish from contaminated waters, as well as through inhalation and skin contact (Carpenter 2006; Chapter 9). A number of epidemiological studies have shown associations between PCBs and breast cancer. In Sweden, a case-control study examined 43 patients who had undergone surgery for invasive breast cancer, and 35 patients who had undergone surgery for benign breast conditions; positive associations were found between breast cancer and elevated levels of PCBs in fatty breast tissue (Liljegren et al. 1998). An Ontario case-control study that compared biopsy tissue from 217 breast cancer cases and 213 benign controls reported similar findings (Aronson et al. 2000). A cohort study found a positive association between PCB exposure and breast cancer risk among women monitored less than three years before diagnosis (Dorgan et al. 1999), and two other cohort studies found significant correlations between increases in breast cancer risk and average levels of PCBs in the blood (Hoyer et al. 2000, 2002). In a nested case-control study of women living on Long Island between 1980 and 1992, O'Leary and colleagues (2004) compared EDC exposure in 105 women diagnosed with breast cancer with 210 controls. Geographic information was gathered regarding the distance between residences and hazardous waste sites containing persistent organic pollutants, including organochlorine pesticides and PCBs. A significant association was found between breast cancer incidence and proximity to PCB-containing sites, and the authors suggested that exposure to pesticides containing PCB-like substances should be more comprehensively investigated with regard to breast cancer risk as a result (O'Leary et al. 2004).

Even with growing evidence of connections between EDC exposure and breast cancer, as in the foregoing example of PCBs, proof of causation remains elusive, as several studies show little or no correlation (Charles et al. 2001; Muscat et al. 2003; Rusiecki et al. 2004). It is interesting to ponder the reasons that some studies show a clear link between contaminants and breast cancer while others do not. The discrepancies between studies may be real, but it is more likely that confounders, or reporting or publication bias (particularly regarding the reporting of only significant results), may be the culprit. Muscat and colleagues 2003 followed women for five years after diagnosis, and their discussion mentions that changes in participant risk factor profiles, as well as their small data set, may have confounded their results. They also mention that while the highest concentration of the pesticide investigated was not significant in cancer recurrence, intermediate levels of some pesticides were associated with an "increased risk [of cancer] relative to the lowest levels" (1477). This was not seen as significant as no clear dose-response was seen by the investigators. Charles and colleagues (2001, 391) comment specifically on why many other studies provide data that suggest differences in EDC levels between cancerous and non-cancerous tissue while their study and another did not: "Two obvious disparities among the studies are the type and number of samples analysed as well as the analytical method employed." Rusiecki and colleagues (2004) report no clear association between PCB exposure and breast cancer, but state that the relatively small number of subjects within the different tumour subtypes they investigated was a limitation. They also note in their discussion of findings that they did in fact find high levels of PCBs in some tumour tissue, but did not report this since it was not statistically significant (Rusiecki et al. 2004). They acknowledge that although their findings were consistent with some studies, other studies found exactly the opposite to be true, and they discuss an interesting reason that this might be the case: "There may be a critical period of [EDC] exposure ... during which the effects of these compounds may be the most significant and lasting. Although serum and adipose tissues reflect a woman's lifetime body burden, we have no way of knowing the timing of exposure. Risk may also vary by genetic characteristics" (Rusiecki et al. 2004, 798).

Early-Onset Puberty

Girls around the world appear to be maturing at a faster pace, with early-onset puberty on the rise (Cesario and Hughes 2007). Early-onset or precocious puberty is the premature appearance of primary sexual characteristics, such as the growth of breasts and pubic hair and the start of menstrual periods. In such circumstances, although a girl's body may have the ability to be sexually active and procreate, this may not correspond with her psychological maturity (see Chapter 7). Studies by the American Academy of Pediatrics indicate that girls who go through early sexual maturity are more likely to engage in risk-taking behaviours, such as drug use and unprotected sex (American Academy of Pediatrics 2006; Zuckerman 2001). The early onset of puberty also places young girls at higher risk of breast cancer later in life (Zuckerman 2001).

In a US survey, the Exposure to Chemical Agents Working Group of the National Children's Study, exposure to EDCs in utero (in the womb) was found to lower the age at which the onset of puberty occurred in offspring, as well as the age at which breast buds developed in girls (Wang, Needham, and Barr 2005). A well-known study carried out between 1994 and 1998 with Puerto Rican girls aged six months to eight years provides strong evidence for the association between exposure to phthalates, a known EDC, and early puberty (Colon et al. 2000; Chapter 7). The researchers analyzed forty-one blood samples from girls diagnosed with early-onset puberty and compared them with thirty-five control samples. High levels of phthalates – on average, 450 parts per billion – were found in 68 percent of the case samples, compared with an average level of only 70 parts per billion in the control group (Colon et al. 2000).

Social Determinants of Health: Are Some Women at Even Higher Risk?

As described in the Introduction and Chapters 1, 2, 3, 7, 9, 10, and 11, social determinants of health are the conditions in which people live and work that affect their opportunities to lead healthy lives (Labonte and Schrecker 2007). Both within and between nations, environmental exposures, such as exposures to EDCs, are linked to persistent social gradients in health (Berkman and Kawachi 2000; Soskolne et al. 2007). It is imperative that when we talk about the environmental risks that affect the health of Canadian women, we consider how social determinants of health can interact to influence the health of the most vulnerable segments of the population. For example, Weck, Paulose, and Flaws (2008) found that a variety of factors can impact pregnancy outcomes when considered in conjunction with toxic exposures, including occupational hazards, psychosocial factors, racialization, chronic stress, and socio-economic status. A woman's socio-economic position has vital effects on her health, regardless of whether she lives in a low-, middle-, or high-income country. Emerging literature now demonstrates the ways in which socio-economic circumstances can modify the effects of air pollution on mortality, with those of lower socio-economic status experiencing greater adverse health effects and premature incidences of death (Laurent et al. 2007). A two-year research project by Toronto-based PollutionWatch indicated that low-income families in Toronto's poorest neighbourhoods face even further health risk because they are breathing contaminated air (PollutionWatch 2008). Families of lower socio-economic status also tend to live in less desirable areas, which often means being in close proximity to industrial areas or point sources of pollution (Welsh 2008).

Following the 2005 United Nations Millennium Development Goals annual conference, O'Neill and colleagues (2007) discussed the importance of linking concerns about equity to concerns about the environment, as the populations most vulnerable to environmental contamination and EDC exposure are those with fewer resources. Unfortunately, research and interventions directed towards these populations are limited (O'Neill et al. 2007). The environmental justice movement has grown out of the need to address and reduce the unfair distribution of environmental burdens, especially on vulnerable populations such as women and children, including those that result from exposures to hazardous substances such as EDCs (Payne-Sturges et al. 2006).

Plastics Production and Waste

Since the Second World War, the use of plastic consumer products has steadily increased worldwide. In an age of multitasking and convenience, the production of items such as plastic containers, wrapping, and packaging has skyrocketed. It is predicted that the manufacture and use of plastics will continue to grow because of product affordability, effectiveness, and lower overall distribution costs for producers. Increases in plastics production and consumer demand for plastic products, combined with insufficient recycling and disposal systems to cope with such demand, present a serious concern: continuing in this direction will result in the volume of plastic waste around the world growing at an astounding pace, with the potential for adverse impacts on the environment (Colborn 2004). For example, research already shows that plastic products used in the health care industry, especially those with single-use applications, have resulted in medical waste becoming one of the world's major sources of environmental contamination (US EPA 1994). A rise in the manufacturing and disposal of medical and domestic single-use plastic products has significant implications for human health, as it will result in the greater release of, and subsequent human exposure to, EDCs.

Plastics, EDCs, and Pathways of Exposure

EDCs are widely found in plasticizers, an additive in plastic products that provides greater product malleability and flexibility. Three main plasticizers are used in the production of plastic consumer goods and medical products: (1) *bisphenol A* (BPA) serves as a raw material during the production of polycarbonates and epoxy resins; (2) *di(2-ethylhexyl) phthalate* (DEHP) is found in polyvinyl chloride, the most common plastic used in producing medical supplies; and (3) *diethylhexyl adipate* is added to plastics to make them flexible (Health Care without Harm 2002; Kamrin 2004; Latini et al. 2003). These three substances represent only a fraction of the thousands of EDC compounds comprising plastic products. People can come into contact with these EDCs through a number of pathways, some of which are described below.

EDC Exposure though Use of Plastic Products

One significant route of EDC exposure is through contact with plastic consumer products (Fernandez et al. 2007). Take the example of BPA. BPA, one of the most common plasticizers used in the plastics industry, is found in a wide range of plastic consumer goods. In 2005, the annual production of BPA worldwide was 3.2 million tons (Fernandez et al. 2007). Automotive applications account for 20 percent of the chemical's use, glazing and sheet form for construction make up another 20 percent, and the use of BPA in optical media, including CDs and DVDs, is on the rise (Fernandez et al. 2007). BPA is also found in a plethora of other consumer products, including thermal papers, plastic water pipes, dental sealants, flame retardant materials, children's toys, medical equipment such as IV bags and lines, paints, solvents, detergents, lubricating oils, cosmetics, and rubber sex toys, as well as in the manufacture of milk and food containers, in baby bottles, and in the interior linings of food cans (Brisken 2008; Health Canada 2010; National Toxicology Program 2008). The scent of a new car? That's the smell of BPA volatilizing from the plastic dashboard. The odour of that new shower curtain? You guessed it: BPA.

The substantial production and consumption of BPA-containing products means that BPA is continuously being released into the environment and leading to human exposures, often as a result of coming into contact with BPA-containing plastic products. Evidence of its harmful effects on human health continues to build. The US Food and Drug Administration initially declared BPA to be safe, but it reopened its examination after a subcommittee of an advisory board found that the agency had ignored valuable studies demonstrating links between BPA and adverse health effects (Rust and Kissinger 2008). In 2008, after months of conferring with researchers and officials, the Canadian government moved to ban polycarbonate baby bottles after officially declaring BPA toxic (Austen 2008; Government of Canada 2010a, 2010b; Layton and Lee 2008). Canada is the first country to do so, and this bold and important first step will no doubt put pressure on other high-income countries to follow suit. In addition, some companies, such as Nalgene (a producer of high-end water bottles), voluntarily removed BPA from their bottles, not only because of safety concerns but also in response to increasing consumer demand for BPA-free bottles.[2] It remains to be seen how other plastic manufacturers will respond.

Exposure through Plastics Disposal

Another significant yet less recognized pathway of EDC exposure is through plastic waste. The two most common means of plastic disposal are incineration and landfill deposition, both of which can present serious health hazards (US EPA 2012b; Zero Waste America, n.d.). When plastic is burned, toxins are released into the air, and plastics left for years in landfills can leach toxic compounds into groundwater, particularly in older sites that lack proper liners. These forms of disposal are most common in low- and middle-income countries that have poorly developed waste management systems (Hosoda 2007).

The Status of Plastics Recycling around the Globe

Many developed and developing countries are dealing with a surplus of plastic waste with only limited recycling capacity to address it as a result of ineffective waste stream techniques, and struggling with manufacturers who do not take responsibility for their actions. In addition, while paper and metal have large and well-established after-markets, which buy recycled goods and turn them into items for consumption, it has been more difficult for governments to develop broad-based techniques for plastics breakdown and reassembly because of the variety of plastic products manufactured and the array of chemical compounds that make up each product (Colborn 2004). For example, the cap of a typical plastic water bottle may be made of a different plastic compound from the body of the bottle. As a result, only a handful of plastics after-markets have been established. Most plastics – sometimes as much as 90 percent, depending on the region – ultimately end up in landfills and incinerators (Themelis and Todd 2004). A number of nations have attempted to address plastics recycling and toxic waste reduction to varying extents, with different degrees of success.

The United States

In 2008, the US Environmental Protection Agency estimated that Americans generated about 250 million tons of trash, an average of 2.1 kilograms of waste per person per day (US EPA 2008a). Twelve percent, or 30 million tons, of this trash is plastic (US EPA 2008a; Killam 2010). Even though much of this waste is put in bins to be recycled by good-hearted consumers, only a small percentage of plastic (about 7 percent) actually goes through the recycling process and is recovered (Killam 2010). This means that 25 percent of landfill weight consists of plastics (Killam 2010), in stark contrast to the large percentage of paper and metal that is recycled, largely a result of thriving after-markets for these materials (Killam 2010; Themelis and Todd 2004).

The majority of recycling systems in the United States follow a multiple-stream model, where consumers separate their recycling into separate containers, waste management companies pick it up, and recovery facilities separate and sort materials. Unfortunately, on arrival at the recovery facility, the majority of materials end up being dumped into one receptacle slated for the landfill (Themelis and Todd 2004).

The associated costs and responsibilities of the multiple-stream model are mostly borne by the consumer: the consumer pays for the product, pays to "recycle" it, and then carries the burden of long-term external costs resulting from the transfer of materials to landfills and their related environmental and health effects (Themelis and Todd 2004). In addition, as a result of limited educational initiatives by government and miscommunication on the part of waste management companies, there is general consumer ignorance of the current reality of recycling. Most consumers continue to buy plastic products in the belief that they will be recycled after use, and remain unaware of the limited amounts of plastic that truly end up in recycling facilities. As a result, public confidence in the recycling process has in fact contributed to the problem of plastics overuse and misuse. The current approach to recycling in the United States is largely ineffective, and economic and market requirements for recycled plastics continue to be unfavourable (Omrani, Zamanzedeh, and Soltandalal 2005). The United States also adds to the burden in developing countries by selling plastic waste and shipping it overseas (Hosoda 2007).

Europe

The packaging industry is the largest user of plastic in Europe, representing 37 percent, or 15 million tons, of the plastics market each year (Euro Publishing Consultancy 2007). Millions of tons of post-consumer plastic waste are produced annually, 16.5 percent of which is actually recycled (Euro Publishing Consultancy 2007). The European Union (EU) accounts for about 25 percent of the world's plastic production; the 36.7 million tons consumed in 2003 averages out to 98 kilograms per person (Nkwachukwu et al. 2013). Of all EU countries, Germany produces the greatest amount of plastic waste, accounting for about 8 percent of global production (Nkwachukwu et al. 2013).

In the face of these statistics, the European Union has taken some important steps in recycling reform. The motivation for more sustainable, successful recycling rates has largely resulted from government legislation. Strict regulations under Europe's End of Life Vehicle Directive promote recycling and provide incentives for environmentally friendly product design (US EPA 2008a, 2012a). These regulations represent an early application by the European Union of extended producer responsibility (EPR) principles for the sustainable life cycle management of products (see the section "Extended Producer

Responsibility: An Answer to the Plastic Waste Dilemma" on page 271). In essence, manufacturers must take more responsibility for the processing of waste from their products. This directive has had a sizable impact, leading to process innovations such as single-stream recycling, a system shown to be more effective in the disassembly and reuse of products at the end of their life cycle (Euro Publishing Consultancy 2007; European Parliament and Council 2000). (Single-stream recycling is discussed later in this chapter.) The European Union has also launched NOVPOL, a research enterprise across several European countries that aims to create a system that can recycle plastic products made up of a mixture of compounds (NOVPOL 2007). If it is successful, recycling plants will be able to increase the rate at which they process plastics, making plastic recycling more economical and efficient (Euro Publishing Consultancy 2007).

The Restriction of Hazardous Substances (RoHS) Directive is another regulatory initiative being adopted. Although it is not specific to chemicals used to make plastics, it has a proven risk assessment structure and strict guidelines that could be applied to EDCs in plastic products, not only by restricting hazardous chemicals used in manufacture but also by requiring compliance measures during plastics recycling (RoHS 2014). These guidelines have the potential to call on producers to limit or discontinue their use of EDCs in plastics production and to develop plans for after-market use, thereby making manufacturers responsible for their merchandise throughout its life cycle.

Initiatives like NOVPOL and RoHS will enable Europe to increase its rate of recycling of plastics while decreasing the amount of hazardous substances being manufactured and utilized, ultimately resulting in the continued demand for improved product life cycle management (Euro Publishing Consultancy 2007). It is hoped that other countries will follow Europe's lead.

Asia

Over the last fifty years, there has been a steep increase in plastics production, especially in Asia. China alone produces more plastic than any other country, at 25 percent of the world's plastic; the rest of Asia produces 16.5 percent (Nkwachukwu et al. 2013). In 2005, over 40 million tons of plastics were consumed annually in China, 20 million tons of which were imported from the United States (Hosoda 2007). China's practice of buying waste from high-income countries that are willing to sell has resulted in its focusing on furthering the productivity and expansion of its plastics industry over the past decade (Hosoda 2007). Even though China has recently begun to shift its attention to the relationship between industrial development and environmental protection, there are few regulations or policies in place to encourage more accepted and effective developments in recycling, and governments have yet to launch any recycling education programs (Euro Publishing Consultancy 2007). With no profit motive and a lack of government intervention, there has been little incentive for Chinese companies to sponsor, share, or develop new, innovative recycling initiatives and technologies, and problems regarding plastics waste and pollution remain critical issues for the industry (Hosoda 2007). In many cases, recycling companies also lack the funds and technical support to increase recycling efforts and provide education about environmental issues (Hosoda 2007).

In India, the use of plastics has become popular for the same reason as elsewhere – low cost, low weight, and high durability (Agarwal et al. 2012). Consumption of plastic products has increased from 900,000 tons in the 1990s to over 3 million tons in 2012, an increase of 300 percent (Agarwal et al. 2012). Once plastic has made its way into the municipal waste system in India, it becomes contaminated and can thus only be incinerated or transferred to landfills (Asnani 2006). The collection, storage, processing, and recycling of plastic waste is not yet organised in Indian cities (Agarwal et al. 2012; Zia, Devadas, and Shukla 2008). As in other developing countries, recycling efforts are carried out mostly by an informal network; this is largely a self-organized activity, driven by economic benefits, and so-called waste pickers are the backbone of this recycling industry (Agarwal et al. 2012). Waste pickers collect clean plastic, scrap metal, paper, and glass and sell them to waste buyers, or *Khabariwala,* who then sell to the next party in the chain, the bulk buyers (Agarwal et al. 2012). As in other countries, plastic waste management planning is a long-term process, a process that the Indian government is still grappling with. No new strategies have been put in place since 1986 (Agarwal et al. 2012).

Africa

In most Sub-Saharan African cities, municipal solid waste (MSW) generation per day per capita is between 0.3 and 0.8 kilograms, far less than half the waste generated per capita in the United States or other developed countries (Din and Cohen 2013). A number of features that define MSW in Africa are specific to that continent. Generally, the level of waste management is very low, differs between cities, and does not include a lot of recycling of plastics (Abel 2007; Oyoo, Leemans, and Mol 2011; Regassa, Sundaraa, and Seboka 2011). There is also a general lack of public awareness and community involvement, lack of mandatory and environmental regulations and enforcement of these regulations, and no master plan designed for each African region (Din and Cohen 2013). High population growth rate and rapid urbanization also make waste management difficult to plan for and implement. Din and Cohen (2013) found in their case study that "the general approach for MSW management in low-income cities has to consider economic evaluation based on the joined phases of collection and transportation and MSW treatment as one project. The first phase of collection and transportation is necessary but cannot stand alone due to economic aspects and environmental regulations" (442).

It is important to note that Africa has become the world's latest destination for obsolete electronic equipment, or e-waste, a huge contributor to plastic waste (Schmidt 2006). Brokers from developed countries who arrange exports often pad shipping containers with useless junk, essentially saddling African importers with plastic electronic garbage, while exploiting women and child labourers (Schmidt 2006).

There are many efforts in developed countries to limit the flow of e-waste to developing countries. The environmental group Basel Action Network (BAN) has recently instituted an "e-Steward" certification program that includes third-party auditing, to replace the voluntary "Electronic Recycler's Pledge of True Stewardship," which had been operating for many years. BAN and Fair Trade Recycling (WR3A)[3] are promoting this new "e-certification program" to help e-waste generators find recyclers who can process their deliveries in an environmentally sustainable way (BAN 2013).

Canada

Canada ranks as one of the biggest producers of solid waste per capita in the world. Statistics Canada data show that Canadians generated almost 1,100 kilograms of MSW per person in 2006, up 8 percent from 2004 (CCME 2009). This represents about 35 million tons, 27 million tons of which were disposed of in landfills and incinerators. Only

7.7 million tons of waste was diverted as recyclables or organics, for a national rate of diversion rate of 22 percent (CCME 2009).

In Canada, responsibility for the management of plastic waste and plastic-related toxic chemicals is shared between federal and provincial jurisdictions. Federal regulation of toxic substances is conducted through the *Canadian Environmental Protection Act, 1999,*[4] with the stated aim of preventing pollution and protecting the environment and human health (Chapter 3). Provincial governments, on the other hand, oversee waste management. Unfortunately, despite efforts by all levels of government over the last three decades, Canada has lagged behind other G8 and Organisation for Economic Co-operation and Development (OECD) countries when it comes to MSW diversion and disposal (CCME 2009). In light of statistics showing 15 to 52 percent of Canadians reported recycling all or some of their recyclable waste in 2007, it is apparent that although many Canadian consumers play their part in recycling, national recycling policies continue to thwart their attempts (Statistics Canada 2010).

Extended Producer Responsibility: An Answer to the Plastic Waste Dilemma

Extended producer responsibility (EPR) is an environmental policy approach in which a manufacturer's responsibility for a product, physically and/or financially, extends beyond production to other stages of a product's life cycle, such as the post-consumer stage (OECD 2001). Although the idea has been around for many years, EPR has not been put into substantial practice in plastics recycling. It has two main features: (1) it shifts responsibility upstream to the producer and away from governments or individuals, and (2) it provides incentives to producers to incorporate environmental considerations in the design of their products (OECD 2001).

In essence, EPR follows the "polluter pays" principle, an internationally recognized policy measure that calls on polluters to be accountable for their pollution, to clean up their practices, and to work towards more sustainable approaches to production (Kibert 2004). Through an improvement in product design or a reduction in the use of hazardous substances, producers create products that are less toxic, more easily dismantled, and reusable or recyclable, and that result in lower waste management costs (Greenpeace 2007).

The following are the key EPR models discussed in the literature (Langrova 2002; Lindhqvist 2000; Milojković and Litovski 2005; Oh and Thompson 2006):

- *Product liability.* The manufacturer is responsible for environmental damage caused by the product, to the extent determined by legislation.
- *Financial responsibility.* The manufacturer covers a number of expenses that are otherwise usually passed on to the consumer, such as the cost of collection, recycling, and/or final disposal.
- *Physical responsibility.* The manufacturer is involved in the physical management of the product, ranging from the development of new recycling technology to the management of "take-back" systems for collecting products and managing them at end of life. Electronic or e-waste is often managed using this type of EPR.
- *Ownership.* The manufacturer retains ownership of products throughout their life cycle, and is therefore responsible for all environmental problems posed by the product. In this case, the producer accepts both physical and economic responsibility, and, in effect, consumers only lease the product.
- *Informative responsibility.* The manufacturer is responsible for providing information about the product and its environmental or health effects at various life cycle stages.

Policy tools rooted in these models include various taxes and product fees (such as virgin material taxes or advance recycling fees) and product take-back mandates (Nnorom and Osibanjo 2008). An analysis of e-waste by Oh and Thompson (2006) found that the most cost-effective EPR practices stem from a combination of tools, thereby exploiting all possible avenues of waste reduction, including curbing production, recycling, material substitution, and product design changes.

Combinations of the foregoing models are currently being considered for legislation in Europe, the United States, and Canada. The effectiveness of these models is yet to be seen. They are being evaluated using the following goals and outcomes (Fishbein 2002):

- focus specifically on the waste generated by end-of-life products
- clearly define the financial responsibility of producers for collection, transport, and recycling of their products
- set meaningful targets for collection and recycling

- differentiate recycling from technologies such as waste-to-energy conversion
- include reporting requirements and enforcement mechanisms
- provide producers with incentives to design their products with reuse or recycling in mind (for example, producing plastics with only one compound, making its after-market easy to target)
- build public awareness and provide incentives for consumers to return used products (for example, a discounts on a new laptop if you return your previous one).

Unfortunately, most EPR programs currently in place around the world remain voluntary (Walls 2006; Widmer et al. 2005). As a result, many initiatives do not meaningfully shift the burden of managing plastic waste and hazardous substances from the consumer to the producer or waste manager. It is increasingly apparent that in order for EPR initiatives to work effectively, they must be incorporated into binding legislation. In addition, some EPR initiatives do not result in actual decreases in environmental exposures. For example, producers who are required to pay a fee for their pollution are not necessarily acting to limit or reduce pollution, and EDCs can still be released into the environment and result in health risks (Fishbein 2002). Most EPR initiatives also fail to address the continued and exponential increase in plastics products worldwide, doing little to curb the processes of mass production and consumption. Finally, EPR has sometimes been confused with product stewardship, a shared approach that mandates producers to cover partial costs at the end of a product's life cycle. Although similar to EPR values, product stewardship does not prevent waste or pollution or place responsibility on any one party. As a result, shared stewardship initiatives continue to shift costs from producers to consumers, fail to provide individual producers with financial responsibility or incentives to reduce waste, and ultimately break the momentum needed to advance waste prevention more broadly (Lifset and Lindhqvist 2008; Marr 2004; McKerlie, Knight, and Thorpe 2006).

Plastic Waste Policy in Canada: Shifting towards Ideas of EPR

The Canadian Council of Ministers of the Environment (CCME) recently explored the potential of EPR as public policy (CCME 2009). They determined that national definitions and principles for EPR would provide a level playing field for the application of waste policy and management initiatives across the country. To this end, the CCME formed an EPR Task Group to establish guidance for such programs and develop an evaluation tool for producers. The aim of the task group is to encourage provincial governments to enact producer responsibility measures by: (1) identifying opportunities to expand and improve EPR-related programs; (2) developing recommendations based on issues that arise from these programs; (3) exploring strategies for new initiatives; and (4) facilitating communication and information exchange among provincial jurisdictions (CCME 2009). By carrying out these activities, the task group hopes to minimize environmental impacts, maximize environmental benefits, promote the transfer of end-of-life responsibility for products to producers, and encourage changes in product design that minimize their environmental footprint. However, although the CCME encourages regional and provincial cooperation in the development of EPR programs, it provides only a supportive role and specific measures remain at the discretion of each jurisdiction, with no hard-and-fast rules and regulations.

At the provincial level, British Columbia has been a leader in legislative action on plastic waste initiatives and recycling measures. As a result of a major shift in the responsibility for waste management from local to provincial authority, current legislation requires all waste management plans to include EPR guidelines (McKerlie, Knight, and Thorpe 2006). Following suit, Ontario, after three decades of the "blue box" program, passed its Waste Diversion Act in 2002.[5] The act gives Waste Diversion Ontario responsibility for developing, implementing, and operating waste diversion programs in the province, and provides a framework for the creation of a provincial single-stream waste management system (Canadian Institute for Environmental Law and Policy 2008a, 2008b). This model, seen as one of the most effective methods of promoting EPR, cites the following general sequence (Eenkema van Dijk 2001):

1 The consumer has only one curbside container for all recyclables.
2 At sorting materials recovery facilities, all recycled material goes into one feeder, where everything is mechanically separated.
3 Groups of materials are sent to sorting rooms, where a crew manually removes large pieces for further separation.
4 The remaining material is sent through another feed to be sorted further. At this phase, 80 percent of recycling has been separated.
5 Any remaining material is sorted manually.

Although a single-stream system reduces overall costs, there is a risk of cross-contamination of recycled items, which causes concern among some after-markets that reuse the material (Hanisch 2000). Additionally, such a system involves a high initial set-up cost. These drawbacks indicate that although this type of recycling has had some success in encouraging an after-market for plastic waste, it isn't necessarily the only answer.

Even with the progress that has been made in recycling programs and practices, some questions still remain regarding the management of plastic waste in Canada:

- Where will the responsibility for waste lie in the future? Criticisms of the Ontario Waste Diversion Act include a lack of successful intervention and unclear program goals (two issues repeated throughout literature on EPR).
- How can a consistent provincial legislative base be established? Mandatory EPR policy has been successful in British Columbia and has shown some promise in Ontario, demonstrating the potential for such programs to be developed in other provinces. If there is to be a reduction in EDC-related releases and exposures resulting from plastics disposal, future provincial policies must include product design guidelines and the development of plastic after-markets.
- How can consumers be properly educated on plastics recycling? The program design principles established by the CCME discuss the transfer of responsibility and cost to waste management authorities, but little is mentioned about building consumer awareness. If more people had a greater understanding of the limitations of plastics recycling, would they change their consumption behaviour by buying less, or would they neglect to recycle altogether? If product manufacture and consumption continue at their current pace, it is plausible that no amount of recycling or the development of EPR programs will be of real assistance. These initiatives must occur in tandem with public education programs that promote reduced consumption of plastics and packaging.
- What about the health-related risks associated with the disposal of plastics, and risks to women's health in particular? Although human health is briefly mentioned in CCME documents, only reductions in toxic releases in relation to an industry's environmental footprint are addressed, not practical considerations regarding human exposure, toxic dose, or sex and gender concerns.

- What is the global burden of plastic waste? It is interesting to note that there is a dramatic difference in the consumption of plastic products worldwide, with a steep gradient between high- and low-income countries. It is estimated that in a lifetime one American citizen has 200 times the negative environmental impact of a person living in a developing country (Wilk 2002). Using this comparison, it is easy to see the negative impacts that developed countries can have on their developing neighbours. Canada and other high-income countries need to keep this in mind when designing and advancing waste management and EDC reduction policies, as well as when devising consumer education programs.

Revisiting the Research and Policy Gaps

Since EDCs are ubiquitous in the environment, it is difficult to find unexposed controls that can be used to compare and assess the effects of exposure, or to make causal links between exposure and illness (Brisken 2008; Foster et al. 2008). As a result, some scientists continue to suggest that endocrine disruption poses minimal risk to human health. However, a growing number of experts argue that EDCs have the potential to cause serious harm to the health of current and future generations, and believe that a framework needs to be developed that will provide useful decision-making tools for communicating scientific uncertainty to the broader public and managing uncertain risks (Martin et al. 2007). The overall weight of evidence points to the need for further study on the relationships between EDC exposure and human health, particularly women's reproductive health; systems of plastics development, production, and processing and their connection to exposure; possible behavioural and educational interventions regarding plastics consumption and recycling; and, perhaps most important, precautionary regulatory actions that might reduce overall exposure, such as policies that remove or limit EDCs in plastics, and EPR measures for waste management systems. The more we learn about the prevalence of EDCs in consumer products and their complex effects on human health, even at low levels of exposure, the more we must consider the importance of taking regulatory action on these substances in the face of scientific uncertainty (Solomon and Schettler 2000).

Taking these points into consideration, it is clear that policy changes are needed that take into account what we know about EDCs in relation to plastics production, risk assessment, and product life cycle management.

The following federal and provincial policy prescriptions offer key recommendations on how governments might effectively manage EDC-containing plastics that threaten Canadian public health.

Prescriptions for Federal Policy

The federal governments in Canada and the United States have already made some important progress with regard to EDC regulation and management. As discussed earlier, the Canadian federal government moved ahead with a bill that banned the use of BPA in plastic baby bottles. Along the same lines, a consumer safety law passed by the US Congress in 2008 took effect in February 2009, banning phthalates in children's toys.[6] These two examples show greater political awareness of, and precautionary action regarding, chemical exposure, and could serve as a springboard for future initiatives, especially related to the disposal of plastic waste.

Along the same lines as the Restriction of Hazardous Substances program in the European Union, new structures could be employed in Canada to identify and control hazardous substances – specifically, to limit EDC exposure from plastics and other products through hazardous chemical restrictions and compliance measures during production and recycling (RoHS 2014). The Consumer Product Safety Act, which came into force on 20 June 2011, may be an avenue for addressing toxic chemicals such as EDCs in Canadian consumer products.[7] Under the act, Health Canada can require manufacturers or importers to provide or obtain safety information – including studies or tests – on products, and prohibit the manufacture, importation, sale, or advertisement of those consumer products that pose an unreasonable danger to the health or safety of Canadians. Through this regulatory framework, the federal government has the potential to ensure greater protection from exposures to EDCs through a reduction in the manufacture, importation, and use of plastics containing these harmful chemicals.

The federal government should also support and fund more epidemiological studies that assess the links between women's reproductive health and EDC exposure related to plastic waste, as well as research involving experimental toxicological models that examine EDC modes of action and exposure pathways (Phillips et al. 2008). Further developments on existing toxicological models could support real-life evaluations of human EDC intake, leading to more accurate exposure limit recommendations and the possible identification of biomarkers for health surveillance (Fernandez et al. 2004; Jonker et al. 2004; Kato et al. 2004; Kusumegi et al. 2004; Rajapakse et al. 2004). Research should also be conducted on potential interventions to reduce exposures throughout the entire life cycle of a plastic product.

In addition, an extensive review of the literature makes it clear that a regular and systematic application of EPR is required in order for plastics recycling processes to consider the life cycle of plastic goods. Federal policy makers should develop an interdisciplinary approach to plastic waste intervention, one that incorporates a holistic view of EPR, including economic, societal, and public health considerations (O'Neill et al. 2007).

Finally, the single-stream waste model, a proven approach in the collection and streaming of plastic waste in Ontario, should be applied at a national level. It is federal government's responsibility to mandate programs across the country that incorporate the single-stream method and to establish after-markets for plastic waste. Recycling cannot be effective unless economic, social, and market requirements for recycled plastics become favourable.

Prescriptions for Provincial Policy

Provincial legislation in Canada should be expanded to include the application of EPR in recycling and waste management initiatives, such as the EPR principles incorporated into the Waste Diversion Act in Ontario. Such legislation should include clear guidelines for producers regarding product design, the streamlining of compounds in plastics production, the implementation of new recycling models, and information on the development of plastics after-markets.

Provincial legislation should also address issues of production and consumption, including more effective educational and outreach initiatives to inform and communicate with the public. Building public awareness of the risks of EDC exposure in relation to plastic waste will help strengthen support for, and development of, Canadian policy on plastics recycling, and put associated health risks in perspective. For example, toxics use reduction strategies could be employed in provinces across the country to increase awareness of EDC exposure, plastics consumption, and recycling. The Internet is another way for policy makers and scientists to bridge the gap between the gathering of information and its dissemination to the public. The Environmental Commissioner of Ontario, a notice-and-comment portal established under Ontario's Environmental Bill of Rights,[8] could be expanded into an interactive learning tool for gathering consumer information from individual Canadians, and HealthCam, an Internet database created in France, also provides a model that could be adapted by provincial governments to build public awareness of EDCs in plastics, human health, and waste management (Environmental Commissioner of Ontario 2012; HealthCam 2010). The HealthCam database uses a variety of media, including the Internet and television, to fill the knowledge gap between scientists and the public on a wide range of health-related topics (HealthCam 2010). It is a bulletin and sounding board through which epidemiologists can discuss the most recent public health issues, such as EDCs, plastic waste, women's health and cancer, and dialogue with consumers in layperson terms. These information portals are becoming the future of public health communication.

In addition, behavioural research models should be developed that examine consumer motivations in relation to both recycling and consumption practices, in order to understand how consumer behaviours affect global markets and economic policies related to waste management (Kaiser, Wolfing, and Fuhrer 1999). Earlier studies have produced mixed findings on whether concern for the environment leads to responsible behaviours like recycling (Mainieri et al. 1997). It has been incorrectly assumed that if people participate in one form of pro-environmental behaviour, such as carpooling, they will also participate in other forms, such as recycling (Mainieri et al. 1997). Even if individuals value the environment and health, barriers might inhibit pro-environmental behaviour, such as the inconvenience and cost of being responsible in both buying and recycling behaviours. Behaviour, social location, education, and accessibility therefore need to be considered in planning provincial educational interventions and making subsequent policy decisions. To this end, good risk communications at the provincial level require the following (Leiss 2001):

- a balanced dialogue among stakeholders on the actual risk factors involved
- interpretation of scientific results in non-expert language
- inclusion of public perceptions.

Behavioural interventions that could stem from these studies and be applied to provincial policy and programs include:

- interventions regarding consumer participation and plastic goods that focus more on consumption behaviour
- interventions regarding exposure to EDCs that focus on all segments of the population, particularly women
- interventions that involve multiple forms of media in order to effectively reach and educate the general population.

Conclusions

Although the health effects of EDC exposure continue to be debated by scientists and policy makers alike, mounting evidence of harms to health resulting from exposure and growing media awareness of chemicals such as BPA have contributed to increasing consumer concern over chemical and product safety. In effect, if producers and waste managers fail to take the lead in addressing EDC exposures from plastics manufacturing and waste management, grassroots efforts may ultimately guide action on issues of EDC exposure, women's health, and recycling.

Nevertheless, individuals cannot make change on their own: government, industry, researchers, and public health professionals have equally important parts to play. Wacholder (2005, 3) states: "We epidemiologists ritualistically declare in the final paragraph of a manuscript that our work will ultimately save lives or at least prevent disease. But ... we seldom actually grapple with prevention." True prevention means a focus on clear, detailed, and binding policy. Policy makers need public health professionals to help guide the writing and implementation of interventions, and health professionals need policy makers to make their research dreams come true. If we are to see true change in the production and consumption of plastics, their proper and effective disposal and recycling, and reduction in the EDCs that they contain and release, experts from all relevant sectors must come together to collaborate, exchange ideas, and integrate their areas of focus. In uniting relevant research and policy making, communities, governments, public health professionals, and industry can develop more effective, sustainable, and safe waste management and recycling processes and policies.

Notes

1 *Toxic Substances Control Act of 1976,* Public Law 94-469, 90 Stat. 2003 (http://www.gpo.gov/fdsys/pkg/STATUTE-90/pdf/STATUTE-90-Pg2003.pdf).

2 http://nalgene.com.
3 http://wr3a.org/.
4 *Canadian Environmental Protection Act, 1999,* S.C. 1999, c. 33 (http://laws-lois.justice.gc.ca/eng/acts/C-15.31/index.html).
5 *Waste Diversion Act, 2002,* S.O. 2002, c. 6 (http://www.e-laws.gov.on.ca/html/statutes/english/elaws_statutes_02w06_e.htm).
6 *Consumer Product Safety Improvement Act of 2008,* Public Law 110-314, 122 Stat. 3017 (http://www.cpsc.gov/cpsia.pdf).
7 *Consumer Product Safety Act,* S.C. 2010, c. 21 (http://laws-lois.justice.gc.ca/eng/acts/C-1.68/index.html).
8 *Environmental Bill of Rights, 1993,* S.O. 1993, c. 28 (https://www.e-laws.gov.on.ca/html/statutes/english/elaws_statutes_93e28_e.htm).

References

Abdelouahab, Nadia, Donna Mergler, Larissa Takser, et al. 2008. "Gender Differences in the Effects of Organochlorines, Mercury, and Lead on Thyroid Hormone Levels in Lakeside Communities of Quebec (Canada)." *Environmental Research* 107: 380-92, doi:10.1016/j.envres.2008.01.006.

Abel, A. 2007. "An Analysis of Solid Waste Generation in a Traditional African City: The Example of Ogbomoso, Nigeria." *Environment and Urbanization* 19 (2): 527-37, doi:10.1177/0956247807082834.

Agarwal, S.P., K.L. Samdani, B.S. Singhvi, et al. 2012. "Plastic Waste Management: An Indian Perspective." *Global Journal of Engineering and Applied Sciences* 2 (1): 56-61.

Arbuckle, T.E. 2006. "Are There Sex and Gender Differences in Acute Exposure to Chemicals in the Same Setting?" *Environmental Research* 101 (2): 195-204. doi:10.1016/j.envres.2005.08.015.

American Academy of Pediatrics. 2006. "Menstruation in Girls and Adolescents: Using the Menstrual Cycle as a Vital Sign." *Pediatrics* 118 (5): 2245-50. http://pediatrics.aappublications.org/content/118/5/2245.full.html.

Aronson, Kristan J., Anthony B. Miller, Christy G. Woolcott, et al. 2000. "Breast Adipose Tissue Concentrations of Polychlorinated Biphenyls and Other Organochlorines and Breast Cancer Risk." *Cancer Epidemiological Biomarkers Preview* 9 (1): 55-63.

Asnani, P.U. 2006. "Solid Waste Management." In *India Infrastructure Report 2006: Urban Infrastructure,* edited by Anupam Ratori, 160-89. Kanpur: Indian Institute of Technology.

Austen, Ian. 2008. "Canada Takes Steps to Ban Most Plastic Baby Bottles." *New York Times,* 19 April. http://www.nytimes.com/2008/04/19/business/worldbusiness/19plastic.html.

Basel Action Network (BAN). 2010. "E-Stewards." http://www.e-stewards.org/about/.

Berkman, Lisa F., and Ichiro Kawachi. 2000. *Social Epidemiology.* New York: Oxford University Press.

Bray, Freddie, Peter McCarron, and D. Maxwell Parkin. 2004. "The Changing Global Patterns of Female Breast Cancer Incidence and Mortality." *Breast Cancer Research* 6 (6): 229-39, doi:10.1186/bcr932.

Brisken, Cathrin. 2008. "Endocrine Disruptors and Breast Cancer." *Chimia* 62 (5): 406-9, doi:10.2533/chimia.2008.406.

Calafat, Antonia M., Zsuzsanna Kuklenyik, John A. Reidy, et al. 2005. "Urinary Concentrations of Bisphenol A and 4-Nonylphenol in a Human Reference Population." *Environmental Health Perspectives* 113 (4): 391-95, doi:10.1289/ehp.7534.

Canadian Institute for Environmental Law and Policy. 2008a. "An Introduction to Ontario's *Waste Diversion Act*." http://www.cielap.org/pdf/WDA_Introduction.pdf.

–. 2008b. "Ontario's *Waste Diversion Act:* Moving beyond Recycling." http://www.cielap.org/pdf/WDA_BeyondRecycling.pdf.

Cantarero, Lourdes, and Isabel Yordi Aguirre. 2010. "Gender Inequities in Environment and Health." In *Environment and Health Risks: A Review of the Influence and Effects of Social Inequities,* edited by the World Health Organization, 217-37. Copenhagen: World Health Organization. http://www.euro.who.int/__data/assets/pdf_file/0003/78069/E93670.pdf.

Carpenter, David O. 2006. "Polychlorinated Biphenyls (PCBs): Routes of Exposure and Effects on Human Health." *Reviews on Environmental Health* 21 (1): 1-23, doi:10.1515/REVEH.2006.21.1.1

Carson, Rachel. 1962. *Silent Spring.* New York: Houghton Mifflin.

Caserta, Donatella, Luca Maranghi, Alberto Mantovani, et al. 2008. "Impact of Endocrine Disruptor Chemicals in Gynaecology." *Human Reproductive Update* 14 (1): 59-72, doi:10.1093/humupd/dmm025.

CCME (Canadian Council of Ministers of the Environment). 2009. *Canada-Wide Action Plan for Extended Producer Responsibility.* http://www.ccme.ca/assets/pdf/epr_cap.pdf.

CCS (Canadian Cancer Society, Steering Committee on Cancer Statistics). 2012. *Canadian Cancer Statistics 2012.* Toronto: Canadian Cancer Society. http://www.canceradvocacy.ca/download/Canadian%20Cancer%20Statistics%202012.pdf.

Cesario, Sandra K., and Lisa A. Hughes. 2007. "Precocious Puberty: A Comprehensive Review of Literature." *Journal of Obstetric, Gynecologic and Neonatal Nursing* 36 (3): 263-74, doi:10.1111/j.1552-6909.2007.00145.x.

Charles, Marie-Jocelyne, Michael J. Schell, Elaine Willman, et al. 2001. "Organochlorines and 8-Hydroxy-2′-Deoxyguanosine (8-OHdG) in Cancerous and Noncancerous Breast Tissue: Do the Data Support the Hypothesis that Oxidative DNA Damage Caused by Organochlorines Affects Breast Cancer?" *Archives of Environmental Contamination and Toxicology* 41 (3): 386-95, doi:10.1007/s002440010264.

Colborn, Theo. 2004. "Neurodevelopment and Endocrine Disruption." *Environmental Health Perspectives* 112 (9): 944-49, doi:10.1289/ehp.6601.

Colborn, Theo, Dianne Dumanoski, and John Peterson Myers. 1996. *Our Stolen Future: Are We Threatening Our Fertility, Intelligence, and Survival? A Scientific Detective Story.* New York: Dutton.

Colon, Ivelisse, Doris Caro, Carlos J. Bourdony, et al. 2000. "Identification of Phthalate Esters in the Serum of Young Puerto Rican Girls with Premature Breast Development." *Environmental Health Perspectives* 108 (9): 895-900, doi:10.1289/ehp.00108895.

Damstra, Terri, Sue Barlow, Aake Bergman, et al., eds. 2002. *Global Assessment of the State-of-the-Science of Endocrine Disruptors.* Geneva: World Health Organization, International Programme on Chemical Safety. http://www.who.int/ipcs/publications/new_issues/endocrine_disruptors/en/.

Daston, George P., Jon C. Cook, and Robert J. Kavlock. 2003. "Uncertainties for Endocrine Disrupters: Our View on Progress." *Toxicological Sciences* 74 (2): 245-52, doi:10.1093/toxsci/kfg105.

Din, Gregory Y., and Emil Cohen. 2013. "Modeling Municipal Solid Waste Management in Africa: Case Study of Matadi, the Democratic Republic of Congo." *Journal of Environmental Protection* 4: 435-45, doi:10.4236/jep.2013.45052.

Dorgan, Joanne F., John W. Brock, Nathaniel Rothman, et al. 1999. "Serum Organochlorine Pesticides and PCBs and Breast Cancer Risk: Results from a Prospective Analysis (USA)." *Cancer Causes and Control* 10 (1): 1-11, doi:10.1023/A:1008824131727.

Doucet, Josee, Brett Tague, Douglas L. Arnold, et al. 2009. "Persistent Organic Pollutant Residues in Human Fetal Liver and Placenta from Greater Montreal, Quebec: A Longitudinal Study from 1998 through 2006." *Environmental Health Perspectives* 117 (4): 605-10, doi:10.1289/ehp.0800205.

Edginton, Andrea N., and Len Ritter. 2009. "Predicting Plasma Concentrations of Bisphenol A in Children Younger than 2 Years of Age after Typical Feeding Schedules, Using a Physiologically Based Toxicokinetic Model." *Environmental Health Perspectives* 117 (4): 645-52, doi:10.1289/ehp.0800073.

Eenkema van Dijk, Eric H. 2001. "Single Stream Recycling: The Future is Now." *Recycling Today,* 3 July. http://www.recyclingtoday.com/Article.aspx?article_id=16870.

Environmental Commissioner of Ontario. 2012. *Ontario's Environmental Bill of Rights and You: A Guide to Exercising Your Right to Participate in Environmental Decision Making in Ontario.* http://www.eco.on.ca/uploads/EBR%20Documents/The%20EBR%20and%20You%202013.pdf.

Euro Publishing Consultancy. 2007. "The Global View of Plastics Recycling Is Far from Uniform." *Euro Publishing Consultancy,* 1 November. http://www.highbeam.com/doc/1G1-172686222.html.

European Parliament and Council. 2000. *Directive 2000/53/EC of the European Parliament and of the Council of 18 September 2000 on End-of-Life Vehicles.* http://eur-lex.europa.eu/LexUriServ/LexUriServ.do?uri=CELEX:32000L0053:EN:NOT.

Fernandez, Mariana F., Ana Rivas, Fatima Olea-Serrano, et al. 2004. "Assessment of Total Effective Xenoestrogen Burden in Adipose Tissue and Identification of Chemicals Responsible for the Combined Estrogenic Effect." *Analytical and Bioanalytical Chemistry* 379 (1): 163-70, doi:10.1007/s00216-004-2558-5.

Fernandez, Mariana F., Juan P. Arrebola, Jalia Taoufiki, et al. 2007. "Bisphenol-A and Chlorinated Derivatives in Adipose Tissue of Women." *Reproductive Toxicology* 24 (2): 259-64, doi:10.1016/j.reprotox.2007.06.007.

Fishbein, Bette K. 2002. *Waste in the Wireless World: The Challenge of Cell Phones.* New York: INFORM.

Foster, Warren G., Michael S. Neal, Myoung-Soek Han, et al. 2008. "Environmental Contaminants and Human Infertility: Hypothesis or Cause for Concern?" *Journal of Toxicology and Environmental Health, Part B: Critical Reviews* 11 (3-4): 162-76, doi:10.1080/10937400701873274.

Government of Canada. 2010a. "Order Adding a Toxic Substance to Schedule 1 to the Canadian Environmental Protection Act, 1999." *Canada Gazette* 144 (21). http://www.gazette.gc.ca/rp-pr/p2/2010/2010-10-13/html/sor-dors194-eng.html.

–. 2010b. "Order Amending Schedule I to the Hazardous Products Act (Bisphenol A)." *Canada Gazette* 144 (7). http://www.gazette.gc.ca/rp-pr/p2/2010/2010-03-31/html/sor-dors53-eng.html.

Greenpeace. 2007. *Individual Producer Responsibility: Helping to Solve the E-waste Problem and to Encourage Eco-Design.* Amsterdam: Greenpeace International. http://www.greenpeace.org/international/Global/international/planet-2/report/2008/2/individual-producer-responsibility.pdf.

Hanisch, Christiane. 2000. "Is Extended Producer Responsibility Effective?" *Environmental Science and Technology* 34 (7): 170A-75A, doi:10.1021/es003229n.

Health Canada. 2010. "Bisphenol A." http://www.hc-sc.gc.ca/fn-an/securit/packag-emball/bpa/index-eng.php.

Health Care without Harm. 2002. *Resource Kit: Implementing the FDA's Public Health Notification on PVC Devices Containing DEHP.* http://www.noharm.org/.

HealthCam. 2010. "Our Goals." http://www.healthcam.org/objectives.html.

Hirabayashi, Yoko, and Tohru Inoue. 2011. "The Low-Dose Issue and Stochastic Responses to Endocrine Disruptors." *Journal of Applied Toxicology* 31 (1): 84-88, doi: 10.1002/jat.1571.

Hosoda, Eiji. 2007. "International Aspects of Recycling of Electrical and Electronic Equipment: Material Circulation in the East Asian Region." *Journal of Material Cycles and Waste Management* 9 (2): 140-50, doi:10.1007/s10163-007-0179-8.

Hoyer, Annette P., Torben Jorgensen, Philippe Grandjean, et al. 2000. "Repeated Measurements of Organochlorine Exposure and Breast Cancer Risk (Denmark)." *Cancer Causes and Control* 11 (2): 177-84, doi:10.1023/A:1008926219539.

Hoyer, Annette P., Anne-Marie Gerdes, Torben Jorgensen, et al. 2002. "Organochlorines, p53 Mutations in Relation to Breast Cancer Risk and Survival: A Danish Cohort-Nested Case-Controls Study." *Breast Cancer Research and Treatment* 71 (1): 59-65, doi:10.1023/A:1013340327099.

Huang, Po-Chin, Pao-Lin Kuo, Yen-Yin Chou, et al. 2009. "Association between Prenatal Exposure to Phthalates and the Health of Newborns." *Environment International* 35 (1): 14-20, doi:10.1016/j.envint.2008.05.012.

Ikezuki, Yumiko, Osamu Tsutsumi, Yasushi Takai, et al. 2002. "Determination of Bisphenol A Concentrations in Human Biological Fluids Reveals Significant Early Prenatal Exposure." *Human Reproduction* 17 (11): 2839-41, doi:10.1093/humrep/17.11.2839.

Jonker, Derek, Andreas P. Freidig, John P. Groten, et al. 2004. "Safety Evaluation of Chemical Mixtures and Combinations of Chemical and Non-Chemical Stressors." *Reviews on Environmental Health* 19 (2): 83-139, doi:10.1515/REVEH.2004.19.2.83.

Kaiser, Florian G., Sybille Wolfing, and Urs Fuhrer. 1999. "Environmental Attitude and Ecological Behaviour." *Journal of Environmental Psychology* 19 (1): 1-19, doi:10.1006/jevp.1998.0107.

Kamrin, Michael A. 2004. "Bisphenol A: A Scientific Evaluation." *Medscape General Medicine* 6 (3): 7. http://www.ncbi.nlm.nih.gov/pmc/articles/PMC1435609/.

Kato, Natsumi, Makoto Shibutani, Hironori Takagi, et al. 2004. "Gene Expression Profile in the Livers of Rats Orally Administered Ethinylestradiol for 28 Days Using a Microarray Technique." *Toxicology* 200 (2-3): 179-92, doi:10.1016/j.tox.2004.03.008.

Kibert, Nicole C. 2004. "Extending Producer Responsibility: A Tool for Achieving Sustainable Development." *Journal of Land Use and Environmental Law* 19 (2): 503-23. http://www.law.fsu.edu/journals/landuse/vol19_2/kibert.pdf.

Killam, Galen. 2010. "Plastic Biodegradation in Landfills." *Green Plastics,* 24 September. http://green-plastics.net/news/45-science/93-plastic-biodegradation-in-landfills.

Krimsky, Sheldon. 2000. *Hormonal Chaos: The Scientific and Social Origins of the Environmental Endocrine Hypothesis*. Baltimore: Johns Hopkins University Press.

Kusumegi, Takahiro, Junko Tanaka, Michihiro Kawano, et al. 2004. "BMP7/ActRIIB Regulates Estrogen-Dependent Apoptosis: New Biomarkers for Environmental Estrogens." *Journal of Biochemical and Molecular Toxicology* 18 (1): 1-11, doi:10.1002/jbt.20004.

Labonte, Ronald, and Ted Schrecker. 2007. "Globalization and Social Determinants of Health: Introduction and Methodological Background (Part 1 of 3)." *Global Health* 3 (5): 1-10, doi:10.1186/1744-8603-3-5.

Langrova, Veronika. 2002. "Comparative Analysis of EPR Programmes for Small Consumer Batteries: Case Study of the Netherlands, Switzerland and Sweden." Master's thesis, International Institute for Industrial Environmental Economics (IIIEE), Lund University.

Latini, Giuseppe, Claudio De Felice, Giuseppe Presta, et al. 2003. "In Utero Exposure to Di-(2-Ethylhexyl) Phthalate and Duration of Human Pregnancy." *Environmental Health Perspectives* 111 (14): 1783-85, doi:10.1289/ehp.6202.

Laurent, Olivier, Denis Bard, Laurent Filleul, et al. 2007. "Effect of Socioeconomic Status on the Relationship between Atmospheric Pollution and Mortality." *Journal of Epidemiology and Community Health* 61 (8): 665-75, doi:10.1136/jech.2006.053611.

Layton, Lyndsey, and Christopher Lee. 2008. *Canada Bans BPA from Baby Bottles.* http://www.washingtonpost.com/wp-dyn/content/article/2008/04/18/AR2008041803036.html.

Leiss, William. 2001. *In the Chamber of Risks: Understanding Risk Controversies.* Montreal and Kingston: McGill-Queen's University Press.

Lifset, Reid, and Thomas Lindhqvist. 2008. "Producer Responsibility at a Turning Point?" *Journal of Industrial Ecology* 12 (2): 144-47, doi:10.1111/j.1530-9290.2008.00028.x.

Liljegren, Groan, Lennart Hardell, Gunilla Lindstrom, et al. 1998. "Case-Control Study on Breast Cancer and Adipose Tissue Concentrations of Congener Specific Polychlorinated Biphenyls, DDE and Hexachlorobenzene." *European Journal of Cancer Prevention* 7 (2): 135-40.

Lindhqvist, Thomas. 2000. "Extended Producer Responsibility in Cleaner Production: Policy Principle to Promote Environmental Improvements of Product Systems." Doctoral dissertation, International Institute for Industrial Environmental Economics (IIIEE), Lund University.

Madigan, M. Patricia, Regina G. Ziegler, Jacques Benichou, et al. 1995. "Proportion of Breast Cancer Cases in the United States Explained by Well-Established Risk Factors." *Journal of the National Cancer Institute* 87 (22): 1681-85, doi:10.1093/jnci/87.22.1681.

Maffini, Maricel V., Beverly S. Rubin, Carlos Sonnenschein, et al. 2006. "Endocrine Disruptors and Reproductive Health: The Case of Bisphenol-A." *Molecular and Cellular Endocrinology* (254-55): 179-86, doi:10.1016/j.mce.2006.04.033.

Mainieri, Tina, Elaine G. Barnett, Trisha R. Valdero, et al. 1997. "Green Buying: The Influence of Environmental Concern on Consumer Behavior." *Journal of Social Psychology* 137 (2): 189-204, doi:10.1080/00224549709595430.

Mantovani, Alberto. 2006. "Risk Assessment of Endocrine Disrupters: The Role of Toxicological Studies." *Annals of the New York Academy of Science* 1076: 239-52, doi:10.1196/annals.1371.063.

Marr, John. 2004. "EPR Background Paper: Learning from Practice." Paper presented at Canada's Third National Extended Producer Responsibility Workshop, Halifax, Nova Scotia, 3-5 March. http://www.gov.ns.ca/nse/waste/epr/docs/BackgroundPaper_English.pdf.

Martin, Olwenn V., John N. Lester, Nikolaos Voulvoulis, et al. 2007. "Human Health and Endocrine Disruption: A Simple Multicriteria Framework for the Qualitative Assessment of End Point Specific Risks in a Context of Scientific Uncertainty." *Toxicological Sciences* 98 (2): 332-47, doi:10.1093/toxsci/kfm008.

McKerlie, Kate, Nancy Knight, and Beverley Thorpe. 2006. "Advancing Extended Producer Responsibility in Canada." *Journal of Cleaner Production* 14 (6-7): 616-28, doi:10.1016/j.jclepro.2005.08.001.

McKinlay, Rebecca, Jane A. Plant, J. Nigel Bell, et al. 2008. "Endocrine Disrupting Pesticides: Implications for Risk Assessment." *Environment International* 34 (2): 168-83, doi:10.1016/j.envint.2007.07.013.

McLachlan, John A., Erica Simpson, and Melvenia Martin. 2005. "Endocrine Disruptors and Female Reproductive Health." *Best Practice and Research, Clinical Endocrinology and Metabolism* 20 (1): 63-75, doi:10.1016/j.beem.2005.09.009.

Milojković, Jelena, and Vančo Litovski. 2005. "Concepts of Computer Take-Back for Sustainable End-of-Life."*Working and Living Environmental Protection* 2 (5): 363-72. http://facta.junis.ni.ac.rs/walep/walep2005n/walep2005-03.pdf.

Muscat, Joshua E., Julie A. Britton, Mirjana V. Djordjevic, et al. 2003. "Adipose Concentrations of Organochlorine Compounds and Breast Cancer Recurrence in Long Island, New York." *Cancer Epidemiology, Biomarkers and Prevention* 12 (12): 1474-78. http://cebp.aacrjournals.org/content/12/12/1474.

National Research Council. 1999. *Exposures: Sources and Dynamics of Horomonally Active Agents in the Environment.* Washington, DC: National Academies Press.

National Toxicology Program. 2008. *NTP-CERHR Monograph on the Potential Human Reproductive and Developmental Effects of Bisphenol A.* NIH Publication 08-5994. Washington, DC: Center for the Evaluation of Risks to Human Reproduction, US Department of Health and Human Services. http://cerhr.niehs.nih.gov/evals/bisphenol/bisphenol.pdf.

Newbold, Retha R., Elizabeth Padilla-Banks, Ryan J. Snyder, et al. 2007. "Developmental Exposure to Endocrine Disruptors and the Obesity Epidemic." *Reproductive Toxicology* 23 (3): 290-96, doi:10.1016/j.reprotox.2006.12.010.

Nkwachukwu, Onwughara Inoocent, Chukwu Henry Chima, Alaekwe Obiora Ikenna, et al. 2013. "Focus on Potential Environmental Issues on Plastic: Towards Sustainable Plastic Recycling in Developing Countries." *International Journal of Industrial Chemistry* 4 (34): 1-13. http://www.industchem.com/content/4/1/34.

Nnorom, Innocent C., and Oladele Osibanjo. 2008. "Overview of Electronic Waste (E-waste) Management Practices and Legislations, and Their Poor Applications in the Developing Countries." *Resources Conservation and Recycling* 52 (6): 843-58, doi: 10.1016/j.resconrec.2008.01.004.

NOVPOL. 2007. "A New Concept for the Recycling of Incompatible Polymers Allowing the Creation of New Polymeric Materials with Enhanced Properties." http://www.ist-world.org/ProjectDetails.aspx?ProjectId=9fd9fb2015ab4913a848abfa227c6026&SourceDatabaseId=7cff9226e582440894200b751bab883f.

O'Leary, Erin S., John E. Vena, Jo L. Freudenheim, et al. 2004. "Pesticide Exposure and Risk of Breast Cancer: A Nested Case-Control Study of Residentially Stable Women Living on Long Island." *Environmental Research* 94 (2): 134-44, doi:10.1016/j.envres.2003.08.001.

O'Neill, Marie S., Anthony McMichael, Joel Schwartz, et al. 2007. "Poverty, Environment, and Health: The Role of Environmental Epidemiology and Environmental Epidemiologists." *Epidemiology* 18 (6): 664-68, doi:10.1097/EDE.0b013e3181570ab9.

OECD (Organisation for Economic Co-operation and Development). 2001. *Fact Sheet: Extended Producer Responsibility: A Guidance Manual for Governments.* Paris: OECD Publications.

Oh, SoonHee, and Shirley Thompson. 2006. "Do Sustainable Computers Result from Design for Environment and Extended Producer Responsibility? Analyzing E-Waste Programs in Europe and Canada." Proceedings of the International Solid Waste Association Annual Congress, Copenhagen, 1-5 October. http://citeseerx.ist.psu.edu/viewdoc/download?doi=10.1.1.88.5038&rep=rep1&type=pdf.

Omrani, Ghasem, M. Zamanzedeh, and M. Soltandalal. 2005. "Plastic Recycling Problems and Its Health Aspects in Tehran." *Journal of Applied Sciences* 5 (2): 351-56. http://scialert.net/qredirect.php?doi=jas.2005.351.356&linkid=pdf.

Oyoo, R., R. Leemans, and A.P.J. Mol. 2011. "Future Projections of Urban Waste Flows and Their Impacts in African Metropolises Cities." *International Journal of Environment Research* 5 (3): 705-24. http://www.ijcr.ir/.

Payne-Sturges, Devon, Gilbert C. Gee, Kirstin Crowder, et al. 2006. "Workshop Summary: Connecting Social and Environmental Factors to Measure and Track Environmental Health Disparities." *Environmental Research* 102 (2): 146-53, doi:10.1016/j.envres.2005.11.001.

Phillips, Karen P., Warren G. Foster, William Leiss, et al. 2008. "Assessing and Managing Risks Arising from Exposure to Endocrine-Active Chemicals." *Journal of Toxicology and Environmental Health, Part B: Critical Reviews* 11 (3-4): 351-72, doi:10.1080/10937400701876657.

PollutionWatch. 2008. *An Examination of Pollution and Poverty in the Great Lakes Basin.* Toronto: Canadian Environmental Law Association and Environmental Defence. http://www.pollutionwatch.org/pub/PW_Pollution_Poverty_Report.pdf.

Rajapakse, Nissanka, Elisabete Silva, Martin Scholze, et al. 2004. "Deviation from Additivity with Estrogenic Mixtures Containing 4-Nonylphenol and 4-Tert-Octylphenol Detected in the E-SCREEN Assay." *Environmental Science and Technology* 38 (23): 6343-52, doi:10.1021/es049681e.

Regassa, N., R.D. Sundaraa, and B.B. Seboka. 2011. "Challenges and Opportunities in Municipal Solid Waste Management: The Case of Addis Ababa City, Central Ethiopia." *Journal of Human Ecology* 33 (3): 179-90.

Ries, Lynn A.G., Carol L. Kosary, Benjamin F. Hankey, et al., eds. 2002. *SEER Cancer Statistics Review, 1973-1999.* Bethesda, MD: National Cancer Institute.

Rogan, Walter J., and Aimin Chen. 2005. "Health Risks and Benefits of Bis(4-Chlorophenyl)-1,1,1-Trichloroethane (DDT)." *Lancet* 366 (9487): 763-73, doi:10.1016/S0140-6736(05)67182-6.

RoHS (Restriction of Hazardous Substances). 2014. "RoHS Compliance FAQ." http://www.rohsguide.com/rohs-faq.htm.

Rusiecki, Jennifer A., Theodore R. Holford, Shelia H. Zahm, et al. 2004. "Polychlorinated Biphenyls and Breast Cancer Risk by Combined Estrogen and Progesterone Receptor Status." *European Journal of Epidemiology* 19 (8): 793-801, doi:10.1023/B:EJEP.0000036580.05471.31.

Rust, Susan, and Meg Kissinger. 2008. "Lawmakers to Seek Ban on BPA: Analysis of Chemical Leaching from Food Containers Gets Action in Washington, Madison." *Milwaukee Journal Sentinel,* 17 November. http://www.jsonline.com/watchdog/watchdogreports/34623239.html.

Salehi, Fariba, Lesley Dunfield, Karen P. Phillips, et al. 2008a. "Risk Factors for Ovarian Cancer: An Overview with Emphasis on Hormonal Factors." *Journal of Toxicology and Environmental Health, Part B: Critical Reviews* 11 (3-4): 301-21, doi:10.1080/10937400701876095.

Salehi, Fariba, Michelle C. Turner, Karen P. Phillips, et al. 2008b. "Review of the Etiology of Breast Cancer with Special Attention to Organochlorines as Potential Endocrine Disruptors." *Journal of Toxicology and Environmental Health, Part B: Critical Reviews* 11 (3-4): 276-300, doi:10.1080/10937400701875923.

Sasco, Annie J. 2003. "Breast Cancer and the Environment." *Hormone Research* 60 (Supp. 3): 50, doi:10.1159/000074500.

Sasco, Annie J., Rudolf Kaaks, and Ruth E. Little. 2003. "Breast Cancer: Occurrence, Risk Factors and Hormone Metabolism." *Expert Review of Anticancer Therapy* 3 (4): 546-62, doi:10.1586/14737140.3.4.546.

Schmidt, Charles W. 2006. "Unfair Trade: E-Waste in Africa." *Environmental Health Perspectives* 114 (4): A232-A235. PMCID: PMC1440802. http://www.ncbi.nlm.nih.gov/pmc/articles/PMC1440802/.

Schönfelder, Gilbert, Burkhard Flick, E. Mayr, et al. 2002. "In Utero Exposure to Low Doses of Bisphenol A Lead to Long-Term Deleterious Effects in the Vagina." *Neoplasia* 4 (2): 98-102, doi:10.1038/sj.neo.7900212.

Sekizawa, Jun. 2008. "Low-Dose Effects of Bisphenol A: A Serious Threat to Human Health?" *Journal of Toxicological Sciences* 33 (4): 389-403, doi:10.2131/jts.33.389.

Shields, Peter G. 2006. "Understanding Population and Individual Risk Assessment: The Case of Polychlorinated Biphenyls." *Cancer Epidemiology, Biomarkers and Prevention* 15 (5): 830-39, doi:10.1158/1055-9965.EPI-06-0222.

Solomon, Gina M., and Ted Schettler. 2000. "Environment and Health: 6. Endocrine Disruption and Potential Human Health Implications." *Canadian Medical Association Journal* 163 (11): 1471-76. http://www.cmaj.ca/content/163/11/1471.full.pdf.

Soskolne, Colin L., Colin D. Butler, Carel Ijsselmuiden, et al. 2007. "Toward a Global Agenda for Research in Environmental Epidemiology." *Epidemiology* 18 (1): 162-66.

Statistics Canada. 2010. *Recycling by Canadian Households, 2007.* Environment Accounts and Statistics Division. Ottawa: Minister of Industry. http://www.statcan.gc.ca/pub/16-001-m/2010013/part-partie1-eng.htm.

Sugiura-Ogasawara, Mayumi, Yasuhiko Ozaki, Shin-ichi Sonta, et al. 2005. "Exposure to Bisphenol A Is Associated with Recurrent Miscarriage." *Human Reproduction* 20 (8): 2325-29, doi:10.1093/humrep/deh888.

Sugiyama, Shin-ichiro, Naoyuki Shimada, Hiroyuki Miyoshi, et al. 2005. "Detection of Thyroid System-Disrupting Chemicals Using In Vitro and In Vivo Screening Assays in *Xenopus laevis*." *Toxicological Sciences* 88 (2): 367-74, doi:10.1093/toxsci/kfi330.

TEDX (The Endocrine Disruption Exchange). 2013. "Endocrine Disruption: Introduction – Overview." http://endocrinedisruption.org/endocrine-disruption/introduction/overview.

Themelis, Nickolas J., and Claire E. Todd. 2004. "Recycling in a Megacity." *Journal of the Air and Waste Management Association* 54 (4): 389-95. http://www.tandfonline.com/doi/pdf/10.1080/10473289.2004.10470921.

US EPA (United States Environmental Protection Agency). 1994. *Volume 1: Estimating Exposures to Dioxin-Like Compounds.* Washington, DC: Office of Health and Environmental Assessment, US EPA. http://cfpub.epa.gov/si/si_public_file_download.cfm?p_download_id=438673.

–. 1997. *EPA Special Report on Endocrine Disruption: Fact Sheet.* Washington, DC: Office of Research and Development, US EPA. http://www.epa.gov/raf/publications/pdfs/endocrine-disruptions-factsheet.PDF.

–. 1998. *Endocrine Disruptor Screening and Testing Advisory Committee (EDSTAC) Final Report.* Washington, DC: Office of Science Coordination and Policy, US EPA. http://www.epa.gov/endo/pubs/edspoverview/finalrpt.htm.

–. 2008a. *Municipal Solid Waste Generation, Recycling, and Disposal in the United States: Facts and Figures for 2008.* Washington, DC: US EPA. http://www.epa.gov/osw/nonhaz/municipal/pubs/msw2008rpt.pdf.

–. 2008b. *Toxicological Profile for 2,2′,4,4′,5-Pentabromodiphenyl Ether (BDE-99).* In support of Summary Information on the Integrated Risk Information System (IRIS). Washington, DC: US EPA. http://www.epa.gov/IRIS/toxreviews/1008tr.pdf.

–. 2008c. *Toxicological Review of Decabromodiphenyl Ether (BDE-209).* In support of Summary Information on the Integrated Risk Information System (IRIS). Washington, DC: US EPA. http://www.epa.gov/iris/toxreviews/0035tr.pdf.

–. 2012a. "Recycling and Reuse: End-of-Life Vehicles and Producer Responsibility." http://www.epa.gov/oswer/international/factsheets/200811_elv_directive.htm.

–. 2012b. "Wastes." http://www.epa.gov/wastes/.

van Larebeke, Nicolas A., Annie J. Sasco, James T. Brophy, et al. 2008. "Sex Ratio Changes as Sentinel Health Events of Endocrine Disruption." *International Journal of Occupational and Environmental Health* 14 (2): 138-43. https://dspace.stir.ac.uk/bitstream/1893/597/1/IJOEH_April08_9Larebeke%20NVL.pdf.

Vandenberg, Laura N., Theo Colborn, Tyrone B. Hayes, et al. 2012. "Hormones and Endocrine-Disrupting Chemicals: Low-Dose Effects and Nonmonotonic Dose Responses." *Endocrine Reviews* 33 (3): 378-455, doi:10.1210/er.2011-1050.

Vom Saal, Frederick S., Bensen T. Akingbemi, Scott M. Belcher, et al. 2007. "Chapel Hill Bisphenol A Expert Panel Consensus Statement: Integration of Mechanisms, Effects in Animals and Potential to Impact Human Health at Current Levels of Exposure." *Reproductive Toxicology* 24 (2): 131-38, doi:10.1016/j.reprotox.2007.07.005.

Wacholder, Sholom. 2005. "The Impact of a Prevention Effort on the Community." *Epidemiology* 16 (1): 1-3, doi:10.1097/01.ede.0000147633.09891.16.

Walls, Margaret. 2006. *Extended Producer Responsibility and Product Design: Economic Theory and Selected Case Studies.* Resources for the Future Discussion Paper 06-08. Paris: OECD. http://www.rff.org/RFF/Documents/RFF-DP-06-08-REV.pdf.

Wang, Richard Y., Larry L. Needham, and Dana B. Barr. 2005. "Effects of Environmental Agents on the Attainment of Puberty: Considerations When Assessing Exposure to Environmental Chemicals in the National Children's Study." *Environmental Health Perspectives* 113 (8): 1100-07, doi:10.1289/ehp.7615.

Weck, Rebekah L., Tessie Paulose, and Jodi A. Flaws. 2008. "Impact of Environmental Factors and Poverty on Pregnancy Outcomes." *Clinical Obstetrics and Gynecology* 51 (2): 349-59, doi:10.1097/GRF.0b013e31816f276e.

Welsh, Moira. 2008. "Poorest Areas Also Most Polluted, Report Shows." *The Star,* 27 November. http://www.thestar.com/news/gta/article/544370-poorest-areas-also-most-polluted-report-shows.

Weltje, Lennart, Frederick S. vom Saal, and Jorg Oehlmann. 2005. "Reproductive Stimulation by Low Doses of Xenoestrogens Contrasts with the View of Hormesis as an Adaptive Response." *Human and Experimental Toxicology* 24 (9): 431-37, doi:10.1191/0960327105ht551oa.

WHO (World Health Organization)/UNEP (United Nations Environment Programme). 2013. *State of the Science of Endocrine Disrupting Chemicals – 2012.* Geneva: WHO/UNEP. http://www.who.int/ceh/publications/endocrine/en/.

Widmer, Rolf, Heidi Oswald-Krapf, Deepali Sinha-Khetriwal, et al. 2005. "Global Perspectives on E-Waste." *Environmental Impact Assessment Review* 25 (5): 436-58, doi:10.1016/j.eiar.2005.04.001.

Wilk, Richard. 2002. "Consumption, Human Needs, and Global Environmental Change." *Global Environmental Change* 12 (1): 5-13, doi:10.1016/S0959-3780(01)00028-0.

Wise, Lauren A., Julie R. Palmer, Kathleen Rowlings, et al. 2005. "Risk of Benign Gynecologic Tumors in Relation to Prenatal Diethylstilbestrol Exposure." *Obstetrics and Gynecology* 105 (1): 167-73, doi:10.1097/01.AOG.0000147839.74848.7c.

Yang, Mihi, Mi Seon Park, and Ho Sun Lee. 2006. "Endocrine Disrupting Chemicals: Human Exposure and Health Risks." *Journal of Environmental Science and Health, Part C: Environmental Carcinogenesis and Ecotoxicology Reviews* 24 (2): 183-224, doi:10.1080/10590500600936474.

Zacharewski, Tim. 1998. "Identification and Assessment of Endocrine Disruptors: Limitations of In Vivo and In Vitro Assays." *Environmental Health Perspectives* 106 (Supp. 2): 577-82.

Zero Waste America. n.d. "Waste and Recycling: Data, Maps, and Graphs." http://zerowasteamerica.org/Statistics.htm.

Zia, H., V. Devadas, and S. Shukla. 2008. "Assessing Informal Waste Recycling in Kanpur City, India." *Management of Environmental Quality: An International Journal* 19 (5): 597-612, doi:10.1108/14777830810894265.

Zuckerman, Diana. 2001. "When Little Girls Become Women: Early Onset of Puberty in Girls." *The Ribbon* 6 (1). National Center for Policy Research for Women and Families, http://www.timeenoughforlove.org/saved/GirlsBecomeWomen.htm.

Chapter 9 Xenoestrogens and Breast Cancer: Chemical Risk, Exposure, and Corporate Power

Sarah Young and Dugald Seely

If past trends are any indication, over 22,000 new cases of breast cancer will be diagnosed this year among Canadian women. Of these, over 5,000 will result in early death (CCS 2012a). The incidence of breast cancer has risen in Canada by over 25 percent in the past twenty-five years, and breast cancer is the leading cause of death in women aged twenty to forty-four (CCS/National Cancer Institute of Canada 2006; see Chapters 8 and 10). Breast cancer comprises over 30 percent of all cancer diagnoses in women (Salehi et al. 2008). Five to 10 percent of these cases result from genetic predisposition, and another 20 percent can be attributed to the following established risk factors: early age at menarche, late age at menopause, having a family member with breast cancer, previous history of breast cancer, not having children, late age of pregnancy, the use of hormone therapy, the use of oral contraceptives, exposure to ionizing radiation, benign breast disease, and mammographically dense breast tissue. However, most newly diagnosed breast cancer cases *cannot* be attributed to these risk factors, meaning that the majority of cases, at least 70 percent, are of unknown origin (Salehi et al. 2008).

In fact, experts now believe that nearly two-thirds of breast cancer cases come from a combination of gene/environment interactions, lifestyle factors (including lack of physical exercise, alcohol intake, and obesity), and exposure to environmental pollutants. What is known for certain is that exposure to estrogen is the most important risk factor for the development of breast cancer (Andreeva, Unger, and Pentz

2007; Maffini et al. 2006). As such, the potential for xenoestrogens – external/exogenous or synthetic chemicals that have an estrogenic effect on the body, such as bisphenol A (BPA), cadmium, and polychlorinated biphenyls (PCBs) – to contribute to breast cancer risk may be vastly underestimated. Now more than ever, there is a clear need to conduct research to identify modifiable risk factors for breast cancer, including occupational and environmental exposures.

The rest of this chapter consists of four main sections. The first describes xenoestrogens and how they may contribute to breast cancer risk. The second explores xenoestrogens within a social determinants of health model, including an explanation of specific xenoestrogens and the studies that may link them to breast cancer risk. The third section looks at the political economy of chemicals and cancer in Canada, and points to the need to refocus research, policy makers, and the public on prevention. The last section provides policy recommendations and outlines the need for research within the Canadian context.

Xenoestrogens and Breast Cancer

In both humans and animals, the mammary gland undergoes continuous morphological and functional changes throughout life, with estrogens playing a key role. A growing body of scientific research has linked estrogen-mimicking chemicals in the environment to increasingly high rates of breast cancer (see also Chapters 6, 8, and 10). Specifically, studies have found that environmental exposures – in combination with genetic predisposition, age at exposure, lifestyle factors, and a women's hormonal milieu – have a cumulative effect on women's total risk for developing breast cancer (Gray, Nudelman, and Engel 2010). Due to damage from environmental carcinogens at critical points during development, genes can undergo mutations, leading to greater cancer susceptibility in the future. Subsequent mutations and a cellular environment promoting cell proliferation increase the chances that breast cancer will develop. In her most recent book, which links environmental toxins and cancer incidence, Devra Davis emphasizes evidence that women who experience higher levels of hormones in the womb or at birth are nearly four times as likely to develop breast cancer as women whose prenatal hormones are at average levels (Davis 2002; Chapters 6, 8, and 10).

Environmental pollutants can be carcinogenic through one of two mechanisms. In the first and most common case, a toxin can be cancer-inducing if it has a mutagenic effect – for example, if it causes molecular changes in the genetic material. Mutagenicity can occur through direct damage to a cell's DNA and activation of cancer-causing genes, or through inactivation of genes that suppress tumours. The other, more subtle way in which a toxin can be carcinogenic is by inducing a hormone-like proliferative effect in susceptible tissues. In this way, toxins specifically known as endocrine-disrupting chemicals (EDCs) promote the development of cancer by causing an increase in cell numbers.[1] Organs thought to be susceptible to cancer induction from EDCs include the prostate, the uterus, the ovaries, and the breasts.

As discussed in other chapters (Introduction and Chapters 3, 8, 10, and 11), EDCs are exogenous compounds (chemicals originating outside the body) that can disrupt normal hormonal function, producing adverse health effects for both adults and their children. Mammary tissue cells express specific receptors, known as alpha-receptors, that can be stimulated by estrogens to induce the proliferation of end buds in the breasts (Kortenkamp et al. 2007). Proliferation of these end buds increases the risk of breast cancer by augmenting the chance of mutation during cell division (Maffini et al. 2006). Xenoestrogens comprise a subclass of EDCs that mimic estrogens and are hypothesized to increase the risk of breast cancer development.

It is now widely accepted that sustained, high levels of estrogens promote hormonally responsive cancers (Chapters 6, 8, and 10). Case-control studies confirm that serum (blood) estrogen levels are higher in breast cancer cases compared with controls (women who do not have breast cancer) (Moore, Chamberlain, and Khuri 2004). There are different types of estrogens with varying degrees of activity. Of the naturally occurring (endogenous) estrogens, estradiol is the most potent; it is ten times more biologically active than estrone, and even more so than estriol. In comparison, diethylstilbestrol (DES) is a xenoestrogen that can be even more biologically active than estradiol (Donovan et al. 2007; Giusti, Iwamoto, and Hatch 1995; Hatch et al. 1998). Considering the inherent risk posed by naturally occurring estrogens as promoters of hormonally responsive cancers like breast cancer, we need to look very closely at the risk posed by xenoestrogens, which are as or even more potent than naturally occurring estrogens.

Routes of Exposure to Xenoestrogens

Hormonally active synthetic chemicals have entered the environment in massive quantities over the past sixty years. In studies measuring concentrations of these chemicals in human tissue, over 95 percent of tissues tested positive for the chemicals, reflecting the ubiquity of xeno-estrogens in and around us (Environmental Defence 2006; Ibarluzea et al. 2004; Maffini et al. 2006). As discussed in other chapters (3, 6, 7, 8, and 10), women working in agriculture or living near industries can be exposed to higher levels of xenoestrogens, as can those using certain kinds of personal care products and plastics (Donovan et al. 2007; Fernandez et al. 2007b), demonstrating that occupation, geographical location, and personal consumption choices play a very important role. The air we breathe, the food we eat, and the water we drink contain a number of hazardous compounds that enter the environment in a variety of ways: they are sprayed directly onto our food in the form of pesticides; they get pumped into the air from diesel exhaust fumes and industrial emissions in the form of polycyclic aromatic hydrocarbons (PAHs); they are used as liners in tin cans, built into plastic water bottles, and found in dental composites in the form of BPA; they are added to personal care products in the form of parabens; they leach into the soil and water from electronic waste in landfills in the form of cadmium; and they are found in the fish we eat in the form of PCBs (Brody et al. 2007b; Clapp, Howe, and Jacobs 2007; Coyle 2004; Davis 2002; Gray, Nudelman, and Engel 2010; Kortenkamp et al. 2007; Rudel et al. 2007).

In the past century, consumer products and industrial production processes have become saturated with synthetic chemicals. Of the more than 80,000 chemicals currently in use in North America, a mere 7 percent have been tested for their potential impact on human health (Gray, Nudelman, and Engel 2010). Of those chemicals tested to date, many have demonstrated both carcinogenic activity and the ability to be hormonally disruptive, adding to a growing body of evidence that some synthetic chemicals play a key role in breast cancer risk (Brody et al. 2007a; Payne, Scholze, and Kortenkamp 2001).

In 2007, Rudel and colleagues (2007) compiled a list of 216 chemicals shown to act as mammary carcinogens in animals, and described human routes of exposure to these chemicals. The list includes 73 chemicals present in consumer products or as contaminants in food, 35 air pollutants, 25 compounds associated with specific occupational exposures, and 29 that are produced at over a million pounds per year in the United States. Numerous xenoestrogenic chemical varieties were named, including 10 pesticides, 6 chlorinated solvents, and 36 industrial chemicals (Rudel et al. 2007). Rudel and colleagues argue that there are relevant links between the carcinogenicity in animals and in humans, pointing out that the International Agency for Research on Cancer (IARC) will deem a chemical carcinogenic to humans if a single animal study shows multiple neoplasms at multiple sites. Nearly all of the 216 chemicals in Rudel and colleagues' list (2007) were mutagenic and caused tumours in multiple organs and species.

The Cocktail Effect and the Importance of Xenoestrogen Mixtures

Although it is difficult to prove a causal link between many single chemicals and cancer, it is likely that the additive and/or synergistic effects of these chemicals increases cancer risk, a phenomenon referred to as "the cocktail effect" (Kortenkamp 2006; see also Chapters 1, 3, 4, 10, and 11). Our bodies are constantly exposed to a cocktail of chemicals through air, food, and water, and, in the case of fat-soluble toxins, accumulation of these compounds often exceeds the body's ability to excrete them. Biomonitoring studies indicate that numerous toxins reside in our bodies and accumulate in fatty organs like the brain and breasts (Betarbet et al. 2000; Gray, Nudelman, and Engel 2010; Ionescu et al. 2006).

Rather than study each chemical one at a time, tightly controlled lab-based studies enable researchers to determine the levels at which there are observable adverse effects from specific groupings of chemicals, such as xenoestrogens (Kortenkamp et al. 2007). According to Payne, Scholze, and Kortenkamp (2001), the effects of these groupings, or mixtures, can be predicted from the potency of individual agents only if the effects of the individual agents and mixtures are analyzed within the same system. There is now a standardized method of assessing the total effective xenoestrogen burden (TEXB) in human adipose tissue and serum, where xenoestrogens are assessed as a group. This methodology, developed by Ibarluzea and colleagues, uses high-performance liquid chromatography (HPLC) to separate environmental xenoestrogens from naturally occurring estrogens. Through the use of this method, numerous organochlorines were detected, including DDT, aldrin, lindane, methoxychlor, and ether (Fernandez et al. 2007a).

These TEXB studies demonstrate that complex interactions between chemicals and hormones can alter the estrogenic environment in mammary tissue and could lead to malignancies. Although it is almost impossible to identify individual amounts of each xenoestrogen in the body, Ibarluzea and colleagues (2004) argue that the ability to separate xenoestrogens from naturally occurring estrogens enables researchers to look at the combined effect of these compounds in adipose tissue. New research opportunities may arise with TEXB considered as a biomarker of exposure – a technique that would demonstrate the role that xenoestrogens can play in breast cancer in a more practical and comprehensive way (Fernandez et al. 2007b).

Xenoestrogens and Social Determinants of Health

In order to determine which groups within society are more vulnerable to exposure than others, a social determinants of health model was created based loosely on the following categories: early life and puberty, food security, social exclusion, education, working conditions, income, housing, and availability of a social safety net (Raphael 2003; see Chapters 1, 2, 3, 6, 7, 8, 10, and 11 for other discussions of social determinants of health). These categories speak directly to breast cancer risk and chemical exposure. Working within this paradigm helps to "bridge the gap between research and public policy" (Raphael 2003, 35), as socio-economic indicators bring the discussion of breast cancer and chemical exposure into a broader framework that includes the political, economic, and environmental contexts. With this idea in mind, we have endeavoured to understand the causes of breast cancer within a wider societal perspective that works to better inform policy directions.

The following subsections explore a few key social determinants of health, and examine literature on the particular human and laboratory evidence connecting xenoestrogens to breast cancer in these contexts. Individual xenoestrogens are provided as examples. We start with early life and puberty exposure, and go on to discuss food security issues, social inequalities, occupational exposures, and geographical location. Specific research and policy implications that emerge for each of these determinants are focused on in depth in the third and fourth sections of this chapter.

Early Life

The timing of exposures to increased levels of natural and chemical xenoestrogens in a woman's body plays a key role in the development of cancer (see discussions of critical windows of vulnerability in Chapters 1-4, 7, and 11). This fact speaks to our first social determinant of health: that of the vulnerabilities inherent to early life. There are two critical periods in a woman's life when she is most prone to increased long-term risk of breast cancer from endogenous and exogenous estrogenic exposures. The first period includes the time in utero and at birth, and the second is during puberty.

Exposures In Utero and at Birth

The first period of susceptibility is in the womb. At this stage, breast tissue is being freshly laid down and increased estrogenic activity can actually increase the likelihood of developing cancerous cells later in life. Recent data indicate that very low levels of exposure in utero and/or from combinations of xenoestrogenic chemicals can increase the lifelong risk of breast cancer (Weiss et al. 1997).

Example: Bisphenol A (BPA)

BPA is widely used in plastics and resins, and is released in large quantities into the atmosphere every year (Chapters 8 and 10; Maffini et al. 2006). It is also commonly found in the lining of tin cans and in baby bottles, wine storage containers, drink containers, water pipes, dental composites, optical lenses, adhesives, compact discs, and paper receipt coatings. In 2003, the worldwide production of BPA exceeded 6 billion pounds (Soto et al. 2008). Continued exposure to BPA and other xenoestrogens occurs in utero and in newborns through breast milk and infant formula (Ye et al. 2006). BPA has been measured in human fetal serum and placental tissue. Animal evidence indicates that it undergoes transgenerational amplification – it increases in effect through generations – and is found in higher concentrations in the mouse fetus than in the maternal mouse (Milligan, Khan, and Nash 1998; Schönfelder et al. 2002). Scientists hypothesize that the accumulation of chemical and natural estrogens in the fetus can alter development and lay the groundwork for disease later in life (Ye et al. 2006).

Tests done on mice injected with BPA in utero demonstrate the dramatic sensitivity of the developing fetus to xenoestrogenic exposure. Impacts that were assessed included higher susceptibility to the naturally occurring estrogen called estradiol, which increased the chances of tumour development, proliferation of cells that line organs and glands, expression of estrogen receptors (ERs) and progesterone receptors (PRs), and decreased vaginal weight (Soto et al. 2008). Translated to humans, BPA may directly alter the growth of mammary tissue in utero, changing both tissue and organ development and possibly leading to breast cancer later in life (Maffini et al. 2006).

Studies demonstrate that seemingly low levels of xenoestrogens like BPA in fatty tissue may have a significant health impact for a number of reasons. Xenoestrogens like BPA may act additively with naturally occurring estrogens to increase overall risk of breast cancer. Moreover, xenoestrogens can bind to plasma-carrier proteins with substantially lower affinities than endogenous estrogens, making them more readily available to target cells like those found in developing mammary glands. Reduced binding of these xenoestrogens results in more freely circulating EDCs, thus causing greater biological activity. Finally, xenoestrogens are known to readily bind to estrogen receptors, thereby causing a larger overall proliferative effect on breast tissue (Silva, Scholze, and Kortenkamp 2007; Soto et al. 2008; vom Saal et al. 2007). It is this proliferative effect that can promote the development of cancer.

Example: Polychlorinated Biphenyls (PCBs)

PCBs belong to a broad family of man-made organic chemicals known as chlorinated hydrocarbons (see Chapter 8). PCBs were domestically manufactured in the United States from 1929 until their manufacture was banned in 1979 (US EPA 2012b). They were used by industry in numerous ways, including in electrical equipment, plasticizers in paints, plastics and rubber products, pigments, dyes, carbonless copy paper, and many other industrial applications (US EPA 2012b). They are still released into the environment today from poorly maintained hazardous waste sites containing residues of these compounds. Once in the environment, PCBs do not readily break down and may persist for long periods, recycling between the air, water, and soil. PCBs are now regularly found in mammals and ocean wildlife in the far north (ATSDR 2000).

People who eat fish may be exposed to elevated levels of PCBs as a result of accumulation in fish tissue. It is important to note that the types of PCBs that tend to bioaccumulate in fish and other animals, and that also bind to sediments, happen to be the most carcinogenic components of PCB mixtures. As a result, people who ingest PCB-contaminated fish may be exposed to PCB mixtures that are even more toxic than those that were released into the environment in the first place (ATSDR 2000). In an Ontario-based case-control study, increased levels of PCBs in breast tissue elevated risk for both pre- and post-menopausal women (Aronson et al. 2000). The IARC (1987) has declared PCBs to be "probably carcinogenic" to humans, and the US National Toxicology Program has stated that it is reasonable to conclude that these compounds are carcinogenic in humans (NIEHS-NTP 2011).

Reminiscent of studies that have examined the bioaccumulation of mercury in fatty tissue and its increase in concentration as it moves up the food chain, PCB levels in human breast milk have been found to exceed safety levels set for cow's milk by the World Health Organization (WHO) (Allen et al. 1997). A study from nearly two decades ago found that Aboriginal women in northern Quebec had high levels of PCBs in their breast milk (Lowell 1990). Wildlife studies further elucidate the impact of environmental toxins on mammals living in the north: polar bears living below the Arctic Circle have had tissue levels of PCBs up to 70 parts per million, and beluga whales have increased numbers of mammary gland cancers concomitant with higher levels of PCBs (Fox 2001). In her 1996 book on breast cancer among Native American women, Colomeda notes that Aboriginals from the north are well aware of the food contamination of wild livestock, referring to caribou meat as "black with little holes in it" and to whales that look "visibly sick." Some Aboriginal women have been told not to breast-feed their children for fear of making them ill (Colomeda 1996).

The fact that northern communities must be wary about breast-feeding points directly to the need to clean up our environment. Not only are wildlife numbers dwindling and food supplies heavily contaminated but women from northern populations, and North American women in general, are beginning to contemplate breast milk contamination. This speaks forcefully for the need to clean up waterways and regulate polluting industries that continue to emit toxic xenoestrogens.

Exposures during Puberty

The second stage of vulnerability to increased estrogen exposures in early life is during puberty. At puberty, mammary tissue starts to undergo cyclic growth. Artificially increased levels of estrogenic activity at puberty may lead to a higher risk of breast cancer later in life. For example, increased risk of breast cancer has been associated with women exposed to the organochlorine pesticide DDT during adolescence (Cohn et al. 2007).

Example: Dichlorodiphenyltrichloroethane (DDT)

DDT was heavily used as a pesticide between 1945 and 1965, before being banned in 1972 because of its potential health risk. Cohn and colleagues (2007) did a prospective study using blood samples taken from young women between 1959 and 1967. They analyzed the serum levels of DDT in those who had breast cancer, and matched those cases with controls. For women exposed to DDT during the early stages of breast development (i.e., puberty), the study found a fivefold increase in breast cancer risk (Cohn et al. 2007). Because levels of DDT in the body decrease over time, and because breast cancer can take decades to appear, this study was significant in being able to compare DDT concentrations in women under twenty with incidences of breast cancer in the same women in their forties. The authors warn that we may see more cases soon, as those who were heavily exposed in the United States during the 1960s and 1970s are only now approaching their fifties and sixties, the age at which women have the greatest risk of developing breast cancer (Cohn et al. 2007).

DDT is just one of many organochlorine pesticides that can harm humans. In their publication on total effective xenoestrogen burden (TEXB), Fernandez and colleagues (2007b) cite twenty studies that associate organochlorine pesticides with breast cancer risk. Pesticides are used extensively in agriculture, homes, and workplaces, and throughout cities on private and public lawns. These substances are immunotoxic (toxic to the immune system), mutagenic (induce mutation), and carcinogenic, and can enhance the metabolism of other exogenous chemicals (Sherman 1994). Among pesticides for which animal studies have demonstrated clear breast cancer risk are atrazine and urethane. Atrazine has been associated with increased mammary tumour development in rats (Rudel et al. 2007). This widely used herbicide is potentially carcinogenic in females living on farms, and the general population can also be exposed to it through inhalation of ambient air, ingestion of drinking water, and ingestion of some foods. Urethane is a solvent used in pesticides, fumigants, and cosmetics; it also occurs in trace doses in foods such as beer, wine, soy sauce, bread, yogurt, and olives partially prepared through the fermentation process.[2] Studies of urethane-fed mice found enhanced development of mammary tumours in both sexes (Rudel et al. 2007). Two different studies have indicated an association between the pesticide dieldrin and an elevated risk of breast cancer in humans; one of these studies involved breast cancer cases among Danish women (Gammon et al. 2002). In addition, two studies have shown links between mirex and breast cancer risk among women who had never produced breast milk (Aronson et al. 2000; Moysich et al. 1998).

Some pesticides, such as DDT, heptachlor, and mirex, have been banned in the past two decades in Western countries. Unfortunately, even those subject to bans persist in the ecosystem and in human tissues for years following their removal from production. The persistence of organochlorine pesticides is well known: they take a long time to break down, are fat-soluble, and bioaccumulate in fatty tissue, ensuring that they remain biologically active for decades (Allen et al. 1997). Since an important predictor of breast cancer risk is one's lifetime exposure to estrogen – in which the periods in utero and during puberty are critical windows of susceptibility – the estrogenic behaviour of persistent organochlorine pesticides may be increasingly linked to greater risk of the disease.

Food Security

Food security is an important social determinant of health that is intrinsically tied to potential toxic exposure. Xenoestrogens and synthetic chemicals are commonly found in our food supplies and water systems. Pesticides are used generously on crops, BPA leaches into food from food wrappers and plastic bottles, PCBs are found in dangerous levels in fish and other wildlife, and cadmium, introduced below, is found in certain agricultural soils.

Although some people are turning to organic produce in order to limit their toxic exposure, many people cannot afford organic foods. Buying organics can almost triple one's household monthly food expenditure over time. Top income earners can make this kind of a choice, but most Canadians would find it much more difficult. Considering the ubiquity of pesticides and other chemicals in our food supply, and the fact that individuals cannot afford to escape this persistent exposure, the responsibility for protecting people from harm needs to shift to industry and government. This can be done by reducing allowance levels for pesticide residues, banning the production and use of BPA and cadmium, cleaning up toxic landfills contaminated with PCBs, and encouraging industry to use less harmful methods of agriculture and to produce less toxic manufactured goods.

Example: Cadmium

Although cadmium is a heavy metal commonly found in the environment, a large body of evidence demonstrates its estrogenic capabilities and possible contribution to breast cancer risk (Garcia-Morales et al. 1994; Ionescu et al. 2006; Johnson et al. 2003; Liu, Yu, and Shaikh 2008; McElroy et al. 2006; McMurray and Tainer 2003; Navarro Silvera and Rohan 2007). A number of studies have found links between cadmium and cancer incidence, including prostrate, pancreatic, and bladder cancer (Kellen et al. 2007; Kriegel et al. 2006; Vinceti et al. 2007). Deemed to be a carcinogen by the IARC, cadmium has been banned in electrical components in Europe (IARC 1997).

Human exposure to environmental cadmium derives from the burning of fossil fuels and municipal waste, the smelting of zinc and copper, and tobacco smoke. Cadmium is used in alloys, nickel-cadmium (NiCd) batteries, plastics, fertilizers, and metal plating processes, and as pigments in glazes and enamel paints. Parts of North America are known to have elevated levels of cadmium in the soil, with soils near industrial sites often having the highest levels. There have been notable instances of toxicity resulting from long-term exposure to cadmium in contaminated water and food such as rice, wheat, and seafood (National Toxicology Program 2002). Once ingested, only a small fraction of cadmium is excreted, resulting in increased body burdens over time and making it notoriously difficult to remove from the body (Kellen et al. 2007).

Cadmium can increase one's overall risk of cancer growth by impairing gene function. From a scientific standpoint, it is thought to induce cancer via genotoxic (DNA-damaging) mechanisms, including the initiation of DNA breaks and the inhibition of DNA repair (Navarro Silvera and Rohan 2007), activation of cancer-causing genes, and the inhibition of programmed cell death (apoptosis), a process that is crucial in developing and maintaining the health of the body (McMurray and Tainer 2003). Smokers have much higher levels of cadmium than non-smokers; in addition, a study comparing accumulated heavy metals (including cadmium) in twenty breast cancer biopsies with eight healthy biopsies indicated that the accumulation of these metals in breast tissue may be closely related to malignant growth processes (Ionescu et al. 2006).

Social Inequalities

Variation in the use of potentially risky products requires us to consider associations between racial, socio-economic, and educational inequalities in Western societies and breast cancer risk. Many academics feel that breast cancer has predominantly been treated as a white woman's disease. As such, prevention tactics, including increased mammography screening, are largely geared towards women with higher income and educational levels, whereas other ethnic subgroups undergo less frequent screening. This disparity is seen as resulting from a combination of cultural preconceptions about mammography as well as barriers in language and education that limit access to preventive care (Andreeva, Unger, and Pentz 2007; Lee-Lin et al. 2007; Smigal et al. 2006). To illustrate, research has demonstrated that on diagnosis, black women's breast cancer is often more advanced and more aggressive in form (Donovan et al. 2007). Crafting an educational campaign that focuses on specific ethnic groups and, in the case of African Americans, that explores potentially unique causes of breast cancer in relation to the use of personal care products, could impact political, economic, and environmental perceptions of breast cancer in society.

Personal care products, including shampoos, creams, conditioners, hairsprays, deodorants, cosmetics, and perfumes, can contain ingredients that are xenoestrogenic and therefore potentially carcinogenic. These ingredients include parabens, phthalates, BPA, and pesticides. Personal care products with estrogenic chemicals may subject breast buds to elevated estrogen exposure during critical windows of vulnerability in utero and during early life. Donovan and colleagues (2007) completed a study exploring the prevalent use of personal care products containing estrogens and xenoestrogens by African American woman and their daughters. They found that African American women use personal care products over ten times more than women of other ethnicities, with more than twenty of the commonly used products containing human placental extract and "estrogen plus" (ATSDR 2000; Donovan et al. 2007). They hypothesized that the persistent, long-term use of specific personal care products may account for a 50 percent higher rate of breast cancer among African American women under the age of thirty, compared with white women in the same age bracket. This pattern of consumption may partly account for the increased mortality rates observed in African American women with breast cancer, although disparities in access to effective health care and treatment options could also play a significant role (Donovan et al. 2007).

The racial, educational, and income inequalities associated with breast cancer risk become even clearer when we look at media representations of such inequalities. In the book *Breast Cancer: Society Shapes an Epidemic,* Kasper and Ferguson (2000) point to the enormous role played by the media in depoliticizing and deconceptualizing breast cancer (Engel et al. 2005). From *Reader's Digest* articles in the 1920s to present-day articles in *Chatelaine* and *Vanity Fair,* stories focus on women's individual successes in overcoming breast cancer. Kasper and Ferguson (2000) argue that in this way, the media shift responsibility from institutions back to the individual. An individualization of breast cancer further inhibits discussions of breast cancer within broader societal frameworks (Kasper and Ferguson 2000).

Occupational Exposure

Occupationally vulnerable groups include women living and/or working on farms or in the automotive industry, health care workers, workers in the aesthetics industry, house cleaners, and electronics factory workers (Engel et al. 2005; Nie et al. 2007; Prince et al. 2006; Tokumaru et al. 2006; Chapters 10 and 11). These groups of women are exposed to an array of chemicals, including pesticides, BPA, PCBs, a wide range of solvents, diesel, and a number of heavy metals.

Agricultural studies have found a link between pesticide exposure and breast cancer, particularly among women farmers or farm tenants exposed to chlorinated (chlordane, aldrin, and lindane) or non-chlorinated (2,4,5-TP and captan) pesticides (Engel et al. 2005). Fernandez and colleagues (2007a) found a nearly twofold increase in the risk of breast cancer among post-menopausal women from Spain, with detectable levels of aldrin and lindane in fatty tissue (Ibarluzea et al. 2004). Brophy and colleagues (2006) conducted two large-scale studies among female farm workers in Ontario, one of which involved over 500 cases and 500 controls in the Windsor area. The studies found that women with breast cancer were three times more likely to have worked in the agricultural industry than controls, and four times more likely to have also worked in either the automotive industry or in health care (Brophy et al. 2006; see Chapters 10 and 11). The authors hypothesize that pesticides used in agriculture, and solvents used in both the automotive and health care industry, may partly explain the increased risk. Timing of exposure is seen to play an important role, as many of the women in the studies worked on farms during their adolescence (Brophy et al. 2006). Brophy and colleagues recently completed a four- and two-year study to determine the role of occupational, environmental, and modifiable lifestyle factors among female farm workers and automotive and health care workers. With over 1,000 cases and 1,000 controls, their observations shed more light on the increased breast cancer risk among these subgroups of women (Brophy et al. 2012).

A number of other occupations are frequently overlooked as areas of risk for women. Work in the aesthetics industry involves exposure to a wide range of chemicals, including solvents, dyes, and varnishes, many of which have been implicated in numerous forms of cancer, including breast, non-Hodgkin's lymphoma, and brain cancer (Clapp, Howe, and Jacobs 2007; Hoy 1995). Airline service is another potentially high-risk occupation, with a meta-analysis of environmental exposures and breast cancer indicating a 41 percent elevated relative risk of breast cancer for female flight attendants (Tokumaru et al. 2006). The authors suggest that a combination of factors, including ionizing radiation, jet fuel, irregular work hours,[3] and pesticide exposure, may all have contributed to increased risk (Tokumaru et al. 2006).

These occupational studies highlight the potential for increased incidence of breast cancer among women working in industries in which chemical exposures are high. Many of these occupations, including agricultural work, factory work, jobs in the beauty and health care industry, and house cleaning, have one thing in common: they are for the most part low-paying jobs. It can be argued that those at the lower end of the income scale are disproportionately exposed to hazardous substances and therefore most in need of protection from exposure (see Chapters 1 and 11). Unfortunately, lower-income workers do not have the resources to adequately protect themselves and fight for their rights. It is the responsibility of government and industry to ensure that identifiable occupational groups do not face an increased risk of disease from their workplaces.

Geographical Location

Studies carried out in North America indicate that women in specific geographical locations may be at greater risk of breast cancer resulting from higher exposures to xenoestrogens. Since the 1950s, over 750 million tons of toxic chemical waste has been discarded in 50,000 hazardous waste sites (HWS) in the United States alone (Hoy 1995). In enumerating these hazardous waste sites, the Environmental Protection Agency (EPA) discovered "strong associations" between excess cancer deaths and proximity to site locations, and concluded that the leaching of pollutants into groundwater posed a major health risk for humans (Hoy 1995). In *The Truth about Breast Cancer* (1995), author Claire Hoy talks about "above average levels" of breast cancer in places with high concentrations of heavy industry, such as Toronto, Montreal, Sydney (Nova Scotia), and Brantford (Ontario). More recent data found on the Canadian Cancer Society's website confirm that large industrialized centres throughout Ontario, Quebec, and the Atlantic Provinces are seeing increased levels of breast and other types of cancer (CCS 2012b).

The northeastern United States is a region with one of the highest rates of breast cancer in North America (Hoy 1995). In the early 1990s, a number of breast cancer coalition groups were created in Long Island, New York, by cancer survivors who had begun to look at the statistics. These groups discovered that in the 1980s there were 110.6 cases of breast cancer for every 100,000 living on the island, as opposed to the national rate of 94.7 (Hoy 1995). This elevated incidence was thought to be associated with the large number of chemical plants in the area. With the aid of some senators, women's groups convinced the local health department to explore the issue further. In 1994, the New York State Department released a study showing that women living near a chemical plant on Long Island had a 62 percent greater risk of developing breast cancer than other women in the region. This risk increased in relation to the total number of plants located near a woman's home (Hoy 1995). Similar to the strong tobacco lobby established in the past, the chemical industry lobby has been able to argue that there remains insufficient proof of a direct link between proximity to industry and incidence of breast cancer. The battle between breast cancer activists and the chemical industry continues in this region.

On a global scale, Canada, the United States, and Great Britain have the highest incidence of breast cancer in the world, whereas Asian countries exhibit the lowest rates (Parkin, Bray, and Devesa 2001).

Until recently, the developing world had a substantially lower incidence of the disease compared with more developed Western countries. Current analyses now show that in countries like China, Japan, and India, incidence is rising dramatically (Parkin, Bray, and Devesa 2001). These countries are rapidly undergoing industrialization, thus increasing the potential for environmental exposures to air and water pollution. In combination with changes in diet and lifestyle, these factors may be contributing to a higher national incidence of breast cancer.

Recent immigrants to the United States demonstrate an increasing incidence of breast cancer the longer they live in the country, especially among second-generation immigrants, whose risk of disease aligns closely with that of non-immigrant women (Andreeva, Unger, and Pentz 2007). Genetics do not change dramatically from one generation to the next; thus, environmental and lifestyle factors largely explain disparities in incidence between women who have remained in their home countries and those who have emigrated to North America and had offspring (Andreeva, Unger, and Pentz 2007; Stanford et al. 1995). Canada lacks a national registry for tracking relationships between ethnicity and cancer rates, making it almost impossible to know which groups of immigrants – or even which groups of Canadian women – are at increased risk of breast cancer. Although the same level of epidemiological evidence is not available for the Canadian population, we can infer that incidences in Canada would follow a similar trend to those in the United States.

Example: Polycyclic Aromatic Hydrocarbons (PAHs)

The PollutionWatch website, a collaborative project of Environmental Defence and the Canadian Environmental Law Association (CELA), tracks air pollution in Canada and clearly shows that large urban centres across the country have "above average levels" of chemical air pollutants (PollutionWatch 2008b). Polycyclic aromatic hydrocarbons (PAHs) are a group of atmospheric pollutants often found in this mix. They enter local environments through a number of processes, including fossil fuel combustion, diesel exhaust, consumption of smoked and grilled foods, and cigarette smoking, and can lead to human exposure. Many PAHs, including benzene and toluene, are known to be carcinogenic and have been classified as a "danger to human health" by Environment Canada (Environment Canada and Health Canada 1994). As with many other lipophilic (fat-soluble) pollutants, they are stored in fatty tissues, including the breasts (Obana,

Hori, and Kunita 1981). PAHs are thought to be xenoestrogenic as a result of their impact on estrogen receptors, certain enzymes, and tumour-suppressing genes (Salehi et al. 2008; Vinceti et al. 2007). Several in vitro, animal, and human studies point to a link between PAHs and increased risk of breast cancer development. For example, in vitro studies have shown PAHs to be strongly mutagenic, making DNA repair difficult and ultimately leading to tumour growth (Jeffy, Chirnomas, and Romagnolo 2002). In addition, several PAHs are potent mammary carcinogens in rodents. PAHs with established carcinogenicity often cause increases in cell cycling and cell proliferation, mechanisms consistent with the development of human breast cancer (Dipple et al. 1999; Jang et al. 2000; Murray et al. 2005). In one human study, elevated PAH levels in fat tissue correlated with a 2.6-fold increase in breast cancer risk, and demonstrated negative effects on DNA repair (Tang et al. 2002). In another study, the link between PAHs and breast cancer risk was found to be exacerbated by alcohol consumption (Terry and Johnson 2003; see Chapter 6). A study of premenopausal breast cancer patients found a trend of increasing risk in estrogen receptor-positive tumours associated with PAH exposure.

Given the ubiquity of PAHs in most urban centres and the recent evidence linking these air pollutants to breast cancer, a growing number of researchers are calling for increased funding for studies that seek to better understand the role of air pollution in breast cancer risk (Brody et al. 2007a; Salehi et al. 2008).

The Burden of Cancer in Canada

In 2006, WHO concluded that 23 percent of mortalities and 24 percent of "health-adjusted life years" (HALYs, or the number of years lost due to disease, disability, or early death) are globally linked to preventable environmental risk factors. Health Canada estimates that 10-15 percent of cancers are linked to environmental exposures (including pollution, UV radiation, occupational exposures, and consumer products), and WHO estimates that as many as 10-23 percent of female cancers in developed countries such as Canada are due to environmental causes (Prüss-Üstün and Corvalán 2006). Boyd and Genius (2008) synthesized the relevant data, focusing specifically on the Canadian context. They found that 3,400 to 10,200 cancer deaths, and 8,000 to 24,000 new cases of cancer, are directly attributable to toxic exposures in our environment. Further, they note that risk factors associated with environmental exposure are largely involuntary, as opposed to modifiable risk factors associated with diet, fitness, smoking, and other personal lifestyle choices. Thus, "this critical distinction puts great onus on governments to reduce risks posed by environmental hazards" (Boyd and Genuis 2008).

Health-Adjusted Life Years and Risk Factors for Cancer in Canada

In 2007, the Public Health Agency of Canada (PHAC) published a paper called *The Population Health Impact of Cancer in Canada, 2001* (Boswell-Purdy et al. 2007). It estimated that one-quarter of HALYs lost to cancer are attributable to smoking, and another quarter are attributable to a combination of alcohol consumption, obesity, physical inactivity, and a lack of fruit and vegetable consumption. Of all the cancers examined, breast cancer was the leading cause of morbidity, more than prostate and colorectal cancers combined (Boswell-Purdy et al. 2007). There was no mention of the risk posed by environmental toxins. While research in this area is considered to be in its early stages and attributable risk is not well quantified, Dr. Samuel Epstein, a specialist in environmental contaminants and cancer, has generated controversy by arguing that environmental exposure accounts for over 80 percent of cancers (Cancer Prevention Coalition 2003; Hoy 1995). It is obvious now that upstream risk factors such as income level can play a role in the prevalence of cancer risk. Ways forward from a research perspective should include the evaluation of these factors in terms of the poorest and the richest quartiles of the population (Boswell-Purdy et al. 2007); estimates of HALYs for each risk factor could provide some indication of the impact of income distribution, particularly with respect to environmental contaminants such as xenoestrogens.

Challenges in Achieving Cancer Prevention

The previous sections developed evidence regarding the xenoestrogenic potential of certain chemicals and their links to increased breast cancer risk among Canadian women. By exploring routes of exposure in early and pubescent life, and in relation to personal care products, food security, occupation, and geographical location, we highlighted the importance of a social determinants of health model in determining appropriate policy changes to bring about a reduction in xenoestrogen exposure and thus reduce total breast cancer risk in Canada. Unfortunately, such policy changes have been difficult to achieve as a result of the political economy of chemicals and cancer, and challenges associated with chemical assessment methodology and research. These are described below.

The Political Economy of Chemicals and Cancer

Along with the release of the film *Pink Ribbons, Inc.* in 2011, there has been a growing awareness of the power, wealth (an estimated $3 trillion base), and influence of the chemical industry and related lobby associations (Din and Pool 2011). The only other industry with this kind of economic clout is the pharmaceutical industry. Interestingly, a number of companies are simultaneously involved in the production of both pharmaceuticals and industrial chemicals. AstraZeneca provides a clear example: not only is it the largest single producer of Tamoxifen, one of the most commonly used treatments for breast cancer, but it is also a major pesticide producer, having manufactured millions of tons of the herbicide acetachlor, a known carcinogen (Sweeney 2006). AstraZeneca is also the largest sponsor of Breast Cancer Awareness month, and is responsible for the marketing and advertising materials of the campaign (Sweeney 2006). AstraZeneca provides a classic example of how a single corporation can simultaneously profit from the manufacture of chemicals that contribute to the incidence of breast cancer, the production of pharmaceutical treatments to battle breast cancer, and the development of a social and political response to the disease through an international campaign. Finally, and without any sense of irony, this same company is also the number one sponsor for promoting educational awareness about breast cancer, including its prevention.

The scenario depicted for AstraZeneca is not illegal, and there is currently no incontrovertible proof linking the chemicals manufactured by the company with higher rates of breast cancer. The situation, however, points to the minimal ethical standards that allow a multinational producer of suspected carcinogens to develop cancer treatments and to sponsor and control public campaigns about cancer.

Major corporate sponsors also tend to be major funders for research, and to date, cancer organizations continue to accept funds from polluting companies. Encana, one of the largest tar sands producers in Canada, came under scrutiny as a key funder of cancer organizations with the airing of the documentary *Between Midnight and the Rooster's*

Crow (Drost 2005). To their credit, organizations such as the Canadian Cancer Society (CCS) no longer accept funds from companies linked to tobacco (Sweeney 2006). Indeed, it may better serve the public and protect the integrity of cancer organizations if they do not accept funds from companies that release carcinogens into the environment.

Another question to consider is, who benefits from the results of studies funded by corporate polluters? If we compare dollars spent on pharmaceutical treatments for breast cancer, and those spent on finding connections between environmental toxins and cancer, the answer is clear. Research on treatment generates a return on investments, whereas research on prevention, an endeavour of far greater benefit to the public, does not. This funding imbalance points to a tension between environmental responsibility and maximizing profits (Hoy 1995). It is important that research on treatment be balanced with the funding of upstream prevention strategies.

In a paper on breast cancer and public education, Sweeney (2006) discusses the lack of funding for examining connections between environmental toxins and breast cancer. As of 2001, the Canadian Institutes of Health Research (CIHR), the Canadian Cancer Society (CCS), and the Canadian Breast Cancer Research Initiative (CBCRI) had each spent less than 3 percent of their annual budget on research on environmental contaminants and toxic chemicals (Sweeney 2006). Comparing the roughly $630,000 spent on environmental contaminants and cancer in 2000-1, with $697,000 in 2008-9, the CCS has changed little in its allocation of funding despite mounting evidence linking cancer to toxins in the environment (Allen 2008; Sweeney 2006). Compare this with the CIHR's annual budget of $960 million in total research funds, and the $1 billion spent on research by the CCS to date. These figures therefore indicate that over 90 percent of research dollars is spent on treatment and basic science. The Canadian government does no better in supporting preventive research and continues to allocate less than 10 percent to fund primary prevention-based research.

A focus on disease prevention is vital if we are to reduce breast cancer among Canadian women. As Sweeney (2006) and others point out, more solutions are available if the focus shifts away from the narrow lens of biology and individual testing and towards an examination of disease from a broader, sociological perspective. By looking at social determinants such as geographic location, socio-economic status, exposures during early life and puberty, and global environmental factors, organizations and governments can better understand the risk factors responsible for breast cancer and work towards prevention and not just treatment.

Unique Challenges in Toxicology Testing

Cancer prevention research faces a number of difficulties in establishing causal relationships between environmental toxins and human health. Although many studies, noted earlier in this chapter, have indicated chemical links to breast cancer risk, others have not found statistical evidence of risk association (Salehi et al. 2008). Contradictions in the evidence point to methodological challenges specific to this type of research. These include the long latency period between exposure to a pollutant and tumour development, as well as the inability to measure previous exposure directly, with many researchers using the sometimes spotty or inconsistent recollections of study participants as a proxy for their exposure (Salehi et al. 2008). In addition, many environmental studies involve small sample sizes, making it hard to find conclusive evidence of toxic exposures. Large-scale studies are resource-intensive and can sometimes be limited by a contaminated background from the outset.

Epidemiological studies that provide the strongest evidence of causality involve clusters of people with related exposures. Clusters of this kind are difficult to identify, and can sometimes be clouded by the desires of industry, and even of the community itself, to remain hidden. Prospective studies, which follow an exposed group over time, are also considered an effective method for measuring risk; however, it can take decades to successfully prove connections to environmental exposures, as health effects from early life exposures may become evident only after many years of latency. Keeping track of a large cohort over such a long period is expensive, requires incredible dedication by researchers, and is hindered by migration, recordkeeping issues, and the inaccuracies bound to occur when studying humans in uncontrolled settings.

A fundamental challenge with toxicological research is the ability to ascertain which compounds have an effect at which dose, and how additive, negative, and potentially synergistic effects are manifested in real-world circumstances (see Chapters 1, 3, 4, 10, and 11). The earlier discussions regarding TEXB studies point to the difficulty in trying to analyze one chemical at a time and determining safe dosage. While controlled laboratory findings may work to exonerate a chemical at low doses of exposure, xenoestrogenic chemicals may have additive or synergistic effects when combined. Along with individual assessment, chemicals should be analyzed in groups based on physiological activity, or mode of action, to better explore their potential impacts on human health (Chapter 3).

These methodological barriers prevent research on xenoestrogens from progressing at a pace that might help reduce breast cancer among Canadian women and women worldwide. Moving away from the traditional toxicological and monocausal approach to disease, and towards more innovative and realistic research methods that include considerations of social determinants, will shed more light on the true role of xenoestrogens' contribution to breast cancer risk (Brody et al. 2007b; Ibarluzea et al. 2004).

It is well understood that we cannot conduct controlled experimental human toxicological studies. The use of animal models and in vitro studies is necessary to directly explore chemical carcinogenicity. The validity of such studies has been confirmed by evidence indicating the presence of pathways and mechanisms in the human body similar to those found in lab-based research (Salehi et al. 2008). The issue of incompatibility between animal- and human-based research has been used successfully by the chemical industry to discredit findings connecting animal and human risk. The double standard is clear when companies test the safety of their products on mice and rats. Although the findings from animal and lab-based experiments have their limitations, they should not be discounted: such preclinical studies can be used to screen for the presence of risk and provide important insight into how chemicals function in the body and affect overall health.

Recommendations for Research, Policy, and Law Reform

Despite some advances in research and policy, Canada still lags in its efforts to address and curb breast cancer risk. As discussed in Chapter 3, tens of thousands of chemicals remain unregulated or weakly regulated in this country. At the same time, there is an increasing mass of studies revealing the adverse effects of exposures to these chemicals on human health. The government has both the opportunity and the duty to continue comprehensive research on the relationship between environmental toxins and breast cancer, and to further reform chemicals regulation to protect human health. Below are some key recommendations for moving Canada forward in these pursuits.

Enact the Precautionary Principle in Research Practices

Despite the challenges facing research on environmental toxins, there is sufficient evidence on the potential harms of xenoestrogen exposure to engage the precautionary principle (as discussed in Chapters 1-5, and 8): initial studies on laboratory animals and breast cell lines support the hypothesis that xenoestrogens cause breast cancer, and over 90 percent of people tested store a cocktail of xenoestrogenic chemicals in their fatty tissue (Environmental Defence 2006; Liu, Yu, and Shaikh 2008; Maffini et al. 2006; Metzdorff et al. 2007; Milligan, Khan, and Nash 1998; Rudel et al. 2007; Silva, Scholze, and Kortenkamp 2007). Additionally, historical trends indicate a 24-67 percent increase in lifetime incidence of breast cancer in women who were born after 1950 and have the greatest genetic disposition to breast cancer development (women with the BRCA1 and/or BRCA2 genes) (Brody, Tickner, and Rudel 2005).

Adopting the precautionary principle broadens current research objectives to include the primary goal of *prevention,* requiring that government act in the face of uncertainty if evidence for a given chemical demonstrates the potential for harm. As Brody, Tickner, and Rudel (2005) argue, prevention-oriented evidence derives from a broadly defined set of methods that include typical hypothesis testing, epidemiological studies, toxicology, exposure assessment, risk assessment, wildlife studies, and human case reports. They point to more recent studies that illustrate such preventive models of research, wherein public involvement, epidemiological studies, and the development and application of new exposure assessment methods are put in practice. They argue that the incorporation of activist groups can help expand research applications beyond traditional circles and increase the public's knowledge of risk factors associated with environmental contaminants. By democratizing scientific inquiry and engaging precautionary approaches, these models provide a new and innovative type of environmental research that can work to inform preventive public health policy (Brody, Tickner, and Rudel 2005, 922).

The democratization of science should also include increased research in the developing world. Dey, Soliman, and Merajver (2009), in their article on the connection between xenoestrogens and breast cancer risk, call for more research on populations in developing countries, especially considering the recent increases in breast cancer rates observed in these nations. Developing countries still employ a number of chemicals that have been banned in the developed world, leaving millions of people involuntarily exposed to harmful substances in their environment. Data collected from El-Shefa Hospital in Gaza, for example, found that breast cancer cases between 1990 and 1994 accounted for 34 percent of all new cases of cancer in women, an increase from 13.8 percent in 1979 (Richter and Safi 1997). A quarter of the population works in agriculture and is exposed to a number of xenoestrogenic pesticides on a regular basis. Current research on xenoestrogens must consider the potential health impacts of exposure on a global scale.

Incorporate the Precautionary Principle in Regulation

The Wingspread Statement on the Precautionary Principle, established by a large group of international scientists and environmental activists in 1998, identifies four central components of precautionary policies: (1) taking preventive action in the face of uncertainty; (2) making those who create risks responsible for studying and preventing them; (3) considering alternatives to potentially harmful activities; and (4) increasing public participation and transparency in decision making (Science and Environmental Health Network 1998). Currently, Canadian and US chemical regulations require substantial evidence of harm before regulatory action is taken, regardless of the availability of alternatives (Chapter 3).

The federal government can and should play a key role in implementing regulations based on the precautionary principle. With the Chemicals Management Plan (CMP) already in place in Canada, we have an opportunity to develop regulatory policies similar to those in Europe, including REACH (Registration, Evaluation, Authorization, and Restriction of Chemical substances), a program that seeks to ban or substitute chemicals that have the potential to negatively impact human health (European Commission 2012; see Chapter 3). The possibility of lowering breast cancer risk should be sufficient incentive for implementing precautionary regulation. The management and/or elimination of chemicals associated with breast cancer risk should be legislated in Canada to the same extent as in other nations leading in this area.[4]

Increase Research in Communities Most at Risk for Chemical Exposures

The 2008 study by PollutionWatch found a direct correlation between highly polluted sites and areas of poverty around the Great Lakes Basin

(PollutionWatch 2008a). Similar studies in the United States in the 1990s led to a presidential executive order requiring all federal agencies to develop strategies to incorporate environmental justice concerns into their activities[5] (Executive Order No. 12898). This included the creation of the Office of Environmental Justice within the EPA and of a national environmental justice advisory council (US EPA 2012a). Now the EPA must consider environmental justice issues in all policy, permitting, and pollution-monitoring decisions.

The PollutionWatch study recommends policy changes in Ontario and across Canada that include: (1) formal recognition that pollution can affect health, and that people living in poverty may be disproportionately affected; (2) further research into the interconnections between health, pollution, and poverty, and the development of poverty reduction plans; and (3) the creation of a clear environmental equity policy framework that considers these connections and is integrated into environmental decision-making processes such as the management of toxic substances (PollutionWatch 2008a).

Improve Data Sets in Cancer Registries

Bryant (2004) points to the lack of information on ethnic origin in Canadian cancer registries (Bryant 2004; A. Ritchie, personal communication, 3 October 2008). With comprehensive registry data, it is difficult to determine which ethnic groups have higher rates of breast cancer than other groups. Researchers need a more sophisticated data set with which to test hypotheses. Similarly, there has been a failure to document lifetime occupational histories and workplace exposures, resulting in an underestimation of occupationally related cancers and a corresponding lack of substantive prevention-based regulations (Brophy et al. 2006). Fostering more research on relationships between ethnicity and cancer rates, and exposures associated with low income and occupations with known exposures and "danger pay," will translate into improved health protection for Canadians from diverse social backgrounds.

Establish a Comprehensive, Long-term Biomonitoring Program

Unlike the United States and Germany, Canada has failed to implement a national program that would detail population-based exposures to environmental pollutants. As discussed earlier, even cases of occupational exposure have no adequate mechanism for systematic data collection. Moreover, provinces have not adequately addressed the particular vulnerabilities of high-risk populations, such as Aboriginal communities. Aamjiwnaang First Nation, mentioned in the Introduction of this volume, and communities in and around the Alberta oil sands present clear examples of high-risk communities that have demonstrated elevated rates of cancer. Precautionary action should be taken in the form of comprehensive environmental cleanup programs and rigorous community health monitoring, in order to continue tracking communities known to be at greater risk of exposure and to build a better record of other communities that might be experiencing similar conditions.

Create a National Environmental Health Strategy

A number of studies and statistics, some of which have been discussed in this chapter, indicate that the environmental burden of disease, including breast cancer, is substantial and continues to escalate. PHAC's National Advisory Committee on SARS and Public Health has argued that the "area of environmental impacts on health has been seriously neglected in Canada and requires urgent investment" (PHAC 2003, 78). Boyd and Genuis (2008) suggest that a National Environmental Health Strategy is required in order for environmental risk factors associated with breast cancer and other diseases to be adequately considered and addressed in a systematic manner.

Enact a "Consumer Right to Know" Law Involving Personal Care Products and Household Cleaning Goods

The amount of estrogenic compounds contained within personal care products is largely unknown, as many companies do not list ingredients on their product labels and are not required to disclose such information to the public, not even in the form of warnings about chemicals that may be harmful to human health. Adequate labelling is necessary in order for people to take effective preventive health measures. Advocates are pushing for manufacturers to fully list the ingredients of their products as a matter of public health policy, stating that the public has a right to know whether products they are using contain compounds that might increase their risk of disease (Donovan et al. 2007). Canadians have a right to know what is in their consumer products, and the federal government needs to regulate industry to ensure greater public knowledge and participation in protecting health.[6] Both California and the European Union have legislation requiring warning labels on a number of consumer goods. This has motivated many companies to remove harmful chemicals from their products (California Office of Environmental Health Hazard Assessment 2010).

Enact "Community Right to Know" Laws

Toronto took an important step in 2010 with the enactment of its Environmental Reporting and Disclosure by-law.[7] As a form of community right-to-know legislation, the bylaw requires all industry in and around the city to quantify and report air and water pollution on an annual basis. This is a huge achievement for the advancement of transparency and citizen's rights in the city, and should be adopted by other municipalities across the country.

Invest in Toxics Use Reduction

Toxics use reduction legislation targets cancer-causing chemicals used or produced by industry. Instead of focusing on which chemicals pose what kind of risk, toxics reduction strategies work to reduce and/or eliminate all hazardous substances emitted into the environment over time. The Ontario Toxics Reduction Act, implemented in 2009, requires major emitters in the province to report and track their toxic releases, and create a plan for reducing the use, emission, or discharge of harmful chemicals.[8] The Ontario act is based on a similar law in the state of Massachusetts that has been in effect for over twenty years, the Massachusetts Toxics Use Reduction Act (1989).[9] A review of the Massachusetts program between 1990 and 1997 reveals not only that the act has been effective in reducing overall toxic emissions and exposures but also that the overall benefits from its enactment over that time period were estimated at around $91 million (Massachusetts Toxics Use Reduction Program 1997). These benefits include less pollution and cleaner environments; safer consumer products and work environments; improvements in public health; more innovative green technologies; lower compliance costs for companies and lower enforcement costs for government agencies; and reductions in the off-site management of hazardous wastes. In other words, the act has aided in cleaning up the environment while simultaneously saving industry and government money. It is also worth mentioning that recent developments in Canada include local and regional bans on cosmetic pesticide use; this is still limited in scope, however.

Invest in Public Education and Knowledge-Transfer Initiatives

Educating the public needs to be a key priority if we are to see a substantial decrease in breast cancer incidence in Canada. According to Sweeney (2006), educational pamphlets created by Canadian cancer organizations about breast cancer are framed in a way that often places the onus on women to avoid the disease. For example, the Canadian Cancer Society's *Seven Steps to Health* recommends eating a healthier diet, not smoking, protecting oneself from the sun, following cancer screening guidelines, visiting a physician at any sign of ill health, and disposing of hazardous materials safely (Sweeney 2006). In addition, educational materials mainly focus on known risk factors for breast cancer that are difficult to modify, such as age at menarche, ability to bear offspring, breastfeeding, age at menopause, and incidence of breast cancer in the family. There is little mention of environmental contaminants or recent research showing links between toxic exposure and cancer, and most women do not understand that known risk factors account for only a small portion of breast cancer cases (Bryant 2004). As a result, women may be misled and end up feeling powerless to avoid cancer risk.

Non-governmental organizations (NGOs), universities, and integrative medicine clinics are attempting to build awareness about chemical exposures and risks through a number of initiatives. The following are but a few examples of the kind of work that should be funded and promoted by the Canadian government:

- *CancerSmart 3.1: The Consumer Guide,* published by the NGO Toxic Free Canada, is an educational tool that provides information about the health effects of exposure to toxins, practical suggestions about alternative, safe practices and products, and the need to urge government to enact policies that reduce the presence of chemicals in the environment and protect people from harmful exposures (Griffin 2011).
- PollutionWatch has created an online database of companies and industries in Canada that pollute our air.

- Environmental Defence's Toxic Nation campaign has published reports outlining the body burden of chemicals in hundreds of Canadian families.
- Along with other environmental groups, Environmental Defence has also been instrumental, and successful, in lobbying the Canadian government for a ban on certain chemicals, such as BPA.

The type of work conducted by these types of NGOs is useful and effective, and provides examples of non-partisan partnerships between government and NGOs to promote public health.

The Canadian College of Naturopathic Medicine (CCNM) has a teaching clinic that has over 25,000 patient visits per year. Their brochure on breast health recommends ways in which women can reduce exposure to xenoestrogens in their daily lives, including avoiding plastics and pesticides and using glass containers whenever possible (CCNM 2009). Community health clinics, hospitals, and health care practitioners can play an important role in educating patients and promoting preventive measures for individuals.

The Centre for Population Health at the University of Ottawa, in conjunction with five other Canadian universities, has created www.emcom.ca, a website where the public can learn more about EDCs and human health issues. A 2008 article highlighted the significant gap in understanding between scientific experts and the public, who need to be informed about the risks of EDCs in a language they can comprehend (Tyshenko et al. 2008). It argues that knowledge translation and transfer can be achieved through non-partisan Internet portals, such as www.emcom.ca, which enable effective risk communication via web-based discussion groups. These discussion groups involve a number of participants from different backgrounds, including scientists, regulators, political leaders, news media, and NGOs. Knowledge transfer through these types of forums can bring together the multiple risk dimensions of science, politics, and society, and move risk communication into the realm of risk management (Tyshenko et al. 2008).

Societal change is extremely difficult to achieve, especially where cultural, industrial, and economic forces are all at play. The British Working Group on the Primary Prevention of Breast Cancer has compiled a list of cultural barriers that work against successful prevention (No More Breast Cancer 2006). These include: societal acceptance that breast cancer is a fact of life; a fixation on treatment and control rather than prevention; a persistent focus on lifestyle factors as key to prevention; the "out of sight, out of mind" mentality resulting from the invisibility of harmful chemicals; procrastination by policy makers in the face of uncertainty; and the overshadowing of preventive measures by vested interests in profit (No More Breast Cancer 2006). Such a list helps us understand some of the challenges we face in attempting to shift action towards the prevention of environmental risks associated with cancer, and should be disseminated to the public more broadly.

Concluding Remarks

Governments worldwide are becoming increasingly aware of the need to look at social dynamics of poor health in order to develop policies that reduce the risk of disease. The social determinants of health model is a powerful tool for promoting knowledge transfer and mobilizing public efforts around health, but it places little emphasis on the environment. The scientific evidence linking xenoestrogen exposure to breast cancer must be viewed through this model in order for policy to shift from treatment towards prevention, especially when considering women's health. In their original 2002 report, *State of the Evidence: The Connection between Breast Cancer and the Environment,* the Breast Cancer Fund and Breast Cancer Action take a strong stand on the need for more research into chemicals and their obvious link to breast cancer (Davis 2002, 190):

> We ignore, at our peril, the increasing evidence that chemicals are contributing to the rising tide of breast cancer. The obligation to understand this evidence and begin to address it through implementation of public policies that put health first rests with all of us. It is in our power to change the course we are on. Now is the time.

Besides governmental action, it is important that researchers also be engaged in the debate. Davis (2002) offers compelling insights into the limitations of earlier research:

> The millions allocated to research have produced what one editor of the *Lancet* describes as a glut of same old, same old studies. Different questions must be asked to break the logjam. For women

> confronting breast cancer today, the central question remains this: What avoidable factors cause nineteen out of twenty cases of the disease?

The bottom line is that we are talking about avoidable environmental risk factors. Rather than attempting to establish the exact percentage of risk associated with xenoestrogenic chemicals, we should be working on mitigating these risks. One's age, one's heredity, and one's ability to have children are, for the most part, out of an individual woman's agency or control. Although these risk factors are important, they account for only a minority of breast cancer cases. Environmental risk factors, on the other hand, may account for a much larger proportion of breast cancer incidence, and risks can potentially be avoided. Even if EDCs such as xenoestrogens accounted for only 20 percent of breast cancer cases, thousands of lives would still be saved by taking them out of the picture (Davis 2007).

In this chapter, we have highlighted specific chemicals with xenoestrogenic properties, examining their biochemical properties, routes and timing of exposure (particularly as they relate to women's health and susceptibility), and additive potential, and have explored the mounting evidence linking these substances to increased risk of breast cancer. Our discussion is meant to contribute to the elmination of exposure to xenoestrogenic chemicals in order to maintain healthy and sustainable people and environments. The possibility of transgenerational exposure underlines our responsibility for the long-term health of our species. We believe that an upstream approach to health, which centres on precaution and prevention in all policy designs and decisions, is ultimately required if Canada is to realize a truly effective and equitable improvement in women's health.

Notes

1 It is important to note that many environmental pollutants are classified as carcinogenic, i.e., cancer causing, if they are mutagenic. A large number of endocrine disruptors are not mutagenic, however, and can be missed in preclinical tests for carcinogenicity. These endocrine disruptors can still promote the development of cancer by increasing cell proliferation.

2 Animal studies cannot give us the whole picture of risk or even pinpoint clear areas that need investigation; however, they do provide stronger evidence than that from cell-based research. Indeed, chemical risk as highlighted by animal research should provide a

yardstick that deserves serious consideration when respecting the application of the precautionary principle.

3 Shift work is emerging as a risk factor for breast cancer in several studies, with particularly serious implications for women in the health care sector, such as nurses.

4 Recently created by the European Union and known as REACH, this governing body has been created as a way of producing effective policy for chemical management. In June 2007, the EU instituted the precautionary principle within REACH that now governs the assessment and application of chemicals in society. Under REACH, the onus falls on chemical companies to provide safety data for all chemicals produced; to include an analysis of alternatives where a suitable alternative exists; to manage any risks associated with the use of these chemicals; and to communicate this information to consumers (Soto et al. 2008).

5 Executive Order 12898 of 11 February 1994, *Federal Register,* 16 February 1994 (http://www.epa.gov/compliance/ej/resources/policy/exec_order_12898.pdf).

6 Devra Davis is the executive director of the Environmental Oncology Centre at the University of Pittsburgh. She is an authority on the topic of chemical exposure and breast cancer risk. Dr. Davis has tried and failed to get companies to reveal what carcinogenic chemicals are present in their products. In a recent interview with the authors, she stated: "There are known animal carcinogens in shampoos, baby bubble bath products, and numerous other personal care products. At this point in time, there is no right to public information about xenoestrogens – there ought to be a right to know whether or not there is a carcinogen in your baby's bubble bath. We need to change the laws in favour of public safety so that manufacturers of these products are forced to be transparent" (D. Davis, personal communication, 2008).

7 *Environmental Reporting and Disclosure* (By-law 1293-2008), Toronto Municipal Code, c. 423 (http://www.toronto.ca/legdocs/municode/1184_423.pdf).

8 *Toxics Reduction Act,* S.O. 2009, c. 19 (http://www.e-laws.gov.on.ca/html/statutes/english/elaws_statutes_09t19_e.htm).

9 *Massachusetts Toxics Use Reduction Act,* 1989, General Laws, c. 21I (http://www.malegislature.gov/Laws/GeneralLaws/PartI/TitleII/Chapter21I).

References

Allen, Beth. 2008. "Environmental Toxin and Cancer Research at the Canadian Cancer Society 2008/2009." Canadian Cancer Society internal report provided in personal communication to Sarah Young.

Allen, Ruth H., Michelle Gottlieb, Eve Clute, et al. 1997. "Breast Cancer and Pesticides in Hawaii: The Need for Further Study." *Environmental Health Perspectives* 105 (S-3): 679-83, doi:10.1289/ehp.97105s3679.

Andreeva, Valentina A., Jennifer B. Unger, and Mary Ann Pentz. 2007. "Breast Cancer among Immigrants: A Systematic Review and New Research Directions." *Journal of Immigrant and Minority Health* 9 (4): 307-22, doi:10.1007/s10903-007-9037-y.

Aronson, Kristan J., Anthony B. Miller, Christy G. Woolcott, et al. 2000. "Breast Adipose Tissue Concentrations of Polychlorinated Biphenyls and Other Organochlorines and Breast Cancer Risk." *Cancer Epidemiology, Biomarkers and Prevention* 9 (1): 55-63. http://cebp.aacrjournals.org/content/9/1/55.full.pdf+html.

ATSDR (Agency for Toxic Substances and Disease Registry). 2000. *Toxicological Profile for Polychlorinated Biphenyls (PCBs).* Atlanta: ATSDR. http://www.atsdr.cdc.gov/toxprofiles/tp17.pdf.

Betarbet, Ranjita, Todd B. Sherer, Gillian MacKenzie, et al. 2000. "Chronic Systemic Pesticide Exposure Reproduces Features of Parkinson's Disease." *Nature Neuroscience* 3 (12): 1301-6, doi:10.1038/81834.

Boswell-Purdy, Jane, William M. Flanagan, Hélène Roberge, et al. 2007. "Population Health Impact of Cancer in Canada, 2001." *Chronic Diseases in Canada* 28 (1-2): 42-55. http://www.ncbi.nlm.nih.gov/pubmed/17953797.

Boyd, David R., and Stephen J. Genuis. 2008. "The Environmental Burden of Disease in Canada: Respiratory Disease, Cardiovascular Disease, Cancer, and Congenital Affliction." *Environmental Research* 106 (2): 240-9, doi:10.1016/j.envres.2007.08.009.

Brophy, James T., Margaret M. Keith, Kevin M. Gorey, et al. 2006. "Occupation and Breast Cancer: A Canadian Case-Control Study." *Annals of the New York Academy of Sciences* 1076: 765-77, doi:10.1196/annals.1371.019.

Brophy, James T., Margaret M. Keith, Andrew Watterson, et al. 2012. "Breast Cancer Risk in Relation to Occupations with Exposure to Carcinogens and Endocrine Disruptors: A Canadian Case-Control Study." *Environmental Journal* 11: 87, doi:10.1186/1476-069X-11-87.

Brody, Julia Green, Kirsten B. Moysich, Olivier Humblet, et al. 2007a. "Environmental Pollutants and Breast Cancer: Epidemiologic Studies." *Cancer* 109 (S-12): 2667-711, doi:10.1002/cncr.22655.

Brody, Julia Green, Ruthann A. Rudel, Karin B. Michels, et al. 2007b. "Environmental Pollutants, Diet, Physical Activity, Body Size, and Breast Cancer: Where Do We Stand in Research to Identify Opportunities for Prevention?" *Cancer* 109 (S-12): 2627-34, doi:10.1002/cncr.22656.

Brody, Julia Green, Joel Tickner, and Ruthann A. Rudel. 2005. "Community-Initiated Breast Cancer and Environment Studies and the Precautionary Principle." *Environmental Health Perspectives* 113 (8): 920-25, doi:10.1289/ehp.7784.

Bryant, Heather. 2004. "Breast Cancer in Canadian Women." *BMC Women's Health* 4 (S-1): S12, doi:10.1186/1472-6874-4-S1-S12.

California Office of Environmental Health Hazard Assessment. 2010. "Proposition 65 in Plain Language!" http://www.oehha.ca.gov/prop65/background/p65plain.html.

Cancer Prevention Coalition. 2003. "Biography of Samuel S. Epstein." http://www.preventcancer.com/about/epstein.htm.

CCNM (Canadian College of Naturopathic Medicine). 2009. Breast Health Brochure provided to patients at the Robert Schad Naturopathic Clinic, Toronto, ON.

CCS (Canadian Cancer Society). 2012. "Breast Cancer Statistics at a Glance." http://www.cancer.ca/.

CCS/National Cancer Institute of Canada. 2006. *Canadian Cancer Statistics 2006.* Toronto: Canadian Cancer Society/National Cancer Institute of Canada.

Clapp, Richard W., Genevieve K. Howe, and Molly M. Jacobs. 2007. "Environmental and Occupational Causes of Cancer: A Call to Act on What We Know." *Biomedicine and Pharmacotherapy* 61 (10): 631-9, doi:10.1016/j.biopha.2007.08.001.

Cohn, Barbara A., Mary S. Wolff, Piera M. Cirillo, et al. 2007. "DDT and Breast Cancer in Young Women: New Data on the Significance of Age at Exposure." *Environmental Health Perspectives* 115 (10): 1406-14, doi:10.1289/ehp.10260.

Colomeda, Lorelei Anne Lambert. 1996. *Through the Northern Looking Glass: Breast Cancer Stories Told by Northern Native Women.* New York: NLN Press.

Coyle, Yvonne Marie. 2004. "The Effect of Environment on Breast Cancer Risk." *Breast Cancer Research and Treatment* 84 (3): 273-88, doi:10.1023/B:BREA.0000019964.33963.09.

Davis, Devra Lee. 2002. *When Smoke Ran Like Water: Tales of Environmental Deception and the Battle against Pollution.* New York: Basic Books.

–. 2007. *The Secret History of the War on Cancer.* New York: Basic Books.

Dey, Subhojit, Amr S. Soliman, and Sofia D. Merajver. 2009. "Xenoestrogens May Be the Cause of High and Increasing Rates of Hormone Receptor Positive Breast Cancer in the World." *Medical Hypotheses* 72 (6): 652-56, doi:10.1016/j.mehy.2008.10.025.

Din, Ravida (producer), and Léa Pool (director). 2011. *Pink Ribbons, Inc.* [motion picture]. Ottawa: National Film Board of Canada.

Dipple, A., Q.A. Khan, J.E. Page, et al. 1999. "DNA Reactions, Mutagenic Action and Stealth Properties of Polycyclic Aromatic Hydrocarbon Carcinogens (Review)." *International Journal of Oncology* 14 (1): 103-14.

Donovan, Maryann, Chandra M. Tiwary, Deborah Axelrod, et al. 2007. "Personal Care Products that Contain Estrogens or Xenoestrogens May Increase Breast Cancer Risk." *Medical Hypotheses* 68 (4): 756-66, doi:10.1016/j.mehy.2006.09.039.

Drost, Nadja, director. 2005. *Between Midnight and the Rooster's Crow* [motion picture]. Brooklyn, NY: Icarus Films.

Engel, Lawrence S., Deirdre A. Hill, Jane A. Hoppin, et al. 2005. "Pesticide Use and Breast Cancer Risk among Farmers' Wives in the Agricultural Health Study." *American Journal of Epidemiology* 161 (2): 121-35, doi:10.1093/aje/kwi022.

Environment Canada and Health Canada. 1994. *Canadian Environmental Protection Act Priority Substances List Assessment Report: Polycyclic Aromatic Hydrocarbons.* Ottawa: National Printers (Ottawa). http://www.hc-sc.gc.ca/ewh-semt/pubs/contaminants/psl1-lsp1/hydrocarb_aromat_polycycl/index-eng.php.

Environmental Defence. 2006. *Polluted Children, Toxic Nation: A Report on Pollution in Canadian Families.* Toronto: Environmental Defence.

European Commission. 2012. "REACH." http://ec.europa.eu/enterprise/sectors/chemicals/reach/index_en.htm.

Fernandez, Mariana F., Jose M. Molina-Molina, Maria-Jose Lopez-Espinosa, et al. 2007a. "Biomonitoring of Environmental Estrogens in Human Tissues." *International Journal of Hygiene and Environmental Health* 210 (3-4): 429-32, doi:10.1016/j.ijheh.2007.01.014.

Fernandez, Mariana F., Loreto Santa-Marina, Jesus M. Ibarluzea, et al. 2007b. "Analysis of Population Characteristics Related to the Total Effective Xenoestrogen Burden: A Biomarker of Xenoestrogen Exposure in Breast Cancer." *European Journal of Cancer* 43 (8): 1290-99, doi:10.1016/j.ejca.2007.03.010.

Fox, Glen A. 2001. "Wildlife as Sentinels of Human Health Effects in the Great Lakes-St. Lawrence Basin." *Environmental Health Perspectives* 109 (S-6): 853-61. http://www.ncbi.nlm.nih.gov/pmc/articles/PMC1240620/pdf/ehp109s-000853.pdf.

Gammon, Marilie D., Mary S. Wolff, Alfred I. Neugut, et al. 2002. "Environmental Toxins and Breast Cancer on Long Island. II. Organochlorine Compound Levels in Blood." *Cancer Epidemiology, Biomarkers and Prevention* 11 (8): 686-97. http://cebp.aacrjournals.org/content/11/8/686.full.pdf+html.

Garcia-Morales, Pilar, Miguel Saceda, Nicholas Kenney, et al. 1994. "Effect of Cadmium on Estrogen Receptor Levels and Estrogen-Induced Responses in Human Breast Cancer Cells." *Journal of Biological Chemistry* 269 (24): 16896-901. http://www.jbc.org/cgi/pmidlookup?view=long&pmid=8207012.

Giusti, Ruthann M., Kumiko Iwamoto, and Elizabeth E. Hatch. 1995. "Diethylstilbestrol Revisited: A Review of the Long-Term Health Effects." *Annals of Internal Medicine* 122 (10): 778-88.

Gray, Janet, Janet Nudelman, and Connie Engel. 2010. *State of the Evidence: The Connection between Breast Cancer and the Environment/From Science to Action.* 6th ed. San Francisco: Breast Cancer Fund.

Hatch, Elizabeth E., Julie R. Palmer, Linda Titus-Ernstoff, et al. 1998. "Cancer Risk in Women Exposed to Diethylstilbestrol In Utero." *Journal of the American Medical Association* 280 (7): 630-34, doi:10.1001/jama.280.7.630.

Hoy, Claire. 1995. *The Truth about Breast Cancer.* Toronto: Stoddart Publishing.

Hyndman, Brian. 2005. *Strategies for the Reduction and Control of Environmental Carcinogens in Canada: What's Happening? What's Missing?* Toronto: Cancer Care Ontario. https://www.cancercare.on.ca/common/pages/UserFile.aspx?fileId=13506.

IARC (International Agency for Research in Cancer). 1987. *IARC Monographs Supplement 7: Polychlorinated Biphenyls.* http://monographs.iarc.fr/ENG/Monographs/suppl7/Suppl7-129.pdf.

–. 1997. "International Agency for Research on Cancer (IARC): Summaries and Evaluations – Cadmium and Cadmium Compounds." http://www.inchem.org/documents/iarc/vol58/mono58-2.html.

Ibarluzea, Jesus M., Mariana F. Fernández, Loreto Santa-Marina, et al. 2004. "Breast Cancer Risk and the Combined Effect of Environmental Estrogens." *Cancer Causes and Control* 15 (6): 591-600, doi:10.1023/B:CACO.0000036167.51236.86.

Ionescu, John G., Jan Novotny, Vera Stejskal, et al. 2006. "Increased Levels of Transition Metals in Breast Cancer Tissue." *Neuroendocrinology Letters* 27 (S-1): 36-39.

Jang, Tae Jung, Myung Soo Kang, Heesoo Kim, et al. 2000. "Increased Expression of Cyclin D1, Cyclin E and p21(Cip1) Associated with Decreased Expression of p27(Kip1) in Chemically Induced Rat Mammary Carcinogenesis." *Japanese Journal of Cancer Research* 91 (12): 1222-32, doi:10.1111/j.1349-7006.2000.tb00908.x.

Jeffy, Brandon D., Ryan B. Chirnomas, and Donato F. Romagnolo. 2002. "Epigenetics of Breast Cancer: Polycyclic Aromatic Hydrocarbons as Risk Factors." *Environmental and Molecular Mutagenesis* 39 (2-3): 235-44, doi:10.1002/em.10051.

Johnson, Michael D., Nicholas Kenney, Adriana Stoica, et al. 2003. "Cadmium Mimics the *In Vivo* Effects of Estrogen in the Uterus and Mammary Gland." *Nature Medicine* 9 (8): 1081-84, doi:10.1038/nm902.

Kasper, Anne S., and Susan J. Ferguson, eds. 2000. *Breast Cancer: Society Shapes an Epidemic.* New York: St. Martin's Press.

Kellen, Eliane, Maurice P. Zeegers, Elly Den Hond, et al. 2007. "Blood Cadmium May Be Associated with Bladder Carcinogenesis: The Belgian Case-Control Study on Bladder Cancer." *Cancer Detection and Prevention* 31 (1): 77-82, doi:10.1016/j.cdp.2006.12.001.

Kortenkamp, Andreas. 2006. "Breast Cancer, Oestrogens and Environmental Pollutants: A Re-evaluation from a Mixture Perspective." *International Journal of Andrology* 29 (1): 193-98, doi:10.1111/j.1365-2605.2005.00613.x.

Kortenkamp, Andreas, Michael Faust, Martin Scholze, et al. 2007. "Low-Level Exposure to Multiple Chemicals: Reason for Human Health Concerns?" *Environmental Health Perspectives* 115 (S-1): 106-14, doi:10.1289/ehp.9358.

Kriegel, Alison M., Amr S. Soliman, Qing Zhang, et al. 2006. "Serum Cadmium Levels in Pancreatic Cancer Patients from the East Nile Delta Region of Egypt." *Environmental Health Perspectives* 114 (1): 113-19, doi:10.1289/ehp.8035.

Lee-Lin, Frances, Usha Menon, Marjorie Pett, et al. 2007. "Breast Cancer Beliefs and Mammography Screening Practices among Chinese American Immigrants." *Journal of Obstetric, Gynecologic and Neonatal Nursing* 36 (3): 212-21, doi:10.1111/j.1552-6909.2007.00141.x.

Liu, Zhiwei, Xinyuan Yu, and Zahir A. Shaikh. 2008. "Rapid Activation of ERK1/2 and AKT in Human Breast Cancer Cells by Cadmium." *Toxicology and Applied Pharmacology* 228 (3): 286-94, doi:10.1016/j.taap.2007.12.017.

Lowell, J. 1990. "PCBs in Inuit Women's Breast Milk." In *Gossip: A Spoken History of Women in the North,* edited by Mary Crnkovich. Ottawa: Canadian Arctic Resources Committee.

Maffini, Maricel V., Beverly S. Rubin, Carlos Sonnenschein, et al. 2006. "Endocrine Disruptors and Reproductive Health: The Case of Bisphenol-A." *Molecular and Cellular Endocrinology* 254-55: 179-86, doi:10.1016/j.mce.2006.04.033.

Massachusetts Toxics Use Reduction Program. 1997. *Evaluating Progress: A Report on the Findings of the Massachusetts Toxics Use Reduction Program Evaluation.* http://www.turi.org/content/download/7141/122648/file/evaluating%20progress%201997.pdf.

McElroy, Jane A., Martin M. Shafer, Amy Trentham-Dietz, et al. 2006. "Cadmium Exposure and Breast Cancer Risk." *Journal of the National Cancer Institute* 98 (12): 869-73, doi:10.1093/jnci/djj233.

McMurray, Cynthia T., and John A. Tainer. 2003. "Cancer, Cadmium and Genome Integrity." *Nature Genetics* 34 (3): 239-41, doi:10.1038/ng0703-239.

Metzdorff, Stine Broeng, Majken Dalgaard, Sofie Christiansen, et al. 2007. "Dysgenesis and Histological Changes of Genitals and Perturbations of Gene Expression in Male Rats after *In Utero* Exposure to Antiandrogen Mixtures." *Toxicological Sciences* 98 (1): 87-98, doi:10.1093/toxsci/kfm079.

Milligan, Stuart R., Omar Khan, and Marnie Nash. 1998. "Competitive Binding of Xenobiotic Oestrogens to Rat Alpha-Fetoprotein and to Sex Steroid Binding Proteins in Human and Rainbow Trout *(Oncorhynchus mykiss)* Plasma." *General and Comparative Endocrinology* 112 (1): 89-95, doi:10.1006/gcen.1998.7146.

Moore, Rhonda J., Robert M. Chamberlain, and Fadlo R. Khuri. 2004. "Apolipoprotein E and the Risk of Breast Cancer in African-American and Non-Hispanic White Women: A Review." *Oncology* 66 (2): 79-93, doi:10.1159/000077433.

Moysich, Kirsten B., Christine B. Ambrosone, John E. Vena, et al. 1998. "Environmental Organochlorine Exposure and Postmenopausal Breast Cancer Risk." *Cancer Epidemiology, Biomarkers and Prevention* 7 (3): 181-88. http://cebp.aacrjournals.org/cgi/pmidlookup?view=long&pmid=9521429.

Murray, Stephen A., Shi Yang, Elizabeth Demicco, et al. 2005. "Increased Expression of MDM2, Cyclin D1, and p27Kip1 in Carcinogen-Induced Rat Mammary Tumors." *Journal of Cellular Biochemistry* 95 (5): 875-84, doi:10.1002/jcb.20414.

National Toxicology Program. 2002. "Cadmium and Cadmium Compounds." *Report on Carcinogens: Carcinogen Profiles* 10: 42-44. http://toxnet.nlm.nih.gov/cgi-bin/sis/search/r?dbs+hsdb:@term+@na+CADMIUM+COMPOUNDS.

Navarro Silvera, Stephanie A., and Thomas E. Rohan. 2007. "Trace Elements and Cancer Risk: A Review of the Epidemiologic Evidence." *Cancer Causes and Control* 18 (1): 7-27, doi:10.1007/s10552-006-0057-z.

Nie, Jing, Jan Beyea, Matthew R. Bonner, et al. 2007. "Exposure to Traffic Emissions throughout Life and Risk of Breast Cancer: The Western New York Exposures and Breast Cancer (WEB) Study." *Cancer Causes and Control* 18 (9): 947-55, doi:10.1007/s10552-007-9036-2.

NIEHS-NTP (National Institute of Environmental Health Sciences-National Toxicology Program). 2011. "Polychlorinated Biphenyls." In *Report on Carcinogens, Twelfth Edition,* 349-52. Research Triangle Park, NC: National Toxicology Program, Department of Health and Human Services. http://ntp.niehs.nih.gov/ntp/roc/twelfth/profiles/PolychlorinatedBiphenyls.pdf.

No More Breast Cancer. 2006. "Primary Prevention." http://www.nomorebreastcancer.org.uk/primary_prevention.html.

Obana, Hirotaka, Shinjiro Hori, and N. Kunita. 1981. "Polycyclic Aromatic Hydrocarbons in Human Fat and Liver." *Bulletin of Environmental Contamination and Toxicology* 27 (1): 23-27, doi:10.1007/BF01610981.

Parkin, D. Max, Fred I. Bray, and S.S. Devesa. 2001. "Cancer Burden in the Year 2000: The Global Picture." *European Journal of Cancer* 37 (S-8): S4-S66, doi:10.1016/S0959-8049(01)00267-2.

Payne, Joachim, Martin Scholze, and Andreas Kortenkamp. 2001. "Mixtures of Four Organochlorines Enhance Human Breast Cancer Cell Proliferation." *Environmental Health Perspectives* 109 (4): 391-97, doi:10.1289/ehp.01109391.

PHAC (Public Health Agency of Canada). 2003. *Learning from SARS: Renewal of Public Health in Canada: A Report of the National Advisory Committee on SARS and Public Health, October 2003.* Ottawa: Health Canada. http://www.phac-aspc.gc.ca/publicat/sars-sras/naylor/index-eng.php.

PollutionWatch. 2008a. *An Examination of Pollution and Poverty in the Great Lakes Basin.* Toronto: PollutionWatch. http://www.pollutionwatch.org/pub/PW_Pollution_Poverty_Report.pdf.

–. 2008b. "Who Is Polluting?" http://www.pollutionwatch.org.

Prince, Mary M., Avima M. Ruder, Misty J. Hein, et al. 2006. "Mortality and Exposure Response among 14,458 Electrical Capacitor Manufacturing Workers Exposed to Polychlorinated Biphenyls (PCBs)." *Environmental Health Perspectives* 114 (10): 1508-14, doi:10.1289/ehp.9175.

Prüss-Üstün, Annette, and Carlos F. Corvalán. 2006. *Preventing Disease through Healthy Environments: Towards an Estimate of the Environmental Burden of Disease.* Geneva: World Health Organization. http://www.who.int/entity/quantifying_ehimpacts/publications/preventingdisease.pdf.

Raphael, Dennis. 2003. "Addressing the Social Determinants of Health in Canada: Bridging the Gap between Research Findings and Public Policy." *Policy Options* (March): 35-40.

Richter, Elihu D., and J. Safi. 1997. "Pesticide Use, Exposure, and Risk: A Joint Israeli-Palestinian Perspective." *Environmental Research* 73 (1-2): 211-18, doi:10.1006/enrs.1997.3717.

Rudel, Ruthann A., Kathleen R. Attfield, Jessica N. Schifano, et al. 2007. "Chemicals Causing Mammary Gland Tumors in Animals Signal New Directions for Epidemiology, Chemicals Testing, and Risk Assessment for Breast Cancer Prevention." *Cancer* 109 (S-12): 2635-66, doi:10.1002/cncr.22653.

Salehi, Fariba, Michelle C. Turner, Karen P. Phillips, et al. 2008. "Review of the Etiology of Breast Cancer with Special Attention to Organochlorines as Potential Endocrine Disruptors." *Journal of Toxicology and Environmental Health, Part B: Critical Reviews* 11 (3-4): 276-300, doi:10.1080/10937400701875923.

Schönfelder, Gilbert, Werner Wittfoht, Hartmut Hopp, et al. 2002. "Parent Bisphenol A Accumulation in the Human Maternal-Fetal-Placental Unit." *Environmental Health Perspectives* 110 (11): A703-A707, doi:10.1289/ehp.021100703.

Science and Environmental Health Network. 1998. "The Wingspread Consensus Statement on the Precautionary Principle." Wingspread Conference on the Precautionary Principle, 26 January 1998, Racine, WI. http://www.sehn.org/wing.html.

Sherman, Janette D. 1994. "Structure-Activity Relationships of Chemicals Causing Endocrine, Reproductive, Neurotoxic, and Oncogenic Effects: A Public Health Problem." *Toxicology and Industrial Health* 10 (3): 163-79.

Silva, Elisabete, Martin Scholze, and Andreas Kortenkamp. 2007. "Activity of Xenoestrogens at Nanomolar Concentrations in the E-Screen Assay." *Environmental Health Perspectives* 115 (S-1): 91-97, doi:10.1289/ehp.9363.

Smigal, Carol, Ahmedin Jemal, Elizabeth Ward, et al. 2006. "Trends in Breast Cancer by Race and Ethnicity: Update 2006." *CA: A Cancer Journal for Clinicians* 56 (3): 168-83, doi:10.3322/canjclin.56.3.168.

Soto, Ana M., Laura N. Vandenberg, Maricel V. Maffini, et al. 2008. "Does Breast Cancer Start in the Womb?" *Basic and Clinical Pharmacology and Toxicology* 102 (2): 125-33, doi:10.1111/j.1742-7843.2007.00165.x.

Stanford, Janet L., Lisa J. Herrinton, Stephen M. Schwartz, et al. 1995. "Breast Cancer Incidence in Asian Migrants to the United States and Their Descendants." *Epidemiology* 6 (2): 181-83, doi:10.1097/00001648-199503000-00017.

Sweeney, Ellen. 2006. "Breast Cancer: The Importance of Prevention in Public Education." *Women's Health and Urban Life* 5 (1): 75-90. https://tspace.library.utoronto.ca/bitstream/1807/9400/1/sweeney.pdf.

Tang, Deliang, Stan Cho, Andrew Rundle, et al. 2002. "Polymorphisms in the DNA Repair Enzyme XPD Are Associated with Increased Levels of PAH-DNA Adducts in a Case-Control Study of Breast Cancer." *Breast Cancer Research and Treatment* 75 (2): 159-66, doi:10.1023/A:1019693504183.

Terry, Paul, and Kenneth C. Johnson. 2003. "Smoking and Breast Cancer." *British Journal of Cancer* 88 (9): 1500, doi:10.1038/sj.bjc.6600933.

Tokumaru, Osamu, Kosuke Haruki, Kira Bacal, et al. 2006. "Incidence of Cancer among Female Flight Attendants: A Meta-Analysis." *Journal of Travel Medicine* 13 (3): 127-32, doi:10.1111/j.1708-8305.2006.00029.x.

Tyshenko, Michael G., Karen P. Phillips, Michael Mehta, et al. 2008. "Risk Communication of Endocrine-Disrupting Chemicals: Improving Knowledge Translation and Transfer." *Journal of Toxicology and Environmental Health, Part B: Critical Reviews* 11 (3-4): 345-50, doi:10.1080/10937400701876293.

US EPA (United States Environmental Protection Agency). 2012a. "Environmental Justice." http://www.epa.gov/environmentaljustice/index.html.

–. 2012b. "Polychlorinated Biphenyls (PCBs): Basic Information." http://www.epa.gov/epawaste/hazard/tsd/pcbs/pubs/about.htm.

Vinceti, Marco, Marianna Venturelli, Chiara Sighinolfi, et al. 2007. "Case-Control Study of Toenail Cadmium and Prostate Cancer Risk in Italy." *Science of the Total Environment* 373 (1): 77-81, doi:10.1016/j.scitotenv.2006.11.005.

Vom Saal, Frederick S., Bensen T. Akingbemi, Scott M. Belcher, et al. 2007. "Chapel Hill Bisphenol A Expert Panel Consensus Statement – Integration of Mechanisms, Effects in Animals and Potential to Impact Human Health at Current Levels of Exposure." *Reproductive Toxicology* 24 (2): 131-38, doi:10.1016/j.reprotox.2007.07.005.

Weiss, Helen A., Nancy A. Potischman, Louise A. Brinton, et al. 1997. "Prenatal and Perinatal Risk Factors for Breast Cancer in Young Women." *Epidemiology* 8 (2): 181-87, doi:10.1097/00001648-199703000-00010.

Ye, Xiaoyun, Zsuzsanna Kuklenyik, Larry L. Needham, et al. 2006. "Measuring Environmental Phenols and Chlorinated Organic Chemicals in Breast Milk Using Automated On-Line Column-Switching-High Performance Liquid Chromatography-Isotope Dilution Tandem Mass Spectrometry." *Journal of Chromatography B: Analytical Technologies in Biomedical and Life Sciences* 831 (1-2): 110-15, doi:10.1016/j.jchromb.2005.11.050.

Part 4: Consumption in the Production Process

Chapter 10 Plastics Industry Workers and Breast Cancer Risk: Are We Heeding the Warnings?

Margaret M. Keith, James T. Brophy, Robert DeMatteo, Michael Gilbertson, Andrew E. Watterson, and Matthias Beck

> I worked at the plastics plant for five years and then developed breast cancer when I was 32. There are six or seven breast cancer [cases] that we know of. They are all younger than 50. (Participant from Focus Group 5)

Are women who work in the plastics injection moulding industry – particularly those producing automotive parts – at greater risk for breast cancer? There are important social and scientific barriers to answering this question, such as gender and class discrimination, and weaknesses in traditional epidemiological and toxicological research approaches. To address these challenges, we implemented a mixed-method research approach to examine the nature and extent of plastics workers' occupational exposures to carcinogenic and endocrine-disrupting chemicals (EDCs). We reviewed the scientific literature as well as findings from primary research that we conducted using epidemiological and qualitative methods. Through the personal stories and observations of workers who participated in focus groups, and the review of a collection of hygiene inspection reports, we learned that: (1) women in the study area (Essex County, Ontario) had held a wide range of jobs in the plastics industry dating back to the 1960s; (2) the majority of automotive plastics manufacturing workers in the study area were women; (3) the work environment was heavily contaminated with dust, vapours, and fumes; (4) there had been a historic failure by government regulators to control exposures; (5) workers received a steady dose of mixtures of chemicals through inhalation, absorption, and ingestion; (6) workers were getting sick; and (7) society was largely unaware of their plight.

The apparent invisibility of blue-collar women raises issues of gender and class bias and discrimination. It has been argued that blue-collar workers in general (Infante 1995) and blue-collar women in particular barely register on society's radar (Firth, Brophy, and Keith 1997; Messing 1998; Chapter 12). We are seemingly blind to the daily miseries and health risks caused by the hot, smelly, smoky vapours permeating plastics workers' occupational environment. Why has there been such concern about consumer exposures to EDCs, such as bisphenol A (BPA) in baby bottles (Smith and Lourie 2009) but negligible attention to the exposures of plastics workers?

Economic Profile of the Plastics Industry

Exploring possible breast cancer risks associated with exposures in the automotive plastics industry may have implications for thousands of women. According to Industry Canada (2010), plastics manufacturing generates $20.7 billion annually and employs about 91,000 people in Canada, primarily in small and medium-sized firms with a low level of unionization. The majority (48 percent) of plastics firms are located in Ontario, where about 51,000 people are employed. The automotive component, which comprises about 18 percent of the overall industry, dominates plastics manufacturing in Essex County, which is generally regarded as the automotive capital of Canada. While plastics manufacturing is on the rise in Canada in general, employment in the manufacturing of plastic parts for the automotive sector has declined from about 18,000 in 2003 to an estimated 13,000 in 2009 (Industry Canada 2010). Plastics still make up a substantial percentage of today's vehicles, which contain on average approximately 150 kilograms of plastic components (American Chemistry Council 2010).

In the course of our study, plastics workers told us that women make up the majority of the workforce in the automotive parts industry in the Essex County area. They also indicated that, because it is a precarious industry, which has seen a steady decline in recent years, there is fear of plant closure and ensuing job loss. This fear has had a chilling effect on efforts to gain occupational health improvements. As a long-term employee of one of the plastics injection moulding plants explained:

> *If we were to tell our company right now that we want to look at local ventilation, we'd be closed down and moved to the United States – just like that. Our plant's not doing well economically. We just can't bring it up. (Participant from Focus Group 1)*

The Plastics Production Process

The array of rapidly changing inventories of ingredients used to produce plastics make it more difficult to identify which specific exposures might increase breast cancer risk. The raw materials used in plastics production are a varied group of petroleum-based substances. There are hundreds of plastic compounds or polymer and co-polymer (combined polymer) resins. In addition, commercial plastics have one or more supplementary agents called additives. These include: plasticizers, flame retardants, heat stabilizers, anti-oxidants, light stabilizers, blowing agents, anti-static agents, initiators, lubricants, curing agents, colourants, fillers and reinforcing agents, solvents, and brighteners. An Ontario Ministry of Labour (MOL) inspection report on worksite conditions at a Windsor automotive plastics plant describes some of the myriad by-products detected through air sampling:

> Fumes were strong and several workers developed symptoms of nausea, dizziness, and headache. Air concentration tests show presence of hydrocarbons and halogenated hydrocarbons: toluene, MEK [methyl ethyl ketone], acetone, alcohol and xylene. (MOL inspection report, 1987)

The production process involves the use of various thermal moulding techniques that typically involve blending and mixing additives; melting resins in the moulding machines; and forcing the melted product under pressure through dies and rollers or into moulds for a desired shape. Injection moulding, the most prevalent process used in the study area, is described as follows:

> In this process, plastics granules [pellets] or powders are heated in a cylinder (known as the barrel), which is separate from the mold. The material is heated until it becomes fluid, while it is conveyed through the barrel by a helical screw and then forced into the mold where it cools and hardens. The mold is then opened mechanically and the formed articles are removed. (Law and Britton 1998)

Why We Need to Explore Possible Links between Work in Plastics and Breast Cancer

As discussed in Chapters 8 and 9, breast cancer is the most commonly diagnosed cancer in women in Canada (Canadian Cancer Society 2011). Many of these cases cannot be explained by genetic susceptibility or the so-called lifestyle risk factors, such as poor diet, excessive consumption of alcohol, and lack of exercise (Health Canada 2001). The incidence of breast cancer rose sharply in the second half of the twentieth century, in the same time frame that saw women entering the workforce in record numbers. This increase also coincided with the dramatic introduction of thousands of new chemicals into industrial production (Epstein, Steinman, and LeVert 1997). To date – perhaps due in part to limited research on occupation-specific breast cancer – there is no firm scientific consensus that workers' exposures in the plastics industry pose an elevated breast cancer risk (IARC 1999). A European study by Lithner, Larsson, and Dave (2011) developed a model for ranking the hazards of chemicals used in the manufacturing of polymers. The highest rankings were assigned to those polymers made from the basic components of plastics called "monomers," which were identified as mutagenic and/or carcinogenic. These evaluations can be used to inform decisions regarding the reduction of exposures and risks, including product substitution and/or proscription of activities involving the most hazardous compounds.

The exclusion of women from many occupational health studies is one of the barriers to the determination of work-related breast cancer risk (Zahm and Blair 2003). Results of occupational cancer research carried out on male cohorts are not necessarily transferable to women, nor can they accurately reflect the hazards facing women workers (Chapters 3 and 11). As suggested throughout this volume (see the Introduction and Chapters 1, 3, 4, 5, 7, and 8), important biological gender differences, such as body size, amount of adipose tissue, reproductive organs, hormones, heart function, and others, can impact the effects of toxic agents on the body (Messing et al. 2003). There are also customary divisions of labour within many work environments, including plastics manufacturing. We learned, for example, that women tend to make up most of the workforce in the machine operating and decorating departments, where more direct exposure tends to occur, while men hold the majority of the skilled trades and maintenance positions. There also appears to be inequality in attention paid to workers' health and safety concerns. We were told that in the study area, a predominantly male workforce occupies the only plastics plant with state-of-the art ventilation.

Another major obstacle to understanding risks to women in blue-collar jobs is the traditional epidemiological methods used to analyze the relationship between occupational exposures and disease. The limited statistical power of many studies is insufficient to detect risk or underlying relationships (Gennaro and Tomatis 2005). When Goldberg and Labrèche (1996) reviewed the numerous studies on the occupational risks for breast cancer, they found that the excess risks that were detected were not generally statistically significant because of the small number of cases available for analysis.

To help to address the gaps in occupational breast cancer research, a multidisciplinary team, made up of the authors of this chapter and co-investigators, carried out a series of epidemiological and qualitative research studies between 1995 and 2010. One of our areas of interest was the automotive plastics industry in the Windsor, Ontario, area.

Biological Plausibility of an Association between Plastics Exposure and Breast Cancer

In attempting to research any associations between occupational exposures to plastics and incidence of breast cancer, we are confronted with two apparently conflicting paradigms. In traditional toxicology, it is assumed that the higher the dose, the greater the effect. However, in the case of substances that function by disrupting the endocrine system, the timing of exposure, even in relation to extremely low concentrations, may be more important (Kortenkamp et al. 2011; Vandenberg et al. 2012; Chapters 1-4, and 7). Despite these difficulties in methodology, we know that several plastics, additives, and related solvents – present in various concentrations and mixtures in the plastics industry – have been identified as mammary carcinogens in traditional animal studies (Rudel et al. 2007). Cancer has been described as a multifactorial process involving initiation, promotion, and progression (Clapp, Jacobs, and Loechler 2008; Trosko and Upham 2005). Only a few chemicals are complete carcinogens that are able to not only initiate cancer cells but also induce all of the "hallmarks" needed for the progression of disease (Hanahan and Weinberg 2000, 2011). The complex mixture of chemicals in the plastics industry work environment likely increases the probability of contributing to all of these hallmarks.

Similarly, based on endocrine disruptor theory, we know that one of the contributors to the development of hormonally related cancers such as breast cancer is exposure to chemicals that mimic estrogen (Crisp et al. 1998; Davis et al. 1993), as discussed in Chapters 7 to 9. Many plastics and related chemicals have been identified as EDCs and, as such, are known to interfere with normal hormone metabolism and functioning (Diamanti-Kandarakis et al. 2009). It is important to emphasize that EDCs can have effects at infinitesimally low levels, sometimes as low as parts per trillion (Myers 2002). To illustrate the potency of these tiny amounts, one part per trillion would be the equivalent of one drop of gin placed in 660 rail tank cars full of tonic, extending along the tracks for six miles (Colborn, Dumanoski, and Myers 1996, 40). These substances may not exhibit the traditional linear dose-response curve. Rather, there is a consensus among endocrine researchers that low doses may exert a greater effect than higher doses (Vandenberg et al. 2012; Diamanti-Kandarakis et al. 2009; Welshons et al. 2003; Chapters 3, 4, and 7).

Combined exposures to different classes of EDCs, such as those found in the plastics industry, may produce additive or even synergistic effects (see Chapters 1, 3, 4, 9, and 11). Studies that investigated the link between endocrine disruptors and breast cancer have produced inconsistent results where exposure to a single chemical was measured, whereas those that considered several concomitant exposures tended to demonstrate a consistent positive link (Ibarluzea et al. 2004; Payne, Scholze, and Kortenkamp 2001). Kortenkamp (2008) has suggested that the knowledge that has been gained about the impacts of low-dose multiple endocrine disruptors in laboratory settings has not yet been fully reproduced in epidemiological studies. This situation may have significant health implications for plastics workers, whose occupational environment is, as one worker called it, "a mix of smoke and smells from heated plastics" (participant from Focus Group 6).

Two Revealing Local Health Surveys of Plastics Workers

Our interest in the possible association between the plastics industry and cancer first arose in 1979. That was the year in which health and safety activists formed a local coalition, Windsor Occupational Safety and Health (WOSH). One of the founding members was Barbara Wimbush, a union health and safety representative from a local plastics injection moulding plant (Brophy and Jackson 1980). Barbara was very concerned about the health problems among her female co-workers, which included "not only headaches and nausea, but also hair and skin discolouration, breathing difficulties, loss of feeling in the hands, and severe nosebleeds" (Brophy and Jackson 1981, 32).

With the help of a physician associated with the Ontario Federation of Labour, WOSH formulated a health questionnaire to document symptoms. Union health and safety representatives distributed the questionnaire to 300 workers in five plants engaged in plastics injection moulding. The 150 returned questionnaires showed a startling picture of ill health among a relatively young worker population. Ninety-four percent said they had symptoms they attributed to their work, such as nosebleeds and nausea. Brophy and Jackson (1981), who publicized the findings, noted that at one plant 71 percent of the respondents said they had experienced chest pains, 88 percent dizziness, and 41 percent blackouts.

These findings, in the midst of increasing public awareness about the health risks in area workplaces, prompted the local Canadian Broadcasting Corporation television station to produce a thirty-minute documentary titled *Dying for Work* (CBC 1981). Two epidemiologists from the US National Institute for Occupational Safety and Health (NIOSH), Dr. Peter Infante and Dr. Joseph Wagner, were interviewed. When questioned about exposure to plastics vapours, particularly vinyl chloride, Dr. Wagner warned:

> When we are dealing with an agent that causes cancer, it would be a gross injustice on my part or anyone who is knowledgeable in the field of cancer research to say that I can tell you what a safe level is. We cannot ... If the group is exposed to vinyl chloride we can expect the typical pattern of cancer that has occurred in populations that have been studied to date. (CBC 1981)

Unfortunately, the auto industry experienced one of its periodic economic downturns just as these revelations were being made. Brophy and Jackson (1981) reported that workers and union officials were hesitant to press for improvements because of the fear of plant closure and job losses.

In 1994, personnel from the Occupational Health Clinics for Ontario Workers (OHCOW) and Windsor Occupational Health Information

Service (WOHIS) – an offshoot of WOSH – collaborated with the Canadian Auto Workers (CAW) union in conducting another survey. Approximately 500 questionnaires were distributed and over 330 were returned (OHCOW and WOHIS 1996). The average age of the respondents was thirty-five years; 68 percent were below the age of forty; 73 percent were women; 72 percent of the workers worked with heated plastics; and 65 percent reported working with solvents. Notable was the prevalence of nosebleeds (26 percent), dizziness (49 percent), nausea or vomiting (34 percent), and unusual tiredness (54 percent). Although the survey was oriented towards acute symptoms, eleven cases of cancer were also identified. The mean age at diagnosis of cancer was thirty-three. In addition, thirty-nine women reported reproductive problems.

Our Subsequent Occupational Breast Cancer Research

Half a decade later, the authors and co-investigators launched a series of studies that explored the hypothesis that occupational exposures might be placing women at a higher risk for breast cancer. In 2000, we undertook a collaborative case-control study called Lifetime Occupational Histories Record (LOHR), funded by the Workplace Safety and Insurance Board and co-sponsored by the Windsor Regional Cancer Centre, University of Windsor, WOHIS, and OHCOW. The results showed that the 564 women with breast cancer who were interviewed for the study were nearly three times more likely to have worked in agriculture than the 599 controls drawn randomly from the population. There was a statistically significant quadrupling of the risk for those who worked in agriculture and subsequently worked in automotive-related manufacturing (Brophy et al. 2006). The latter finding is particularly relevant to our plastics manufacturing hypothesis. We were intrigued by the observation that many of the women were very young when they began working on the farm. The findings of an increased risk for breast cancer in this cohort suggest that the timing of exposures may be an important factor and that there may have been a compounding of their risk as a result of subsequent exposures. In other words, we hypothesize that agents or conditions in the farming environment may influence breast development, particularly in biologically vulnerable individuals such as children or adolescents, and predispose them to increased risk from subsequent occupational exposures to carcinogens in automotive-related industries.

While the findings from the LOHR study were provocative, they lacked the specificity needed to identify particular causative exposures or conditions, that is, modifiable risk factors. With the goal of exploring actual exposures in more detail, we launched two integrated follow-up studies: a comprehensive case-control study, Lifetime Histories Breast Cancer Research (LHBCR), and a qualitative study, the Exposure Exploration Study, designed to provide us with a deeper understanding of the historical working conditions experienced by women employed in key occupations, including the plastics industry. The studies were again conducted with the cooperation of community partners. Both were funded by the Canadian Breast Cancer Foundation – Ontario Region.

In 2003, we began data collection for the LHBCR case-control study utilizing a comprehensive interviewer-administered questionnaire that we adapted from previously tested occupational cancer and/or breast cancer survey instruments (Institute for Survey Research 1998; Siemiatycki 1991; Stewart et al. 1996). The interviews, each of which took up to two hours to complete, gathered information about the subjects' socio-economic status, lifestyle, reproductive history, family history, parental occupations, residency, and work history. The questionnaire also included many open-ended questions about working conditions and occupational exposures. During the six years of data collection, we interviewed 1,006 women with breast cancer and 1,146 randomly selected controls from the community. A total of 184 of the study participants worked in the plastics industry at some point in their lives. The study found that the risk for breast cancer more than doubled among women who had worked in automotive plastics manufacturing for ten years and who had been assessed as having been highly exposed to EDCs and/or carcinogens (OR = 2.68; 95 percent CI 1.47-4.88). The risk rose to almost fivefold for pre-menopausal women (OR = 4.76; 95 percent CI 1.58-14.4). The risk for women who worked in food canning, where it is plausible that they were exposed to BPA from can linings, also more than doubled (OR = 2.35; 95 percent CI 1.00-5.53). Their risk for pre-menopausal breast cancer rose to more than fivefold (OR = 5.70; 95 percent CI 1.03-31.5) (Brophy et al. 2012).

Although the case-control study participants were able to provide us with some exposure information, there were limitations to their recollections. For example, some were unable to recall precise details or lacked knowledge about the chemicals being used or their concentrations. Furthermore, we had no air sampling data with which to make direct exposure estimates. We therefore had to use indirect measures in our effort to reconstruct the history of the subjects' likely exposures. To accomplish this, we undertook the Exposure Exploration Study in 2007, which utilized a qualitative research methodology. This decision was based on the observation that, for purposes of exploring day-to-day experiences in workplace and community settings, such an approach can produce a richness and depth of understanding that cannot be achieved with conventional epidemiology alone (Brown 2003).

Because cancer can have a latency period between exposure and diagnosis, ranging anywhere from several years to several decades, it was essential for us to understand the working conditions dating back thirty or more years. With the help of local unions and the CAW national office, we recruited twenty-seven individuals ranging in age from thirty-four to sixty-two from thirteen different local plastics plants. We held six separate focus groups over a two-year period. One of the key data-gathering techniques employed was *hazard mapping* (Keith et al. 2001): we asked participants to draw their workplaces on large pieces of paper and to include workflow, position of their machines or task areas, nearby processes that might contribute to their exposures, sources of ventilation, and other relevant details. Like other mapping proponents, we have found that graphic representations help participants to recall details and more clearly describe conditions, even those in the distant past (Keith and Brophy 2004). The focus groups were audio-recorded and transcribed. The transcripts revealed a breadth of experiential accounts that painted a vivid picture of work in the automotive plastics industry and of the timing, degree, and types of exposures the workers encountered. For example, in describing conditions as they were a decade ago, a worker told us:

> *There would be lots of fumes. The safety alarms were shut off. We ran polyvinyl chloride, lead, and chromium and silica – all designated substances. And we had no control programme ... That is why we wanted ventilation. We were told, "You don't work in a flower shop." (Participant from Focus Group 2)*

Several participants would often tell similar stories, which served to bolster our confidence in the accuracy of various recollections. We have selected a number of the transcribed comments that best reflect important themes. These excerpts, which provide snapshots of individual personal experiences, are attributed to participants by their group numbers rather than by names or other identifiers. We have supplemented the information provided by the participants with a review of relevant scientific literature, as well as several government and industrial hygiene inspection reports (DeMatteo et al. 2012).

We also reviewed a small collection of material safety data sheets (MSDSs), which were provided to us by union representatives. Unfortunately, in Canada the law requires that MSDSs be provided only for currently used materials. There is no legal requirement to preserve older MSDSs or maintain them in a registry. As a result, most of the older MSDSs had been discarded, making it difficult to know whether a present-day health problem might be related to a past exposure. A union health and safety representative told us that "a person was hired to shred all the pre-1986 MSDSs" (participant from Focus Group 4). Another worker noted: "In three years they [MSDSs] are gone and then you never see them again ... you can be exposed to something 15 years ago and now how can you make a link?" (participant from Focus Group 1). We were told of an instance in which no MSDS was provided for a particular ingredient added to a material as a filler because, as the workers were informed by their supervisor, "it was a 'trade secret'" (participant from Focus Group 4). In addition, many available MSDSs were criticized by workers for being incomprehensible or incomplete.

The assessment of precise exposure and associated risk was also hampered by the reliance of government inspectors on air sampling, a system that is flawed at best. Critics contend that legislated exposure limits are set at levels that are economically achievable by industry rather than those that will protect human health (Castleman and Ziem 1988; Roach and Rappaport 1990; Chapter 3). Moreover, these limits do not take into account the possible effects at very low doses that are characteristic of endocrine disruptors (Vandenberg et al. 2012). It is not surprising that in both the scientific literature and personal experience, workers' complaints are seldom validated by meter readings. As one worker reported:

> *Committees have had the ministry come in and do testing and it's never over the exposure limits ... We would run ABS [acrylonitrile-butadiene-styrene] ... and there were a lot of issues with people suffering from symptoms and the test results always came back under what was allowed. (Participant from Focus Group 1)*

Air monitoring can also underestimate true body burdens, which are the measurable concentrations of chemicals in the body. For example, a study we reviewed regarding acrylonitrile, a common plastics component, indicated that levels measured in air samples taken by industrial hygienists were below occupational exposure limits. However, blood and urine concentrations were significantly higher in exposed workers than concentrations measured in referent groups (Houthuijs et al. 1982). There were similar findings in other reviewed studies measuring styrene (Brugnone et al. 1993), phthalates (Hines et al. 2008), brominated flame retardants (Thuresson, Bergman, and Jakobsson 2005), and BPA (He et al. 2009). In many cases, the levels found in workers were far higher than those found in the general population. It is important to note that these general population levels were found to produce dramatic adverse health effects, including breast tumours in laboratory animals.

Accurate exposure assessment for individual plastics workers was further complicated by the "toxic soup" found in most plastics industry environments, as well as by the multiple job responsibilities typical of small plastics operations. Focus group participants informed us that it was a common practice to move from process to process in the course of their work, thereby increasing the opportunities for exposure to an array of substances. On any given day, there might be several types of materials being used. We were told:

> *We rotate throughout our entire plant ... we are pretty much being exposed to different materials every day ... like one machine was ABS, another machine was nylon and they were ten feet away from each other. (Participant from Focus Group 1)*

> *Different jobs would run simultaneously. There were side-by-side machines using different pellets for different products. (Participant from Focus Group 6)*

> *PVC [polyvinyl chloride] was probably the worst type of material. We did have other material as well. We used ABS, TPO [thermoplastic olefins], and acrylic. (Participant from Focus Group 1)*

Resins are brought into the plant in the form of pellets, bulk powders, granules, and liquids or syrups. Workers can be exposed during the manual handling of drums or bags, including opening, pouring, scooping, and closing drums, or the manual blending, mixing, and pouring done in preparation for thermal processing. As the melted resin is forced into moulds or extruded through dies or rollers, various gases and vapours are released from the moulding compound and are often vented directly into the work environment. In addition, drilling, grinding, sanding, and buffing all produce significant concentrations of dust. As two government inspection reports of plastics grinding areas noted:

> Problem with grinder grinding parts ... The product being ground was polypropylene modified with talc, rubber, and aluminum. (MOL inspection report, n.d.)

> Dust from regrind machine was being blown into wire bins and escaping in surrounding area. Large quantities of dust in area observed by inspector. (MOL inspection report, n.d.)

Prepping, painting, and decorating can expose workers to solvents, paints, glues, and bonding agents. As workers recalled:

> *While on assembly near decorating, the parts were frequently spray-painted with gray paint by an automated sprayer. Since we were close by, we would also get a good dose of spray paint all over us. It was everywhere. We would look like the "Tin Man" in the Wizard of Oz. (Participant from Focus Group 5)*

> *We used to have some glue, some nasty glue ... I remember smelling it and one lady, she would just shake; it was from that stuff ... there was no ventilation, it was terrible. I couldn't stand it. I would have to go outside; you would get high. (Participant from Focus Group 1)*

Inadequate engineering controls can contribute to these exposures. For example, some inspection reports that we reviewed noted a problem with ventilation of the gluing spray booth at one of the automotive plastics plants. It was revealed that for over ten years the plant had exhausted contaminated air from the gluing booths into the general workplace. As the reports note:

> Exposures to volatile organic compounds from spray glue operation are high and workers will require respiratory protection

> when working in the east booth. Currently the exit for the booth exhaust fan is inside the plant and air is re-circulated. (Industrial hygiene report, 1990)

> Inspector responds to work complaint of illness from spray booth exhaust and issues an order to exhaust booth air to the outside as required by Industrial Regulation. (MOL inspection report, 1990)

In the purging (removing polymer from moulding machines), cleaning, and maintenance of the various moulding machines, there can be significant exposures to emissions of various resins and solvents. Many workers attested that purging contributed significantly to the environmental concentration of thermal degradation products:

> *You would pull the screw out ... and this would just shoot the molten plastic there and then it would drop on the floor to smoke. (Participant from Focus Group 6)*

> *When they purged the machines, hot stinky gunk would sit there and off-gas. (Participant from Focus Group 3)*

> *In the winter some days are worse than others. Depends if they just purged and they don't take the purge away or if they have gone from white to black and it takes forever to purge; that stuff is going nowhere or they don't bother to turn that one fan on that we actually have. (Participant from Focus Group 1)*

Exposures were compounded when equipment failed to operate correctly, which workers told us was a common occurrence:

> *Last Thursday I had some smoke on my machine. I was told it was probably some residue burning off; this lasted all day. They bypass the safeties on the machines; it is very common for them to do. (Participant from Focus Group 5)*

> *I looked behind the mold and I could see a big cloud of smoke and there was a fire and ... the smoke is clearing and here is one of our workers standing in the middle of it; you couldn't even see her and it was just plastic burning. (Participant from Focus Group 1)*

> *The PVC was over-processed so it would decompose; my grinder would smoke; it was emitting all this stuff and I would be standing there continually working and I didn't know any better. (Participant from Focus Group 1)*

> *The MSDS would indicate that something should be run at 400 degrees; we would run it at 475 degrees. A lot of time the water is off and the "thermolators" [temperature control units] break down and that affects the temperature. (Participant from Focus Group 2)*

> *PVC only ran on one machine but when it overheated it was bad. You had to go inside the machine to pull out the parts. (Participant from Focus Group 2)*

> *It's wear and tear on the nozzle ... the plastic is not shooting through ... and it's burning, it's clogged up and oozing out ... that plastic is heating up ... the MSDS says it's great at 300 degrees, but what happens when it gets to 500 or 600 degrees? (Participant from Focus Group 1)*

Government inspectors also noted excessive emissions related to machine malfunctions:

> At one ABS injection molding machine, the injection barrel cover flaps near the exhaust ventilation were both open and could not be closed ... In the operation of injection molding equipment and maintenance of the dies, plastic fumes and smoke are often produced when the plastic is overheated and begins to breakdown [sic] chemically. (MOL inspection report, 1989)

> Smoke from excess plastics built up at the feed end of molding machine and fumes from normal operation. Use of acrylics, ABS, PP [polypropylene], PE [polyethylene] in injection molding. Fumes from use of welding torch to cut away PE, PP build up on nozzles which last about an hour and occur about 4 hours per week. Fifteen molding machines with 20 workers on 3 eight-hour shifts, three times per week. (MOL inspection report, 1978)

Our research revealed that although union activism has resulted in improvements in some plants, in many others the conditions remain much as they were decades ago. Government inspectors would occasionally write up recommendations or orders to improve specific problem areas, but overall the air quality – and therefore the airborne exposures – continue to be of considerable concern to the workers. A machine operator complained to us: "Nothing has really changed. They just added a few ceiling fans – no local ventilation whatsoever … when it gets bad they just open the bay door" (participant from Focus Group 5). This observation was substantiated by an inspection report from 2004 that stated: "Mold injection units are not equipped with local exhaust ventilation."

From an industrial hygiene perspective, complaints of odours, respiratory problems, and skin irritations are considered to be an indication that workers are breathing in or handling hazardous substances. As one worker recalled:

> *You can smell it and by the end of the night your head is just banging, your throat is dry and burning, your eyes are watering and you are doing this for seven hours. (Participant from Focus Group 5)*

Another worker commented:

> *I think almost everyone in our plant had their upper respiratory [systems] compromised. There is no doubt in my mind, whether they get bronchitis or asthma or they get colds they can't get rid of all winter long; nose bleeds too. (Participant from Focus Group 1)*

A hygiene inspector noted a worker's health symptoms in his report (undated):

> Worker complained that every time she puts a part in the grinder, dust would come flying out at her which caused her to get itchy. She also has respiratory problems and could feel dust going up her nose and in her lungs.

Why Exposures in Plastics Production Are of Concern in Relation to Breast Cancer

The workers' and inspectors' descriptions of exposures in the plastics industry work environment become all the more disturbing in view of the epidemiological and toxicological literature suggesting a link between exposures and breast cancer. While epidemiological research into the possible association between breast cancer and plastics manufacturing remains relatively scant, the following review indicates that evidence is beginning to mount:

- The earliest epidemiological evidence of an association dates back to 1977, when elevated breast cancer mortality was found among PVC-fabricating workers (Chiazze, Nichols, and Wong 1977). Subsequent analyses of the data found that the risk was statistically non-significant (Chiazze et al. 1980; Chiazze and Ference 1981). The authors found that the study had insufficient statistical power to confirm the association because of the small number of subjects.
- A more recent study did identify a statistically significant increase (almost double) in the risk of breast cancer in plastics and rubber industry workers; the same study showed an elevated risk among workers exposed to organic solvents and benzene (Petralia et al. 1998).
- Two other studies reported an increased breast cancer risk approaching statistical significance among rubber and plastics workers (Gardner et al. 2002; Ji et al. 2008). Adding weight to this association is a study that demonstrated a more than quadrupling of breast cancer risk among male workers in the rubber and plastics industries, although the findings fall just short of statistical significance (Ewertz et al. 2001). This finding is especially noteworthy considering that breast cancer among men is a rare occurrence and is not complicated by traditional hormonal risk factors.
- Labrèche and colleagues (2010) linked elevated risk of breast cancer with odds ratios (OR) of 7.69 for occupational exposure to acrylic fibres and 1.99 for nylon fibres when exposures occurred before age thirty-six. They found that exposure to acrylic and rayon fibres and monoaromatic hydrocarbons doubled the OR of estrogen/progesterone-positive tumours. A threefold increase was linked to exposure to polycyclic aromatic hydrocarbons (PAHs). This study is particularly relevant because modern textiles typically consist of synthetic fibres made from acrylic, nylon, rayon, and polyesters. These are essentially plastic resins treated with additives such as plasticizers and flame retardants, many of which are recognized mammary carcinogens and endocrine disruptors (Rudel et al. 2007). The finding of an increased risk related to exposure before age thirty-six is also notable because breast tissue is particularly vulnerable before it reaches maturation through the completion of the first

full-term pregnancy (Brody et al. 2007; Clark, Levine, and Snedeker 1997; Cohn et al. 2007; Fenton 2006).

- Villeneuve and colleagues (2011) found an almost twofold increase in breast cancer risk among French plastics and rubber product makers.

Turning to toxicology, there is a further body of evidence that may be relevant to our breast cancer hypothesis. The following review (which should not be considered comprehensive) examines some of the many cancer-causing and endocrine-disrupting agents that are typically found in plastics manufacturing. Hazardous monomers are frequently present in common polymers (Burgess 1982). A recent study ranked fifty-five polymers used in plastics production. The highest ranking was applied to fifteen substances that were carcinogenic and/or mutagenic, including polyvinyl chloride, styrene-acrylonitrile, and acrylonitrile-butadiene-styrene (ABS) (Lithner, Larsson, and Dave 2011).

- Vinyl chloride monomers can be released during PVC production. Vinyl chloride is classified by the International Agency for Research on Cancer (IARC 2011) as carcinogenic to humans (Group 1). It has also been shown to be a mammary carcinogen in animals (Rudel et al. 2007).
- Although styrene use has declined over the years, studies done in the 1990s showed significant exposure among plastics workers (Jensen et al. 1990). Plastics workers had higher bioaccumulation of styrene compared with the general population (Galassi et al. 1993). Styrene is classified by the IARC (2011) as possibly carcinogenic to humans (Group 2B), and is shown to cause mammary gland tumours in animal studies (Rudel et al. 2007). It also acts as an EDC (Gray et al. 2009).
- Thermal degradation of ABS during normal plastics processing results in acrylonitrile monomer in the air. As previously mentioned, measurements of urinary concentrations indicate a significantly higher body burden in plastics workers compared with unexposed control groups (Houthuijs et al. 1982). Acrylonitrile is classified by the IARC (2011) as possibly carcinogenic to humans (Group 2B) and has been shown to be a mammary carcinogen in animals (Rudel et al. 2007). In addition, it is linked to genital abnormalities in children born to exposed mothers, and may have xenoestrogenic potential (Czeizel, Hegedüs, and Tímár 1999). Acrylonitrile is also linked to an increase in lymphocyte (white blood cell) counts, severe liver damage, lung cancer, and increased genetic aberrations in exposed

workers during chromatid and chromosome exchange (Major et al. 1998; Scélo et al. 2004). The presence of acrylonitrile and co-contaminants is documented in a 1995 MOL laboratory report: "Finding of lab test of ABS pellets used in production in response to worker complaints about ABS injection molding: acrylonitrile as well as benzene, styrene, acetaldehyde, xylene, and toluene were released."

Numerous additives are used in the production of plastics. The following are of concern due to their carcinogenic and/or endocrine-disrupting potential:

- Phthalates are a broad class of substances used to make plastics soft and pliable (see Chapter 7). They can be released into the air by high-temperature processing or malfunctions. The potential estrogenic action of di(2-ethylhexyl) phthalate (DEHP) used to plasticize PVC may be linked to incidence of male breast cancer, testicular cancer, and adverse pregnancy outcomes among PVC-fabricating workers (Ahlborg, Bjerkedal, and Egenaes 1987; Ewertz et al. 2001; Hardell, Ohlson, and Fredrikson 1997). A recent study of male PVC workers in Taiwan found adverse effects on semen quality among men with the highest concentrations of DEHP (Huang et al. 2011). A study of a phthalate-exposed population in Northern Mexico revealed elevated breast cancer risk among women (López-Carrillo et al. 2010). Several focus group participants observed comparable health phenomena: "I know in our plant we had three guys with testicular cancer" (participant from Focus Group 1); "Many men and women had reproductive problems like sterility in both as well as lots of miscarriages and some kids were born with developmental problems" (participant from Focus Group 5); "We had lots of cancers in our plant ... 15 women and 2 men, all under 50 years old. And we also had one guy with breast cancer, which seemed odd. I never knew men could get breast cancer" (participant from Focus Group 5).
- Bisphenol A (BPA) is a monomer used to manufacture polycarbonate plastic, the resin used in linings for most food and beverage cans and in dental sealants, and an additive in many other consumer products, including automotive parts. Animal studies show a number of effects in offspring of BPA-exposed mice that are known risk factors for breast cancer, such as abnormal development of mammary glands

(Gray et al. 2009). Human BPA studies identify adverse effects in women with a high BPA body burden, including chromosomal abnormalities, abnormal karyotype (sets of chromosomes) in fetuses, recurrent miscarriages, polycystic ovary syndrome (a condition in which women experience an imbalance in female sex hormones, leading to changes in development, menstruation, and pregnancy), high male hormone levels, obesity, and endometrial hyperplasia (increase in the number of cells in the lining of the uterus) (Sugiura-Ogasawara et al. 2005; Takeuchi and Tsutsumi 2002; Takeuchi et al. 2004; Yamada et al. 2002; see Chapters 8 and 9).

- Various metal compounds are used as stabilizers and colourants in polymers. These can include: inorganic lead compounds, cadmium, organic tin compounds, barium, calcium, zinc carboxylates, and antimony compounds. Lead compounds used in PVC stabilization are classified by the IARC (2011) as possibly carcinogenic to humans (Group 2B). Lead is also considered to be an EDC (Telisman et al. 2000). Reproductive effects from exposure can occur in men and women. In reference to an MSDS for one of the pellets being moulded, a plastics plant machine operator explained: "This product has lead in it … It's in the pellet. It could have been used anywhere, perhaps five machines at a time" (participant from Focus Group 1). A decorator told us: "In the paint departments we used red paint that had lead in it" (participant from Focus Group 5). When cadmium is used as a pigment in thermoplastics, the injection moulding process can result in measurable air concentrations (Bonilla and Milbrath 1994). Cadmium is classified by the IARC (2010) as a human carcinogen (Group 1). It also functions as a xenoestrogen (Johnson et al. 2003).
- Flame retardants are present in many of the materials used by automotive plastics industry workers. As a union health and safety representative recounted, automotive parts are "required by law to include flame retardant." There are two categories of flame retardants: inorganic and organic. Inorganic flame retardants include metal oxides, hydroxide, and basic carbonates. Organic flame retardants "include phosphate esters and halogenated materials, especially organobromine compounds. Polybrominated biphenyls (PBB) have been shown to be strongly estrogenic and in some instances have been classified by the IARC (2011) as possibly carcinogenic to humans (Group 2B). The primary flame retardants that have been used in

plastics production are organohalogen- and organophosphorus-containing compounds (Green 1987). According to a union health and safety representative, "'Tris' was the original flame retardant used in the plants but it was replaced after it was phased out" (participant from Focus Group 4). Tris(2-chloroethyl) phosphate (TCEP), a phosphororganic compound, has been shown to be environmentally persistent and was added to the European Candidate List of Substances of Very High Concern for Authorization (European Chemicals Agency 2010) because of its potential for being "toxic to reproduction." The metal antimony trioxide is used as a flame retardant in some plastics. It has been shown to cause respiratory cancer in female rats, and the few epidemiological studies that have been done on this substance have indicated possible negative reproductive effects (CAREX Canada 2011). CAREX Canada ranks antimony trioxide as an immediate high priority for review as it relates to occupational settings. It has been classified by the IARC (2011) as possibly carcinogenic to humans (Group 2B).

In addition to the many carcinogenic and/or endocrine-disrupting chemicals found primarily in plastics production, several hazardous compounds used in the plastics industry environment are common to most manufacturing jobs:

- Polycyclic aromatic hydrocarbons are emitted by machining, fuel combustion, and other decomposition processes. PAHs have been identified as mammary carcinogens in animal testing (Rudel et al. 2007; Chapter 9). Benzo[*a*]pyrene, one of the PAHs produced by incomplete combustion, has been identified by the IARC (2011) as carcinogenic to humans (Group 1).
- Numerous solvents are used in plastics production. Chlorinated hydrocarbons (which includes benzene), methyl ethyl ketone, and toluene are used in the painting, gluing, and decorating of plastics products (Law and Britton 1998). Benzene, methylene chloride, toluene, and several other organic solvents have been found to cause mammary tumours in animals (Rudel et al. 2007). Labrèche and Goldberg (1997) hypothesize that organic solvents may initiate or promote the breast cancer process through genotoxic or related mechanisms. Many solvents are also considered to be EDCs (Diamanti-Kandarakis et al. 2009).

- Formaldehyde has been classified by the IARC (2011) as a human carcinogen (Group 1). It was linked to an increase in breast cancer risk in a 1995 study of industrial workers, and similar results were obtained in other international studies (Gray et al. 2009). The presence of formaldehyde in the plastics injection moulding setting was noted in a MOL inspection report in 1982: "Concerning formaldehyde fumes in injection molding operations during purging with higher temperatures: The report confirms that formaldehyde is a problem."
- In several studies, night workers were found to have elevated rates of breast cancer (IARC 2007; Megdala et al. 2005; Schernhammer et al. 2006). Light at night is thought to suppress the hormone melatonin, which regulates circadian rhythms (Griffin 2009). The IARC (2007) has concluded that "shift work that involves circadian disruption is probably carcinogenic to humans" (i.e., Group 2A). The Danish workers' compensation system now recognizes the association between work at night and breast cancer (Chustecka 2009).

Conclusion

At the beginning of this chapter, we posed the question: "Are women who work in the plastics injection moulding industry – particularly those producing automotive parts – at greater risk for breast cancer?" Our subsequent discussion revealed several key facts. First, many indicators suggest that this marginalized group of blue-collar workers is being regularly exposed to agents that could increase their risk of breast cancer. Second, epidemiological studies and experiential evidence indicate that plastics workers are developing the disease at an elevated rate. Finally, we as a society are not doing very much about it. Although our scientific understanding of the links between exposures and disease has progressed, conditions in the plastics manufacturing environment – at least within the automotive parts sector in our study area – appear to have improved very little over the past decades.

Advances in endocrine disruptor theory and new results from experimentation undertaken at extremely low doses appear to indicate not only the inadequacy of existing testing protocols but also the need to re-evaluate all of the regulatory standards and guidelines for occupational health. Although allowable exposure limits may not be regularly exceeded in the plastics industry environment on a chemical-by-chemical basis, the danger associated with complex mixtures has not been adequately evaluated. As a participant from Focus Group 5 asked: "What is the synergistic effect of everything being mixed together?" Moreover, new theories about the development of cancer suggest that the presence of such mixtures can increase women's risk for breast cancer. Some workers are being exposed to substances that have been shown to not only initiate the multi-stage cancer process but also carry it forward through the promotion and progression of the disease.

Despite the strong evidence that women in the plastics industry environment are exposed to agents that can cause breast cancer, to our knowledge no plastics worker in Canada has ever received worker's compensation following a diagnosis, nor have *any* breast cancers been recognized as work-related (Griffin 2009). Furthermore, no public inquiries or commissions have been convened to examine the risks to women posed by exposures in the plastics industry, nor have there been any focused institutionalized research initiatives, prevention campaigns, educational programs, or regulatory changes.

As researchers and as a society, our challenge lies in overcoming the inherent limitations of the dominant scientific paradigm for establishing causation. As noted in Chapters 1, 3, and 7, we should not indiscriminately interpret a lack of definitive research findings as evidence that a chemical agent or mixture is *safe.* The absence of evidence may in reality be due to the absence of studies or the failure to apply suitable research methods (Gennaro and Tomatis 2005). Nor should researchers depend solely on technical measurements and professional judgment in assessing exposures without the rich insights and experiences of the exposed population (Watterson 1999).

Whether due to gender bias or the ideology of scientists and research funders, there has been very little research into the preventable environmental causes of breast cancer (Wilkinson 2007). Because of the inherent limitations of epidemiological and toxicological approaches, it is necessary that we employ innovative research methods to fully understand the occupational contributions to breast cancer incidence. In a recent article in *Science, Technology, and Human Values,* Brown and colleagues (2006, 530) asserted that activism can play an important role in research aimed at breast cancer prevention:

> New methods of knowledge construction that account for lay perspectives are pivotal in creating new means to study and understand environmental causes of disease. While scientists involved with breast cancer advocates strongly uphold scientific

> rigor, the activist agenda also has facilitated new methodologies and proposed a new set of norms for standards of proof through the precautionary principle.

From a public policy standpoint, the question arises as to what we should do in the face of scientific uncertainty or, as is often the case, incomplete knowledge (Scott 2005). Currently, in a perversion of our system of justice, the assumption of innocence is extended to workplace and environmental pollutants (Law Reform Commission of Canada 1986). By this standard, plastics production is considered to be innocent until proven guilty beyond a reasonable doubt. Yet, "uncertainty" should not be seen solely as a scientific matter, but, rather, as a question of political perspective, power, and the extent to which our society values human and environmental health. In 1995, Robert Proctor, author of *Cancer Wars: How Politics Shapes What We Know and Don't Know about Cancer,* argued that incomplete knowledge is not an excuse for failing to reduce hazards: "To do otherwise, to wait and see, to delay in the face of good but partial evidence, is tantamount to experimenting on humans" (261).

Recognition of the risks borne by workers and the requisite prevention strategies are long overdue (Watterson, Gorman, and O'Neill 2008). This is true for women and occupational breast cancer risks, as Sandra Steingraber (1997, 685), a biologist, author, and cancer survivor, explained:

> When we in the activist community look at breast cancer research, we are interested not only in results, sample size, elegance of experimental design, or validity of the conclusion based on the data. We are also interested in – saving women's lives.

Acknowledgments

We acknowledge the contributions of the co-investigators for the Lifetime Histories Breast Cancer Research case-control study and the qualitative Exposure Exploration Study; research assistants Jane McArthur, Kathy Mayville, and Daniel Holland; CAW representatives Deb Fields, Sari Sairanen, and Colette Hooson, who provided information and facilitated many of the contacts with workers in the plastics industry; and Dale DeMatteo and Robert Park, who reviewed a draft of this chapter. We are especially grateful to the focus group participants, who generously and courageously provided us with their invaluable first-hand knowledge of exposure conditions in the plastics industry work environment.

The Canadian Breast Cancer Foundation – Ontario Region provided funding for the case-control study and related qualitative research. The Breast Cancer Society of Canada and the Windsor Essex County Cancer Centre Foundation provided additional funding for the case-control study. Several CAW locals also made donations. The University of Windsor hosted the research and provided ethical approval. The case-control study was conducted in partnership with the Windsor Regional Cancer Centre (Windsor Regional Hospital), which provided additional ethical approval. The Occupational Health Clinics for Ontario Workers co-sponsored the studies. The National Network on Environments and Women's Health provided support for a follow-up workshop and presentations.

References

Ahlborg, Gunnar, Tor Bjerkedal, and John Egenaes. 1987. "Delivery Outcome among Women Employed in the Plastics Industry in Sweden and Norway." *American Journal of Industrial Medicine* 12 (5): 507-17, doi:10.1002/ajim.4700120505.

American Chemistry Council. 2010. "Plastics in Automotive Applications." http://plastics.americanchemistry.com/Market-Teams/Automotive.

Bonilla, Jose V., and Randy A. Milbrath. 1994. "Cadmium in Plastic Processing Fumes from Injection Molding." *American Industrial Hygiene Association Journal* 55 (11): 1069-71, doi:10.1080/15428119491018358.

Brody, Julia Green, Kirsten B. Moysich, Olivier Humblet, et al. 2007. "Environmental Pollutants and Breast Cancer: Epidemiologic Studies." *Cancer* 109 (Supp. 12): 2667-711, doi:10.1002/cncr.22655.

Brophy, Jim, and John Jackson. 1980. "The Windsor Occupational Safety and Health Council: The Story of W.O.S.H." *Canadian Dimension* 14 (7): 39-45.

–. 1981. "Fighting for Health and Safety: Windsor, Ontario." *Radical America* 15 (5): 27-38.

Brophy, James T., Margaret M. Keith, Kevin M. Gorey, et al. 2006. "Occupation and Breast Cancer: A Canadian Case-Control Study." *Annals of the New York Academy of Sciences* 1076: 765-77, doi:10.1196/annals.1371.019.

Brophy, James T., Margaret M. Keith, Andrew Watterson, et al. 2012. "Breast Cancer Risk in Relation to Occupations with Exposure to Carcinogens and Endocrine Disruptors: A Canadian Case-Control Study." *Environmental Health* 11 (87): 1-17, doi:10.1186/1476-069X-11-87.

Brown, Phil. 2003. "Qualitative Methods in Environmental Health Research." *Environmental Health Perspectives* 111 (14): 1789-98, doi:10.1289/ehp.6196.

Brown, Phil, Sabrina McCormick, Brian Mayer, et al. 2006. "'A Lab of Our Own': Environmental Causation of Breast Cancer and Challenges to the Dominant Epidemiological Paradigm." *Science, Technology, and Human Values* 31 (5): 499-536, doi: 10.1177/0162243906289610.

Brugnone, F., L. Perbellini, G.Z. Wang, et al. 1993. "Blood Styrene Concentrations in a 'Normal' Population and in Exposed Workers 16 Hours after the End of the Workshift." *International Archives of Occupational and Environmental Health* 65 (2): 125-30.

Burgess, R.H., ed. 1982. *Manufacturing and Processing of PVC*. London: Applied Sciences Publishers.

Canadian Cancer Society. 2011. *Canadian Cancer Statistics 2011*. Toronto: Canadian Cancer Society's Steering Committee on Cancer Statistics.

CAREX Canada. 2011. *"Antimony Trioxide." http://www.carexcanada.ca/en/antimony_trioxide/.*

Castleman, B., and G.E. Ziem. 1988. "Corporate Influence on Threshold Limit Values." *American Journal of Industrial Medicine* 13 (5): 531-59.

CBC (Canadian Broadcasting Corporation). 1981. *Dying for Work* [documentary]. Windsor: CBC Television.

Chiazze, Leonard Jr., and Lorraine D. Ference. 1981. "Mortality among PVC-Fabricating Employees." *Environmental Health Perspectives* 41: 137-43.

Chiazze, Leonard Jr., William E. Nichols, and Otto Wong. 1977. "Mortality among Employees of PVC Fabricators." *Journal of Occupational Medicine* 19 (9): 623-28.

Chiazze, Leonard Jr., Otto Wong, William E. Nichols, et al. 1980. "Breast Cancer Mortality among PVC-Fabricators." *Journal of Occupational Medicine* 22 (10): 677-79.

Chustecka, Zosia. 2009. "Denmark Pays Compensation for Breast Cancer after Night-Shift Work." *Medscape Medical News,* 23 March. http:www.medscape.com/view article/590022.

Clapp, Richard W., Molly M. Jacobs, and Edward L. Loechler. 2008. "Environmental and Occupational Causes of Cancer: New Evidence 2005-2007." *Reviews of Environmental Health* 23 (1): 1-37.

Clark, Rachel Ann, Roy Levine, and Suzanne Snedeker. 1997. "Breast Cancer and Environmental Risk Factors: Biology of Breast Cancer." Fact Sheet #5. New York: Cornell University Program on Breast Cancer and Environmental Risk Factors in New York State. http://envirocancer.cornell.edu/factsheet/General/fs5.biology.pdf.

Cohn, Barbara A., Mary S. Wolff, Piera M. Cirillo, et al. 2007. "DDT and Breast Cancer in Young Women: New Data on the Significance of Age at Exposure." *Environmental Health Perspectives* 115 (10): 1406-14, doi:10.1289/ehp.10260.

Colborn, Theo, Dianne Dumanoski, and John Peterson Myers. 1996. *Our Stolen Future: Are We Threatening Our Fertility, Intelligence, and Survival? A Scientific Detective Story.* New York: Dutton.

Crisp, Thomas M., Eric D. Clegg, Ralph L. Cooper, et al. 1998. "Environmental Endocrine Disruption: An Effects Assessment and Analysis." *Environmental Health Perspectives* 106 (Supp. 1): 11-56, doi:10.1289/ehp.98106s111.

Czeizel, Andrew E., Susan Hegedüs, and László Tímár. 1999. "Congenital Abnormalities and Indicators of Germinal Mutations in the Vicinity of an Acrylonitrile Producing Factory." *Mutation Research/Fundamental and Molecular Mechanisms of Mutagenesis* 427 (2): 105-23, doi:10.1016/S0027-5107(99)00090-1.

Davis, Devra Lee, H. Leon Bradlow, Mary Wolff, et al. 1993. "Medical Hypothesis: Xenoestrogens as Preventable Causes of Breast Cancer." *Environmental Health Perspectives* 101 (5): 372-77, doi:10.1289/ehp.93101372.

DeMatteo, Robert, Margaret M. Keith, James T. Brophy, et al. 2012. "Chemical Exposures of Women Workers in the Plastics Industry with Particular Reference to Breast Cancer and Reproductive Hazards." *New Solutions* 22 (4): 427-48, doi:http://dx.doi.org/10.2190/NS.22.4.d http://baywood.com

Diamanti-Kandarakis, Evanthia, Jean-Pierre Bourguignon, Linda C. Giudice, et al. 2009. "Endocrine-Disrupting Chemicals: An Endocrine Society Scientific Statement." *Endocrine Reviews* 30 (4): 293-342, doi:10.1210/er.2009-0002.

Epstein, S., D. Steinman, and S. LeVert. 1997. *The Breast Cancer Prevention Program.* New York: Macmillan.

European Chemicals Agency. 2010. "Candidate List of Substances of Very High Concern for Authorisation." http://echa.europa.eu/chem_data/authorisation_process/candidate_list_table_en.asp.

Ewertz, Marianne, Lars Holmberg, Steiner Tretli, et al. 2001. "Risk Factors for Male Breast Cancer: A Case-Control Study from Scandinavia." *Acta Oncologica* 40 (4): 467-71.

Fenton, Suzanne E. 2006. "Endocrine-Disrupting Compounds and Mammary Gland Development: Early Exposure and Later Life Consequences." *Endocrinology* 147 (6): S18-S24, doi:10.1210/en.2005-1131.

Firth, Matthew, James Brophy, and Margaret Keith. 1997. *Workplace Roulette: Gambling with Cancer.* Toronto: Between the Lines.

Galassi, Claudia, Manolis Kogevinas, Gilles Ferro, et al. 1993. "Biological Monitoring of Styrene in the Reinforced Plastics Industry in Emilia Romagna, Italy." *International Archives of Occupational and Environmental Health* 65 (2): 89-95, doi:10.1007/BF00405725.

Gardner, Kathleen M., Xiao Ou Shu, Fan Jin, et al. 2002. "Occupations and Breast Cancer Risk among Chinese Women in Urban Shanghai." *American Journal of Industrial Medicine* 42 (4): 296-308, doi:10.1002/ajim.10112.

Gennaro, Valerio, and Lorenzo Tomatis. 2005. "Business Bias: How Epidemiologic Studies May Underestimate or Fail to Detect Increased Risks of Cancer and Other Diseases." *International Journal of Occupational and Environmental Health* 11 (4): 356-59.

Goldberg, Mark S., and France Labrèche. 1996. "Occupational Risk Factors for Female Breast Cancer: A Review." *Occupational and Environmental Medicine* 53 (3): 145-56, doi:10.1136/oem.53.3.145.

Gray, Janet, Nany Evans, Brynn Taylor, et al. 2009. "State of the Evidence: The Connection between Breast Cancer and the Environment." *International Journal of Occupational and Environmental Health* 15 (1): 43-78.

Green, Joseph. 1987. "Halogen and Phosphorus Containing Flame Retardants." In *Handbook for Fillers for Plastics,* edited by Harry S. Katz and John V. Milewski, 313-63. New York: Van Nostrand Reinhold.

Griffin, S. 2009. *Environmental Exposure: The CancerSmart Guide to Breast Cancer Prevention.* Vancouver: Toxic Free Canada.

–. 2011. *CancerSmart 3.1: The Consumer Guide.* Vancouver: Toxic Free Canada.

Hanahan, Douglas, and Robert A. Weinberg. 2000. "The Hallmarks of Cancer." *Cell* 100 (1): 57-70, doi:10.1016/S0092-8674(00)81683-9.

–. 2011. "Hallmarks of Cancer: The Next Generation." *Cell* 144 (5): 646-74.

Hardell, Lennart, Carl-Göran Ohlson, Mats Fredrikson. 1997. "Occupational Exposure to Polyvinyl Chloride as a Risk Factor for Testicular Cancer Evaluated in a Case Control Study." *International Journal of Cancer* 73 (6): 828-30.

He, Yonghua, Maohua Miao, Chunhua Wu, et al. 2009. "Occupational Exposure Levels of Bisphenol A among Chinese Workers." *Journal of Occupational Health* 51 (5): 432-36, doi:10.1539/joh.O9006.

Health Canada. 2001. *Summary Report: Review of Lifestyle and Environmental Risk Factors for Breast Cancer.* Ottawa: Minister of Public Works and Government Services Canada.

Hines, Cynthia J., Nancy B. Nilsen Hopf, James A. Deddens, et al. 2008. "Urinary Phthalate Metabolite Concentrations among Workers in Selected Industries: A Pilot Biomonitoring Study." *Annals of Occupational Hygiene* 53 (1): 1-17, doi:10.1093/annhyg/men066.

Houthuijs, Danny, Bregt Remijn, Han Willems, et al. 1982. "Biological Monitoring of Acrylonitrile Exposure." *American Journal of Industrial Medicine* 3 (3): 313-20, doi:10.1002/ajim.4700030306.

Huang, Li-Ping, Ching-Chang Lee, Ping-Chi Hsu, et al. 2011. "The Association between Semen Quality in Workers and the Concentration of Di(2-Ethylhexyl) Phthalate in Polyvinyl Chloride Pellet Plant Air." *Fertility and Sterility* 96 (1): 90-94.

IARC (International Agency for Research on Cancer). 1999. *Some Monomers, Plastics and Synthetic Elastomers, and Acrolein.* IARC Monographs Volume 19. Lyon: World Health Organization.

–. 2007. "IARC Monographs Programme Finds Cancer Hazards Associated with Shiftwork, Painting and Firefighting." Press Release 180. Lyon: World Health Organization. http://www.iarc.fr/en/media-centre/pr/2007/pr180.html.

–. 2011. *Agents Classified by the* IARC Monographs, *Volumes 1-109.* Lyon: World Health Organization. http://monographs.iarc.fr/ENG/Classification/ClassificationsAlphaOrder.pdf.

Ibarluzea, Jesús M., Mariana F. Fernández, Loreto Santa-Marina, et al. 2004. "Breast Cancer Risk and the Combined Effect of Environmental Estrogens." *Cancer Causes and Control* 15: 591-600.

Industry Canada. 2010. "Canadian Synthetic Resins Industry." Resource Processing Branch, Government of Canada. http://www.ic.gc.ca/eic/site/plastics-plastiques.nsf/eng/pl01384.html.

Infante, Peter F. 1995. "Cancer and Blue-Collar Workers: Who Cares?" *New Solutions* 5 (2): 52-57.

Institute for Survey Research. 1998. *The Breast Cancer Comprehensive Questionnaire.* Prepared for the National Action Plan on Breast Cancer of the US Public Health Service Office on Women's Health. Philadelphia: National Cancer Institute and Temple University.

Jensen, Allan Astrup, Niels Oluf Breum, Jens Bacher, et al. 1990. "Occupational Exposures to Styrene in Denmark 1955-88." *American Journal of Industrial Medicine* 17 (5): 593-606, doi:10.1002/ajim.4700170505.

Ji, Bu-Tian, Aaron Blair, Xiao-Ou Shu, et al. 2008. "Occupation and Breast Cancer Risk among Shanghai Women in a Population-Based Cohort Study." *American Journal of Industrial Medicine* 51 (2): 100-10, doi:10.1002/ajim.20507.

Johnson, Michael D., Nicholas Kenney, Adriana Stoica, et al. 2003. "Cadmium Mimics the *In Vivo* Effects of Estrogen in the Uterus and Mammary Gland." *Nature Medicine* 9 (8): 1081-84, doi:10.1038/nm902.

Keith, Margaret M., and James T. Brophy. 2004. "Participatory Mapping of Occupational Hazards and Disease among Asbestos-Exposed Workers from a Foundry and Insulation Complex in Canada." *International Journal of Occupational and Environmental Health* 10 (2): 144-53.

Keith, Margaret M., Beverley Cann, James T. Brophy, et al. 2001. "Identifying and Prioritizing Gaming Workers' Health and Safety Concerns Using Mapping for Data Collection." *American Journal of Industrial Medicine* 39 (1): 42-51, doi: 10.1002/1097-0274(200101)39:1<42::AID-AJIM4>3.0.CO;2-I.

Kortenkamp, Andreas. 2008. "Low Dose Mixture Effects of Endocrine Disruptors: Implications for Risk Assessment and Epidemiology." *International Journal of Andrology* 31 (2): 233-40, doi:10.1111/j.1365-2605.2007.00862.x.

Kortenkamp, Andreas, Olwenn Martin, Michael Faust, et al. 2011. "State of the Art Assessment of Endocrine Disrupters: Final Report." http://ec.europa.eu/environment/chemicals/endocrine/pdf/sota_edc_final_report.pdf.

Labrèche, France P., and Mark S. Goldberg. 1997. "Exposure to Organic Solvents and Breast Cancer in Women: A Hypothesis." *American Journal of Industrial Medicine* 32: 1-14.

Labrèche, France P., Mark S. Goldberg, Marie-France Valois, et al. 2010. "Postmenopausal Breast Cancer and Occupational Exposures." *Occupational and Environmental Medicine* 67: 263-69, doi:10.1136/oem.2009.049817.

Law, P.K., and T.J. Britton. 1998. "Plastics Industry." In *Encyclopaedia of Occupational Health and Safety,* 4th ed. vol. 3, edited by the International Labour Organization, 77.22-77.29.

Law Reform Commission of Canada. 1986. *Working Paper 53: Workplace Pollution*. Ottawa: Law Reform Commission.

Lithner, Delilah, Åke Larsson, and Göran Dave. 2011. "Environmental and Health Hazard Ranking and Assessment of Plastic Polymers Based on Chemical Composition." *Science of the Total Environment* 409 (18): 3309-24, doi:10.1016/j.scitotenv.2011.04.038.

López-Carrillo, Lizbeth, Raúl U. Hernández-Ramírez, Antonia M. Calafat, et al. 2010. "Exposure to Phthalates and Breast Cancer Risk in Northern Mexico." *Environmental Health Perspectives* 118 (4): 539-44, doi:10.1289/ehp.0901091.

Major, J., Aranka Hudák, Gabriella Kiss, et al. 1998. "Follow-up Biological and Genotoxicological Monitoring of Acrylonitrile- and Dimethylformamide-Exposed Viscose Rayon Plant Workers." *Environmental and Molecular Mutagenesis* 31 (4): 301-10, doi:10.1002/(SICI)1098-2280(1998)31:4<301::AID-EM1>3.0.CO;2-L.

Megdala, Sarah P., Candyce H. Kroenkeb, Francine Ladenb, et al. 2005. "Night Work and Breast Cancer Risk: A Systematic Review and Meta-Analysis." *European Journal of Cancer* 41 (13): 2023-32, doi:10.1016/j.ejca.2005.05.010.

Messing, Karen. 1998. *One-Eyed Science: Occupational Health and Women Workers*. Philadelphia: Temple University Press.

Messing, Karen, Laura Punnett, Meg Bond, et al. 2003. "Be the Fairest of Them All: Challenges and Recommendations for the Treatment of Gender in Occupational Health Research." *American Journal of Industrial Medicine* 43 (6): 618-29.

Myers, John Peterson. 2002. "Scientific Heaven, Regulatory Hell." Comments at the International Scientific Symposium on Endocrine Disruption, 28 November, Hiroshima, Japan. http://www.ourstolenfuture.org/commentary/JPM/2002-1128sciheaven.htm.

OHCOW (Occupational Health Clinics for Ontario Workers) and (WOHIS) Windsor Occupational Health Information Service. 1996. "Health Hazards in the Plastics Product Industry." Unpublished report available upon request.

Payne, Joachim, Martin Scholze, and Andreas Kortenkamp. 2001. "Mixtures of Four Organochlorines Enhance Human Breast Cancer Cell Proliferation." *Environmental Health Perspectives* 109 (4): 391-97.

Petralia, Sandra A., Wong-Ho Chow, Joseph McLaughlin, et al. 1998. "Occupational Risk Factors for Breast Cancer among Women in Shanghai." *American Journal of Industrial Medicine* 34 (5): 477-83, doi:10.1002/(SICI)1097-0274(199811)34:5<477::AID-AJIM8>3.0.CO;2-N.

Proctor, Robert N. 1995. *Cancer Wars: How Politics Shapes What We Know and Don't Know about Cancer*. New York: Basic Books.

Roach, S.A., and S.M. Rappaport. 1990. "But They Are Not Thresholds: A Critical Analysis of the Documentation of Threshold Limit Values." *American Journal of Industrial Medicine* 17 (6): 727-53, doi:10.1002/ajim.4700170607.

Rudel, Ruthann A., Kathleen R. Attfield, Jessica N. Schifano, et al. 2007. "Chemicals Causing Mammary Gland Tumors in Animals Signal New Directions for Epidemiology, Chemicals Testing, and Risk Assessment for Breast Cancer Prevention." *Cancer* 109 (Supp. 12): 2635-66.

Scélo, Ghislaine, Vali Constantinescu, Irma Csiki, et al. 2004. "Occupational Exposure to Vinyl Chloride, Acrylonitrile, and Styrene and Lung Cancer Risk (Europe)." *Cancer Cases and Control* 15 (5): 445-52, doi:10.1023/B:CACO.0000036444.11655.be.

Schernhammer, Eva S., Candyce H. Kroenke, Francine Laden, et al. 2006. "Night Work and Risk of Breast Cancer." *Epidemiology* 17 (1) 108-11, doi:10.1097/01.ede.0000190539.03500.c1.

Scott, Dayna Nadine. 2005. "Shifting the Burden of Proof: The Precautionary Principle and Its Potential for the Democratization of Risk." In *Law and Risk,* edited by the Law Commission of Canada, 50-86. Vancouver: UBC Press.

Siemiatycki, Jack. 1991. *Risk Factors for Cancer in the Workplace.* Boca Raton, FL: CRC Press.

Smith, Rick, and Bruce Lourie. 2009. *Slow Death by Rubber Duck: How the Toxic Chemistry of Everyday Life Affects Our Health.* Toronto: Knopf Canada.

Steingraber, Sandra. 1997. "Mechanisms, Proof, and Unmet Needs: The Perspective of a Cancer Activist." *Environmental Health Perspectives* 105 (Supp. 3): 685-87.

Stewart, Patricia A., Walter F. Stewart, Ellen F. Heineman., et al. 1996. "A Novel Approach to Data Collection in a Case-Control Study of Cancer and Occupational Exposures." *International Journal of Epidemiology* 25 (4): 744-52, doi:10.1093/ije/25.4.744.

Sugiura-Ogasawara, Mayumi, Yasuhiko Ozaki, Shin-ichi Sonta, et al. 2005. "Exposure to Bisphenol A Is Associated with Recurrent Miscarriage." *Human Reproduction* 20 (8): 2325-29, doi:10.1093/humrep/deh888.

Takeuchi, Toru, and Osamu Tsutsumi. 2002. "Serum Bisphenol A Concentrations Showed Gender Differences, Possibly Linked to Androgen Levels." *Biochemical and Biophysical Research Communications* 291 (1): 76-78, doi:10.1006/bbrc.2002.6407.

Takeuchi, Toru, Osamu Tsutsumi, Yumiko Ikezuki, et al. 2004. "Positive Relationship between Androgen and the Endocrine Disruptor, Bisphenol A, in Normal Women and Women with Ovarian Dysfunction." *Endocrine Journal* 51 (2): 165-69.

Telisman, Spomenka, Petar Cvitkovic, Jasna Jurasovic, et al. 2000. "Semen Quality and Reproductive Endocrine Function in Relation to Biomarkers of Lead, Cadmium, Zinc, and Copper in Men." *Environmental Health Perspectives* 108 (1): 45-53.

Thuresson, Kaj, Åke Bergman, and Kristina Jakobsson. 2005. "Occupational Exposure to Commercial Decabromodiphenyl Ether in Workers Manufacturing or Handling Flame-Retarded Rubber." *Environmental Science and Technology* 39 (7): 1980-86, doi:10.1021/es048511z.

Trosko, James E., and Brad L. Upham. 2005. "The Emperor Wears No Clothes in the Field of Carcinogen Risk Assessment: Ignored Concepts in Cancer Risk Assessment." *Mutagenesis* 20 (2): 81-92, doi:10.1093/mutage/gei017.

Vandenberg, Laura N., Theo Colborn, Tyrone B. Hayes, et al. 2012. "Hormones and Endocrine-Disrupting Chemicals: Low-Dose Effects and Nonmonotonic Dose Responses." *Endocrine Reviews* 33 (3): 378-455.

Villeneuve, Sara, Joelle Févotte, Antoinette Anger, et al. 2011. "Breast Cancer Risk by Occupation and Industry: Analysis of the CECILE Study, a Population-Based Case-Control Study in France." *American Journal of Industrial Medicine* 54 (7): 499-509.

Watterson, Andrew. 1999. "Why We Still Have 'Old' Epidemics and 'Endemics' in Occupational Health: Policy and Practice Failures and Some Possible Solutions." In *Health and Work: Critical Perspectives,* edited by Norma Daykin and Lesley Doyal, 107-26. London: Macmillan.

Watterson, Andrew, Thomas Gorman, and Rory O'Neill. 2008. "Occupational Cancer Prevention in Scotland: A Missing Public Health Priority." *European Journal of Oncology* 13 (3): 161-69.

Welshons, Wade V., Kristina A. Thayer, Barbara M. Judy, et al. 2003. "Large Effects from Small Exposures. I. Mechanisms for Endocrine-Disrupting Chemicals with Estrogenic Activity." *Environmental Health Perspectives* 111 (8): 994-1006, doi:10.1289/ehp.5494.

Wilkinson, S. 2007. "Breast Cancer: Lived Experience and Feminist Action." In *Women's Health in Canada: Critical Perspectives on Theory and Policy,* edited by Marina Morrow,

Olena Hankivsky, and Colleen Varcoe, 408-33. Toronto: University of Toronto Press.

Yamada, Hideto, Itsuko Furutaa, Emi H. Katoa, et al. 2002. "Maternal Serum and Amniotic Fluid Bisphenol A Concentrations in the Early Second Trimester." *Reproductive Toxicology* 16 (6): 735-39, doi:10.1016/S0890-6238(02)00051-5.

Zahm, Shelia Hoar, and Aaron Blair. 2003. "Occupational Cancer among Women: Where Have We Been and Where Are We Going?" *American Journal of Industrial Medicine* 44 (6): 565-75, doi:10.1002/ajim.10270.

Chapter 11 Power and Control at the Production-Consumption Nexus: Migrant Women Farmworkers and Pesticides

Adrian A. Smith and Alexandra Stiver

Recently, a great deal of interest has been focused on women's exposures to chemical toxicants in everyday consumer and household goods (see, for example, David Suzuki Foundation 2010; Deacon 2011; Environmental Working Group 2008; Smith and Lourie 2009). As Keith and colleagues affirm in Chapter 10, however, occupational exposure to chemicals has received far less critical attention. In the debate over exposure to toxics, women as paid workers have been all but ignored. This chapter examines the production-consumption nexus in contemporary agricultural production in Ontario, in order to evaluate the impact of chemical exposures, such as exposures to pesticides, on the health of migrant women. The dearth of empirical data documenting such an impact forces us to approach the problem in a different way: we draw from a critical and informed reading of several areas of research, including environmental justice and occupational health and safety. What we find is the migrant agricultural population appears to be disproportionately vulnerable to pesticide exposure in the production process compared with non-migrant agricultural workers – a population that is already greatly at risk.

The exposure of migrant agricultural workers to pesticides is exacerbated by Canada's temporary labour migration regulatory framework. Taking this as a basis for analysis, we surmise that there are gendered impacts of toxic chemical exposures that, when accounted for, may further compound existing legal regulatory dynamics. As a result, migrant women workers may constitute a segment of the population that is at even greater risk of experiencing serious environmental and occupational illness and injury.

In three sections, the analysis constructs the case for taking seriously the health impacts of pesticide use on agricultural workers – migrant and non-migrant generally, and women migrant agricultural workers in particular. The first section theorizes the nexus between production and consumption to find a highly interdependent relationship reliant on labour exploitation. The second section explores the gender dimensions of workers' exposures to pesticides and the long-term health implications. The final section situates the analysis within the constraints on agricultural workers' power and control in labour processes. The degree of power and control currently available to workers is mediated through overlapping processes and dynamics of racialization and racism, legal regulation of citizenship status, and – in a crucially important but underappreciated respect – gendering (but see Preibisch and Encalada 2011). Taken together, these processes and dynamics produce an environment in which migrant agricultural workers, especially women, are more vulnerable to risk of disease and injury on the job.

The Production-Consumption Nexus

Agricultural workers' exposures to pesticides engage environmental justice concerns. A key tenet of environmental justice is that "communities must have control over their environment" (Arcury, Quandt, and Russell 2002, 233). From this perspective, the degree of control a person or a community can exercise with respect to a given risk is critical in evaluating its "fairness" (Scott 2008). With regard to agricultural work, several studies have demonstrated that migrant workers often perceive protective measures for pesticide use as outside their control (Austin et al. 2001; Vaughan 1993a, 1993b). The ability of agricultural workers to engage in safe practices with respect to pesticide use, as discussed below, is known to depend on issues of English language proficiency, power relationships within the labour process, and access to protective equipment (Arcury, Quandt, and Russell 2002). As Levenstein and Wooding (2000) assert, the "social relations of production determine the health and well-being of workers": the risk of disease and injury for migrant women agricultural workers is shaped entirely by choices about the production process made by others outside of their ambit of control.

This chapter aims ultimately to explore why and how certain women may be at greater risk from chemical exposures. Through an examination of the production-consumption nexus in capitalist relations, our approach perceives women as both consumers and producers of food. As noted by others in this volume (see, for example, Chapters 1, 2, 4, and 6-9), several social gradients – such as socio-economic location (including level of educational attainment) – contribute to the capacities of women consumers to avoid chemical exposures in their food choices. We know much less about women's exposures to chemicals as agricultural workers, and even less about their relative capacities to prevent those exposures.

The use of pesticides in agriculture threatens the health of both consumers, through food residues, and producers, through accumulation in the bodies of workers (see Chapters 5, 8, and 9). We deliberately shift the focus from the impacts of chemicals in products on the end-user (consumer) to the impacts of chemicals in the production process. Judith Helfand made a similar move in her 2002 documentary *Blue Vinyl*. Helfand raised concerns about the toxicity of vinyl, not only because we are surrounded by vinyl products on a daily basis but also because workers whose labour and bodies are essential to the production of vinyl – and to its availability as a commodity – suffer devastating health effects (Gold and Helfand 2002). Similarly, Deborah Barndt's book *Tangled Routes: Women, Work, and Globalization on the Tomato Trail* (2002) contextualizes the production-consumption nexus by highlighting the path of the tomato, and the role of labouring women along that path, from growth to sorting, packing, sale, and ultimately consumption.

Since the impact of pesticides is observable in both the production and consumption of food, a more rigorous exploration of the relationship between these spheres is required. Sidney Mintz's classic anthropological text, *Sweetness and Power* (1986), can act as a point of departure. In examining the impact of sugar in the historical development of global capitalism, Mintz articulates a sophisticated and integrated approach to production and consumption. Seemingly adhering to Marx's dictum that production is also consumption, Mintz explains how sugar commodity production, shaped by overseas export policies in the Caribbean and defined by its reliance on slave labour, stemmed from the shifting consumption habits of the English working class. In linking the emergence of modern sugar production with the growth in sugar consumption, Mintz situates labour exploitation as a crucial dimension of both.

In underlining the reciprocal relationship of production and consumption in global capitalism and the centrality of labour exploitation "at both ends of the commodity chain," Mintz opens the door for further work that better accounts for "the fundamental significance of gender to the development of capitalism" (Carney 2008). Thus we seek to clarify the relationship of women workers to global capitalist development by drawing "attention to the ways women are unevenly incorporated into commodity production as workers and consumers as well as the implication of workforces dependent on female labour for the reproduction of households" (Carney 2008, 128). Following Mintz's seminal approach, we treat production and consumption as coterminous and highly interdependent but reliant on labour exploitation, which, under capitalism, obtains from the sphere of production. Generally speaking, the production-consumption nexus is embodied within women workers, as producers and consumers and as paid and unpaid labour. Women are workers and members of households in which they disproportionately bear the burden of unpaid labour, especially the wide and varying yet ongoing demands of social reproduction, which encapsulates "the biological reproduction of the labour force, both generationally and on a daily basis," including the provision of food, shelter, and clothing (Katz 2001, 710). In this respect, a pivotal distinction is drawn between the productive and reproductive economies (see also Chapters 1, 2, and 10). Demarcated by the existence or absence of compensation, the productive-reproductive distinction represents a gendered fault line within capitalism.

Political contestation, not human nature, best explains the gendered dimensions of the productive-reproductive distinction. Although the defining boundary is "drawn differently in different contexts and cultures," and "it varies even within societies considerably over time," the productive and reproductive spheres are "inextricably intertwined." That said, a key dimension of capitalist economies is the subordination of reproduction to production or accumulation (Picchio 1992), despite critical recognition that production cannot occur without reproductive activity (and vice versa). Despite mounting rates of participation in the productive economy, women continue to bear the brunt of the devaluing of reproduction.

These insights parallel recent historical developments in the context of global agricultural production and food commodity chains, where the use of harmful chemicals and other highly industrialized farming practices reveals a blatant disregard for human health and capabilities, the disproportionate demands of social reproduction imposed on women, and the direct relationship between paid and unpaid labour.

In the post-Second World War period – especially in the neoliberal moment – agricultural crop production has undergone considerable restructuring. Increasing corporate consolidation in the global food chain has contributed to the industrialization of agriculture (Chapter 5). Dramatic shifts in cultivation methods have occurred, resulting from the development of more sophisticated forms of irrigation and an increase in the use of pesticides (Bolaria 1992). These intensive farming practices have led to "efficiencies" in the process and a greater total output, the economic benefits of which have accrued to growers and their financial lenders. "Cheap food" – a public policy that has been in place from the mid-twentieth century onward – is also widely accepted as beneficial to Canadians (and Americans) as consumers, but from our perspective, this ignores the human health and ecological consequences of intensive farming and devalues the role of workers as producers and reproducers (Albritton 2009).

Migrant Agricultural Work in Ontario

Migrant agricultural workers migrate to Ontario primarily from the Caribbean and Mexico, as well as more recently from Thailand, the Philippines, and Guatemala. They labour in fields and greenhouses across the province, planting, picking, and packing crops for local and wider consumption. The Seasonal Agricultural Workers Program (SAWP) and the Temporary Foreign Workers Program (TFWP) Low-Skill Pilot Project are federal migration schemes overseen by Human Resources and Skills Development Canada (HRSDC) and Citizenship and Immigration Canada (CIC) to facilitate steady or reliable sources of relatively cheap labour from abroad.

This chapter focuses specifically on the SAWP, although it has more general applicability to the newly emergent TFWP; as such, the latter program's provisions are referenced where possible. A sizeable literature critically evaluates the SAWP's structural features and impact, especially from the standpoint of migrant workers (Satzewich 1991; Basok 2002; McLaughlin 2009). The SAWP was initiated in 1966 as the primary means of recruiting foreign labourers to do agricultural work in Canada. As a result of the program, over 27,000 workers now enter the country each year (HRSDC 2012). In any given year, at least

85 percent of the SAWP workforce is employed in Ontario (Ferguson 2004). Only about 3 percent of the total pool of SAWP workers were women in 2005; the ratio of women to men workers was 1:44 (Preibisch 2007, 9). Key features of the SAWP include the recruitment of workers to spend up to eight months in Canada tied to specific growers, with little real possibility of changing employers, as well as the requirement imposed on growers to provide housing accommodations for workers, which typically occurs on (or relatively near) the farm.

Pesticides, Gender, and Human Health

Migrant agricultural workers toil in fields and greenhouses on a daily basis for all but a few months of the year. Pesticides are used aggressively in a number of their tasks, applied to crops as a solid powder, liquid, or vapour. These chemicals can have direct effects on the working environments of farmworkers.[1] Pesticide exposure can result from inhalation, skin contact (dermal absorption), and ingestion. As a result, those working directly with pesticides – as applicators, crop pickers, or other agricultural workers – have an inherently elevated risk of exposure.

To take greenhouse tomato crop production as an example, specific tasks for a migrant worker might include the following: loading and mixing of pesticides, such as endosulfan, for use; preliminary ground-bloom application of pesticide to the tomato crop; and manual spraying of pesticide to the crop. In this work, farmworkers are repeatedly and routinely exposed to pesticides. Even when not directly applying pesticides, workers have great difficulty avoiding exposure to these chemicals. Accordingly, as Salvatore and colleagues (2009) note, the pesticide exposures of agricultural workers are a serious occupational health concern.

The negative health effects associated with pesticide exposures are numerous and well documented (Alavanja, Hoppin, and Kamel 2004; Frank et al. 2004; Sanborn et al. 2004; Villarejo and McCurdy 2008; Wilk 1986; Chapters 5, 8, 9, and 10). They include both acute effects such as "rashes, headaches, nausea and vomiting, disorientation, shock, respiratory failure, coma, and, in severe cases, death," and chronic effects like cancer or neurological and reproductive health issues (Arcury, Quandt, and Russell 2002, 233). Further, the World Health Organization categorizes pesticides as hormone-mimicking, endocrine-disrupting chemicals (EDCs), a "diverse group of industrial and agricultural chemicals ... that have the capacity to mimic or obstruct hormone function" in the body "by fooling it into accepting new instructions" that distort and disrupt processes of development, growth, and reproduction (Damstra et al. 2002; Krimsky 2000, 2). The association between public health risk and pesticide use is, of course, not a new idea. Rachel Carson's seminal work *Silent Spring* (1962), expressed concern about the dangers of pesticides, the potential for cancer in human populations, and the lack of regulatory response (see Chapter 8).

Women have unique vulnerabilities to pesticide exposure compared with men. As EDCs, pesticides can impact various parts of the body in different ways, and will have distinct effects depending on a worker's sex and the timing of exposure (see Part 3 of this book). Women who perform agricultural work are particularly prone to exposure during certain times of development, known as critical windows of vulnerability (see Chapters 1-4, 7, and 9).[2] Women can experience greater susceptibility to exposure, even at very low doses, during these periods, with endocrine disruption leading to interference in fetal development or menopause. As is becoming clear through an array of studies, low doses of EDC exposure can prove even more harmful when possible additive, cumulative, and synergistic effects are considered (see Chapters 1, 3, 4, 9, and 10).

Although all women are susceptible to the risks of pesticide exposure, research suggests that certain subgroups of women can be disproportionately affected. For example, even though there is no universally accepted causal relationship between pesticides and breast cancer rates, evidence of this association has been accumulating through scientific studies. When taking into account the sheer volume of toxicants used in agriculture, it is clear that a greater consideration of pesticide exposure and its effects on the health of women agricultural workers is warranted (Woodruff, Kyle, and Bois 1994). Certain studies have already begun to explore women agricultural workers as a group of women directly exposed to pesticides (Brophy et al. 2002, 2006, 2012). However, there is limited empirical data on the specific impact of pesticide use on women migrant workers in agriculture, a population at even greater risk of exposure and subsequent harms to human health (but see Edmunds et al. 2011).

It is difficult to accurately assess risk based exclusively on occupation or, in Keith and colleagues' terms (Chapter 10), by using occupational categories as "surrogates for exposure" (Brophy et al. 2006, 773). Even establishing definite associations between pesticides and suspected chronic health effects such as cancer is challenging, especially since some government surveys of pesticides exclude work environments such as greenhouses, where concentrations of, and exposures to, pesticides are especially high (Chapter 10). In fact, very little conclusive work is available on women's occupational exposure. This lack of research does little to alleviate the burdens of risk (Preibisch 2010).

For migrant workers, the situation appears stark. According to Das and colleagues (2001), there are several reasons that so little is known or documented about pesticide exposure and migrant agricultural worker health. First, the application of pesticides on crops typically occurs in combination. Hundreds of active ingredients are registered as pesticides, and thousands of products circulating in the marketplace fall into each pesticide "family" (e.g., herbicides, insecticides, fungicides). Poor and inconsistent data undermine efforts to capture a holistic understanding of the health effects of multiple exposures. In addition, pesticide formulations usually contain "inert" ingredients for which toxicity testing is unavailable or incomplete, and for which manufacturers are usually not required to disclose information. To complicate matters further, workers might work with several different crops within short periods of time, making it difficult to identify individual pesticide exposures (Brophy et al. 2012; Das et al. 2001).

Second, processes of migration may lead to uneven access to health care or undermine the quality of care received. Migration processes contribute to a discontinuity in the provision of care, which proves most disruptive in the treatment of chronic conditions (Das et al. 2001). Social inequalities compound the problem, as does tenuous immigration status (discussed below), which in some cases may lead workers to avoid seeking medical treatment in hospitals and clinics or to be denied such assistance altogether. These challenges not only affect individual workers but also make it difficult for those documenting or attempting to prove the association between exposure and illness or injury. Thus, "making links between specific classes of pesticides and chronic effects is limited by exposures to multiple agents, inadequate exposure assessment, and difficulty in long-term follow-up" (Sever, Arbuckle, and Sweeney 1997, 306).

These methodological challenges in toxicology and epidemiology have real consequences for workers in the workplace. Existing regulatory structures are completely dependent on the technical advice that health risk assessments provide. This is as true for the occupational safety standards incorporated into provincial law as it is for the federal regulatory process that approves pesticide use in Canada. These risk assessment processes fail to consider the variations in vulnerability of different workers to chemical exposure, and they also grossly underestimate the disparities of power that operate in workplaces to ensure that workers bear a disproportionate risk of exposure and potential harms to health.

Occupational Health and Safety in Agricultural Production

Here we situate the significant and continuous pesticide exposures experienced by labourers in agricultural production within the regulatory framework governing the occupational health and safety of migrant agricultural workers. In June 2006, the Ontario government extended protective coverage, under the Occupational Health and Safety Act (OHSA) and Ontario Regulation 414/05, to paid farmworkers in provincial farming operations.[3] However, the Expert Advisory Panel on Occupational Health and Safety that reported in 2010 called for greater attention to the hazards of farm work, without making any mention of the health risks associated with pesticide exposures for farmworkers (Dean 2010, recommendation 32).

Emphasizing prevention of workplace accidents, injuries, and diseases, the OHSA employs an "internal responsibility system" to hold all parties in the workplace jointly responsible for guaranteeing health and safety (Smith 2000; Swinton 1983). Under this system, workers are provided with three core statutory rights. First, OHSA provides a right to participate in the resolution of health and safety issues. Distinct from the OHSA (s. 9[1]), which imposes differential participatory rights based on the size of workplaces, the right to participation in farming operations applies to workplaces of twenty or more workers conducting regular employment in such areas as mushroom and greenhouse farming (O. Reg. 414/05, s. 2[2]). Applicable farming operations must establish a health and safety committee. Second, workers enjoy the right to refuse work they deem unsafe or dangerous (OHSA, s. 43). Third, workers are granted the right to know about potential hazards to which they may be exposed.

In addition to these core rights, OHSA offers protection to workers against retaliation or reprisals when they attempt to invoke statutorily protected rights (OHSA, s. 50; see recommendation 33 in Dean 2010). OHSA also requires employers to report an occupational illness contracted by a current or former worker, or a claim made by such a worker, to the Workplace Safety and Insurance Board (WSIB), the

Ministry of Labour, and the respective joint health and safety committee or representative, as the case may be (OHSA, s. 52[3]).[4]

In all of the stated ways, agricultural workers now enjoy the same formal statutory rights and protections as other Ontario workers protected under OHSA. Important criticisms have been levelled at the overall effectiveness of the internal responsibility system (see Smith 2000) and occupational health and safety regulation historically (Tucker 1990). There is a dearth of empirical data to document the early impact of OHSA on farming practices and working conditions. In practice, however, any latent potential within the OHSA appears stifled within the context of temporary labour migration.

Notwithstanding the widened coverage of protections under OHSA, agricultural workers are not likely to enjoy meaningful improvements in working conditions. As Eric Tucker (2006) asserts, the end of legislative exclusion cannot itself free agricultural workers from "the vicious circle of precariousness and powerlessness" (275) in which they are situated. Tucker (2006, 276) calls for the devotion of "sufficient resources" to statutory enforcement, the development of "a scheme of collective representation that is responsive to their conditions," and a strengthened commitment to collectively organizing agricultural workers.

For collective representation especially, the Supreme Court of Canada's recent denial of meaningful freedom of association rights to agricultural workers in Ontario poses a significant impediment to confronting precariousness in agriculture.[5] Ontario labour law continues to limit agricultural workers' access to forms of collective representation and action prevalent in the industrial unionism model as set out in the Ontario Labour Relations Act.[6] For instance, rather than collective bargaining, arbitration, grievance procedures, and the right to strike, agricultural workers are granted the right to form a voluntary association that does not enjoy privileged status in terms of collective bargaining. Thus, in instances in which agricultural workers manage to organize employee associations, employers have no formal obligation to recognize and bargain with them. Despite critical interventions that question the general effectiveness and feasibility of the industrial unionism model (Panitch and Swartz 2008), particularly in agricultural work (see Industrial Accident Victims Group of Ontario and Justice for Migrant Workers 2009), the denial of meaningful collective bargaining is an affront to efforts aimed at removing insecurities prevalent within the agricultural labour market.

The failure to end the "the vicious circle of precariousness and powerlessness" through an enhancement of collective representation for agricultural workers severely restricts the constitutionally entrenched right of freedom of association. It also narrows the possibilities for collective, worker-driven enforcement of statutory protections. Further, as the cycle continues, and indeed intensifies, workers are faced with troubling health effects. Precarious employment poses an elevated threat to workers' mental and physical health (Lewchuk et al. 2006). The lack of access to formal workplace complaint processes and other minimum statutory protections renders precariously employed workers more susceptible to poor health outcomes (de Wolf 2006; Lewchuk et al. 2006). Not only does labour market insecurity detract from efforts to build healthy and safe work environments, but, in fact, poor health and safety outcomes are a feature of precarious employment. As a precarious form of employment, agricultural work threatens the health and well-being of workers so engaged. Thus, the unwillingness to give substance and meaning to the free associational rights of agricultural workers diminishes their health and safety outcomes and contributes to the viciousness with which they experience precariousness.

For migrant agricultural workers, Canada's temporary labour migration programs also intensify the viciousness of precariousness and powerlessness. Here, precarious citizenship, immigration, or migration status (Das Gupta 2006; Goldring, Berinstein and Bernhard 2007, 2009) reflect tight political and legal constraints on the spatial mobility of migrant workers. This interferes with workers' geographic mobility and labour market circulation within Canada, and with their capacity to confront unhealthy and unsafe working conditions (Smith 2005). Unlike workers with citizen or permanent resident status, SAWP workers are tied to a specific employer and are not permitted to remain in Canada following the expiration of their employment contract (Smith 2005). Most stay in the country for a significant portion of the year. Employers are granted the authority to repatriate workers on a moment's notice without justification, and at the workers' expense (Smith 2005).

Under the TFWP, a worker is permitted to work in Canada for a maximum of four years (Citizenship and Immigration Canada 2011). Although such workers are permitted to apply for permanent residence, little credit is given for their service and crucial importance to agricultural production in Canada. Following a four-year term, most

TFWP workers are required to undergo a four-year "cooling-off" period prior to reapplying for a work permit. In both temporary migration programs, therefore, once "mobilized" to cross national borders, workers are "immobilized" by constraints on their labour market and geographic mobility – which, it should be noted, provides historical continuity with nineteenth century "unfree" labour migration regimes (see, for example, Mohapatra 2004).

The immobilization of migrant agricultural workers produces uneven access to statutory protections. Such political and legal constraints undermine the protective coverage afforded to agricultural workers through minimum standards legislation. The core OHSA rights, including anti-reprisal protection, are rendered wholly ineffective and inaccessible for migrant workers. In the face of this immobilization, how do temporary migrant agricultural workers exercise their OHSA right to refuse unsafe work or to access alternative employment due to unsafe working conditions? It becomes evident that immobilization serves to discourage worker assertions of statutory rights.

It is our claim that Canada's temporary labour migration framework gives rise to especially acute concerns with respect to chemical exposures. This is evident in the general design and structure of the temporary migration programs. Migrant workers may be at greater risk due to compounded exposures and proximity to pesticides. These elevated exposures occur not only as a result of labour in the field but also due to the fact that migrant worker housing accommodations are typically located on farm property (Harrison 2011). Because workers live on or near their workplaces, physical proximity can compound both susceptibility to exposure and degree of risk, both through inhalation and through clothing and footwear brought into the home (Brophy et al. 2012; Ferguson 2004). In this respect, bystander exposure may add to already high levels of occupational exposure, as workers end up being in contact with hazardous chemicals around the clock.[7] Risk could be intensified by other elements of proximity, such as the fact that workers' sources of water might be from local wells contaminated with pesticides, or that they may consume foods carrying pesticide residues (Brophy et al. 2012). As well, pesticide drift, defined as "the airborne movement of agricultural pesticides," is now recognized by activists around the world as a cause for concern (Harrison 2011, 2-3).

In the context of occupational health and safety protections, it is worthwhile to examine the reasons that workers may not be able to engage in the safe use of pesticides. In particular, several studies have demonstrated that migrant workers often perceive protective measures for pesticide use as outside of their control (Austin et al. 2001; Vaughan 1993a). Arcury, Quandt, and Russell (2002, 234) found that the ability of farmworkers to engage in safe pesticide use depended on three critical factors: workers' "ability to communicate with their employer, power relationships at work, and the availability of protective equipment."

With respect to workplace communication, English language comprehension has a direct impact on workers' right to know about occupational hazards, and on their ability to invoke the right to refuse unsafe work. Migrant agricultural workers who are not proficient in English may face heightened risks of chemical exposure where employers provide limited safety measures, such as a lack of multilingual signage, warnings, or instructions. A lack of English language proficiency may also act as a disincentive to the invocation of rights. For instance, workers may not understand – or may misunderstand – instructions and caution signs, or they may lack the skill set to clearly articulate concerns and to appreciate supports and protections in place (Sargeant and Tucker 2009). Further, language proficiencies may be a gendered phenomenon. The survey of Green and colleagues (2008) identified male migrant workers as possessing stronger English language skills than female migrant workers (Sargeant and Tucker 2009). These linguistic disparities may inhibit women from completing job tasks, and may make it more difficult for them to obtain agricultural employment where employers exploit communication challenges in their hiring process (Sargeant and Tucker 2009).

With respect to power relationships, the lack of employment security and the tenuous immigration status of migrant agricultural labour produce troubling impacts on women workers. Within the SAWP, the low numbers of women have been used by employers to threaten women, suggesting that they can easily be replaced with men if they question their working or living conditions (Preibisch 2007). This reality discourages women from asserting their rights or articulating complaints. This gender discrimination extends to other aspects of their jobs, as there are "few opportunities for female workers, and women are heavily controlled and disciplined in various ways by employers" (Justicia/Justice for Migrant Workers 2005, 2).

Women can be viewed as "sexually available" within the group, and documented behaviour ranges from male attention to sexual harassment, originating both from fellow workers and from employers

(Preibisch 2007). Women experience gender discrimination as there are "few opportunities for female workers, and women are heavily controlled and disciplined in various ways by employers" (Justicia/ Justice for Migrant Workers 2005). Employers also inappropriately concern themselves with the social and the sexual activities of women workers. As a result, not only are women more likely than men to face informal controls over their mobility but their sense of power and control in the workplace – such as with respect to demanding or employing protective equipment in their use of pesticides – is likely to be even less than that felt by their male counterparts. For similar reasons, women's health-related medical concerns often go unaddressed (Preibisch 2007).

Migrant workers are also hesitant to take time off for sickness or medical treatment, for fear of losing wages or employment altogether (Sargeant and Tucker 2009). Economic compulsion to maintain employment may prove especially intense in households with single-income earners. Most women arriving as SAWP workers are single heads of households, resulting not only in a sense of anxiety related to separation from their children but in an emotional burden related to the notion that as mothers they have neglected their responsibilities in that respect. This can lead to an overwhelming sense of responsibility to work hard, or even a tendency to impose on themselves greater pressure to maximize their pay over the course of the season (Preibisch 2007). To this extent, the economic pressure to maintain employment might be particularly strong and enticing for women migrant workers, despite potential health risks.

In other respects, the low or marginal power and control workers exercise within the labour process is underscored by empirical studies. Farmworker behaviour with respect to preventive safety measures for the use of pesticides does not change when those workers are educated as to the risks of pesticides. Studies reveal that although worker perceptions of risks might change, the reality of their workplace conditions, and the inherent precariousness of their employment and immigration status, will deter workers from taking the necessary safety measures (Arcury, Quandt, and Russell 2002). To paraphrase the words of famed occupational health and safety organizer Bob Sass, the right to know is about information, not power: migrant workers' knowledge of their legal rights does not, in itself, constitute power (Smith 2000; Smith 2005). OHSA regulation alone cannot overcome the disparities in power relations.

Analyses of occupational risks associated with pesticide use often rely on assumptions about the type of protective equipment that can be accessed by workers. For example, governmental risk assessments of pesticides used in agriculture assume that employers make such equipment available. Yet, even if workers are presented with extensive information on pesticides as workplace hazards, and are provided with protective equipment, the prominence of pesticides in agricultural production still renders agricultural workers susceptible to ongoing exposures. As workers often lack any meaningful way to contest contemporary production methods, they are beholden to the overwhelming authority of capitalist owners. For migrant workers, the situation worsens as the legal regulatory framework governing temporary labour migration renders workers a "captive" labour force (Basok 2002). Thus, agricultural workers, and the subcategory of migrant workers especially, are caught in a dismal plight.

Gendering the Agricultural Production-Consumption Nexus

Following Sidney Mintz's classic account, as discussed at the outset, the analysis takes labour exploitation as a central and overlapping feature of the coterminous relationship between production and consumption. In the context of temporary labour migration in agriculture, this understanding encourages analytical reflection attentive not merely to consumption of commodities but, in fact, to industrial capitalist production. We place primary emphasis on occupational hazards and risks arising from the use of pesticides in contemporary agricultural production. Having noted how power relations disrupt the latent potential that may exist within OHSA, the analysis also stresses how the legal framework governing temporary labour migration further undermines any remaining regulatory promise.

Migrant agricultural labour represents a form of "unfree" labour incorporation in which labour exploitation is produced, deepened, and intensified through processes of racialization and racism and through precarious citizenship status. Immigration law assumes a pivotal role in legitimizing differential treatment between, on the one hand, citizen and near-citizen workers and, on the other, migrant or non-citizen workers. The tenuous nature of their citizenship status impedes migrant worker access to statutory rights and protections, including those relating to occupational health and safety. The physical proximity to pesticide use in the labour process and in the location of their housing accommodations may place migrant agricultural workers at greater risk of continuous chemical exposure and subsequent health effects. Because precarious citizenship functions to immobilize migrant agricultural workers, they experience pesticide exposure differently from non-migrant agricultural workers.

Notwithstanding the importance of attending to the production-consumption nexus within capitalist agricultural production, a key analytical task is to pay heed to the gendered nature of capitalist development. In setting about to clarify and account for the experience of women workers within global capitalist development, we find that women assume the roles of not only producer and consumer but also paid and unpaid labour. The latter roles in particular highlight a gendered fault line within capitalism – the productive and reproductive distinction.

Social reproduction represents a pivotal, if overlooked, dimension of temporary labour migration programs. Although temporary labour migration is increasingly understood within social reproduction terms, especially as it relates to the global phenomenon of care work, agricultural labour migration has yet to undergo a sustained feminist political economic critique grounded in social reproduction. That said, we must not lose sight of the complex challenges within the production sphere as well. While agricultural workers constitute a population at risk of serious environmental and occupational illness and injury, migrant agricultural workers constitute a population at even greater risk, and women migrant workers are particularly vulnerable. Women who perform agricultural work may be uniquely susceptible to pesticide exposures because of critical windows of vulnerability in their biological development (Chapters 1-4, 7, and 9). This susceptibility to exposure may put women migrant workers at greater health and safety risks than acknowledged by policy makers and researchers.

Our aim is not to essentialize the experiences of women within temporary labour migration programs, but, rather, to take seriously the multifaceted production-consumption and productive-reproductive dimensions of their involvement, especially as these forms of precarious employment proliferate globally. The unique positioning of women as producers (paid and unpaid) and as consumers invites a more careful revisiting of the impact of chemical exposures. With few exceptions, women do not receive meaningful consideration within wider discussions of migrant agricultural labour.

This reluctance to engage with the gendered nature of temporary labour migration in agriculture proves quite myopic. For instance, this is especially worrisome as concerns specific to women have arisen within the context of the SAWP, including stigmatization, sexual harassment, and restrictions and limits to mobility based on gender (Encalada Grez 2011; Preibisch 2007). As Susan Buckingham and Rakibe Kulcur have argued, environmental injustices that uniquely affect women often seem invisible to those not actively studying and searching for them (Buckingham and Kulcur 2009, 664). With limited research pertaining specifically to agricultural work and women's health, risks to the women migrant population continue to be minimized or ignored. Despite these constraints, we must acknowledge the unique ways in which agricultural pesticide exposure plays out on the female body, as well as the distinct and discriminatory effects migrant women experience as a result of the current legal regulatory context of temporary migrant work. As a result, the relatively small number of women affected by these specific circumstances ought not to detract from the significance of the problem.

Conclusion

In drawing attention to the production-consumption nexus, this chapter emphasizes the importance of addressing the occupational pesticide exposure inherent in migrant agricultural labour, particularly the exposures experienced by women workers. We contend that, as the introduction to this book makes clear, environmental harms are central – and not incidental – to industrial production. The structural transformation of agricultural production, which has reaped "efficiencies" by increasing production output that benefits growers and by increasing reliance on chemical inputs, has had negative and lasting implications for labourers, especially migrant women agricultural workers, who act as both producers and consumers of pesticide-doused crops.

This chapter textures the discussion by providing an illustrative example of agricultural workers under Canada's temporary labour migration programs, and specifically the disproportionate harms migrant workers face on the job from pesticide exposures. Existing data suggest that the nature of our social, physical, and built environments contribute to health disparities, and the relationship between pesticide exposure, migration, and health is no exception. It is precisely at the interconnected axes of the biological/scientific and the legal/regulatory that we find explanations for health disparities. The biological effects of pesticides coupled with the politico-legal framework of temporary labour migration render working conditions in agricultural production unhealthy and unsafe. Because women workers in these migrant worker programs inhabit the space at the intersection of these threads, they represent a population in need of sustained and critical attention. In this vein, the dearth of comprehensive and large-scale research on the health risks facing women agricultural workers in Canada is especially disturbing.

Although the relationship between exposures to pesticides and specific health outcomes, such as breast cancer, remains unconfirmed and inconclusive, evidence linking agricultural pesticides to a range of health harms is sufficient to require immediate and concrete action (Clapp, Jacobs, and Loechler 2008). There is an urgent need to protect migrant workers' health and to reduce existing and potential hazards. Implementation of a safe substitution principle, restrictions on the use of pesticides, greater enforcement, especially through meaningful collective bargaining, and citizenship status on arrival are merely the core conditions under which workers can pursue a truly transformative agenda. Waiting to identify firm or conclusive links in advance could prove a risky and irresponsible strategy in light of the potential impacts on human health.

Finally, this chapter highlights the need for further study in the face of increasing use of migrant labour, including within agriculture. Women engaged in agricultural production represent a challenging population to study, as factors – from residence, to occupation, to specific tasks, to broader environment – are difficult to tease apart. That said, we have attempted to situate (with an acknowledged level of tentativeness) research findings within the legal regulatory framework governing temporary labour migration and occupational health and safety, in order to articulate concerns about occupational exposure. We are guided by the belief that "health disparities result not only from individual factors but also from factors operating at multiple levels" (Gee and Payne-Sturges 2004, 1645). Our analysis suggests that women as migrant agricultural workers are disproportionately impacted by pesticide exposure, and that focused research is required across a number of interconnected disciplines, including gender, pesticide exposure, agricultural migrant work, and human health, in order to develop a richer perspective on women's environmental health and targeted action plans that confront women's myriad susceptibilities to chemicals in their environment.

Notes

1 Pesticide use in Canada is heavy. Only six of the twenty-eight OECD nations score worse than Canada in pesticide use per capita (Boyd 2001, 25). Data suggest that between 1970 and 1995, there has been an over 400 percent increase in pesticide use across the country. Despite this figure, Boyd (2001, 25) argues that a precise reading on the presence of pesticides in Canada is challenging to establish, with even Environment Canada lamenting the lack of detail regarding production, use, effect, and potency of pesticides.

2 We acknowledge that the "windows of vulnerability" argument for heightened susceptibility to exposures in women migrant agricultural workers is at the intersection of biological and sociological explanations. That women's bodies are more "vulnerable" to chemical pollution is a contested claim. As Buckingham and Kulcur (2009, 665) note, "it is the continued structuring of females as 'different' to a male norm which results in a biological manifestation of a socio-environmental problem." That said, our contention is not meant to encourage an essentialist analytical approach to the impact of agricultural production on women workers' health. Rather, we aim to open space for taking these health concerns seriously within the critique of industrial capitalist production and the supporting regulatory frameworks.

3 *Occupational Health and Safety Act,* R.S.O. 1990, c. 0.1 (http://www.e-laws.gov.on.ca/html/statutes/english/elaws_statutes_90o01_e.htm); O. Reg. 414/05, *Occupational Health and Safety Act,* R.S.O. 1990, c. 0.1 (http://www.e-laws.gov.on.ca/html/regs/english/elaws_regs_050414_e.htm).

4 OHSA defines an occupational illness as "a condition that results from exposure in a workplace to a physical, chemical or biological agent to the extent that the normal physiological mechanisms are affected and the health of the worker is impaired ... and includes an occupational disease for which a worker is entitled to benefits under the Workplace Safety and Insurance Act, 1997" (OHSA, s. 1[1]).

5 *Ontario (Attorney General) v. Fraser,* 2011 SCC 20 (http://scc.lexum.org/en/2011/2011scc20/2011scc20.pdf).

6 *Labour Relations Act,* S.O. 1995, c. 1, Schedule A (http://www.e-laws.gov.on.ca/html/statutes/english/elaws_statutes_95l01_e.htm); see *Agricultural Employees Protection Act,* S.O. 2002, c. 16 (http://www.e-laws.gov.on.ca/html/statutes/english/elaws_statutes_02a16_e.htm).

7 The notion that residential setting affects vulnerability is interesting, and provides further proof of the importance of considering multiple routes or means of exposure, and of why occupational exposure cannot be evaluated in isolation. Chapter 10 uses the example of farm women – a group that shares elements of risk with seasonal agricultural women workers – to illustrate that even the science of routes of exposure is not immune to gendered analysis. In the home, where women have unique domestic duties, there can be additional risks, as agricultural chemicals can drift from the intended targets and result in the contamination of living spaces.

References

Alavanja, Michael, Jane Hoppin, and Freya Kamel. 2004. "Health Effects of Chronic Pesticide Exposure: Cancer and Neurotoxicity." *Annual Review of Public Health* 25: 155-97, doi:10.1146/annurev.publhealth.25.101802.123020.

Albritton, Robert. 2009. *Let Them Eat Junk: How Capitalism Creates Hunger and Obesity.* Winnipeg: Arbeiter Ring Press.

Arcury, Thomas A., Sara A. Quandt, and Gregory B. Russell. 2002. "Pesticide Safety among Farmworkers: Perceived Risk and Perceived Control as Factors Reflecting Environmental Justice." *Environmental Health Perspectives* 110 (S-2): 233-40, doi: 10.1289/ehp.02110s2233.

Austin Colin, Thomas A. Arcury, Sara A. Quandt, et al. 2001. "Training Farmworkers about Pesticide Safety: Issues of Control." *Journal of Health Care for the Poor and Underserved* 12 (2): 236-49, doi:10.1353/hpu.2010.0744.

Barndt, Deborah. 2002. *Tangled Routes: Women, Work, and Globalization on the Tomato Trail.* New York: Rowman and Littlefield.

Basok, Tanya. 2002. *Tortillas and Tomatoes: Mexican Transmigrant Harvesters in Canada.* Montreal and Kingston: McGill-Queen's University Press.

Bolaria, Singh B. 1992. "Farm Labour, Work Conditions, and Health Risks." In *Rural Sociology in Canada,* edited by David Hay and Gurcharn Basran, 228-45. Toronto: Oxford University Press.

Boyd, David R. 2001. *Canada vs. the OECD: An Environmental Comparison.* Victoria: Eco-Research Chair of Environmental Law and Policy, University of Victoria. http://www.environmentalindicators.com/htdocs/PDF/CanadavsOECD.pdf.

Brophy, James T., Margaret M. Keith, Kevin M. Gorey, et al. 2002. "Occupational Histories of Cancer Patients in a Canadian Cancer Treatment Centre and the Generated Hypothesis regarding Breast Cancer and Farming." *International Journal of Occupational and Environmental Health* 8 (4): 346-53.

–. 2006. "Occupation and Breast Cancer: A Canadian Case-Control Study." *New York Academy of Sciences* 1076: 765-77, doi:10.1196/annals.1371.019.

Brophy, James T., Margaret M. Keith, Andrew Watterson, et al. 2012. "Farm Work in Ontario and Breast Cancer Risk." In *Rural Women's Health,* edited by Beverly D. Lelpert, Belinda Leach, and Wilfreda E. Thurston. Toronto: University of Toronto Press.

Buckingham, Susan, and Rakibe Kulcur. 2009. "Gendered Geographies of Environmental Injustice." *Antipode* 41 (4): 659-83, doi:10.1111/j.1467-8330.2009.00693.x.

Carney, Judith. 2008. "Reconsidering Sweetness and Power through a Gendered Lens." *Food and Foodways* 16 (2): 127-34, doi:10.1080/07409710802085999.

Carson, Rachel. 1962. *Silent Spring.* Boston: Houghton Mifflin.

Citizenship and Immigration Canada. 2011. *Temporary Foreign Worker Program: Operational Instructions for the Implementation of the Immigration and Refugee Protection Regulatory Amendments.* Operational Bulletin 275-C. http://www.cic.gc.ca/english/resources/manuals/bulletins/2011/ob275C.asp.

Clapp, Richard W., Molly M. Jacobs, and Edward L. Loechler. 2008. "Environmental and Occupational Causes of Cancer: New Evidence, 2005-2007." *Review of Environmental Health* 23 (1): 1-37, doi:10.1515/REVEH.2008.23.1.1.

Damstra, Terri, Sue Barlow, Aake Bergman, et al., eds. 2002. *Global Assessment of the State-of-the-Science of Endocrine Disruptors.* Geneva: World Health Organization, International Programme on Chemical Safety. http://www.who.int/ipcs/publications/new_issues/endocrine_disruptors/en/.

Das, Rupali, Andrea Steege, Sherry Baron, et al. 2001. "Pesticide-Related Illness among Migrant Farm Workers in the United States." *International Journal of Occupational and Environmental Health* 7 (4): 303-12. http://www.cdph.ca.gov/programs/ohsep/Documents/migrantfarmworkers.pdf.

Das Gupta, Tania. 2006. "Racism/Anti-Racism, Precarious Employment, and Unions." In *Precarious Employment: Understanding Labour Market Insecurity in Canada,* edited by Leah F. Vosko. Montreal and Kingston: McGill-Queen's University Press.

David Suzuki Foundation. 2010. *What's Inside? That Counts: A Survey of Toxic Ingredients in Our Cosmetics.* Vancouver: David Suzuki Foundation.

de Wolff, Alice. 2006. "Privatizing Public Employment Assistance and Precarious Employment in Toronto." In *Precarious Employment: Understanding Labour Market Insecurity in Canada,* edited by Leah F. Vosko. Montreal and Kingston: McGill-Queen's University Press.

Deacon, G. 2011. *There's Lead in Your Lipstick.* Canada: Penguin.

Dean, Tony. 2010. *Expert Advisory Panel on Occupational Health and Safety: Report and Recommendations to the Minister of Labour.* Toronto: Ontario Ministry of Labour.

Edmunds, Kathryn, Helene Berman, Tanya Basok, et al. 2011. "The Health of Women Temporary Agricultural Workers in Canada: A Critical Review of the Literature." *Canadian Journal of Nursing Research* 43 (4): 68-91.

Encalada Grez, Evelyn. 2008. "Migrant Workers Reap Bitter Harvest in Ontario." *Toronto Star,* 28 October. http://www.thestar.com/comment/article/525483.

–. 2011. "Vulnerabilities of Female Migrant Farm Workers from Latin America and the Caribbean in Canada." *FOCAL: The Canadian Foundation for the Americas.* Policy Brief.

Environmental Working Group. 2008. "Reducing Your Exposure to PBDEs in Your Home." http://www.ewg.org/pbdefree.

Ferguson, Sue. 2004. "Conditions Tough for Canada's Migrant Workers: An Unregulated Can of Worms." *Maclean's,* 11 October.

Frank, Arthur, Robert McKnight, Steven Kirbhorn, and Paul Gunderson. 2004. "Issues of Agricultural Safety and Health." *Annual Review of Public Health* 25: 225-45, doi:10.1146/annurev.publhealth.25.101802.123007.

Gee, Gilbert C., and Devon C. Payne-Sturges. 2004. "Environmental Health Disparities: A Framework Integrating Psychosocial and Environmental Concepts." *Environmental Health Perspectives* 112 (17): 1645-50, doi:10.1289/ehp.7074.

Gold, Daniel B., and J. Helfand, directors. 2002. *Blue Vinyl* [motion picture]. New Video Group.

Goldring, Luin, Carolina Berinstein, and Judith Bernhard. 2007. "Institutionalizing Precarious Immigration Status in Canada." *Early Childhood Education Publications and Research.* Paper 4. http://digitalcommons.ryerson.ca/ece/4.

–. 2009. "Institutionalizing Precarious Migratory Status in Canada." *Citizenship Studies* 13 (3): 239-65, doi:10.1080/13621020902850643.

Green, Anne E., David Owen, Paul Jones, et al. 2008. *Migrant Workers in the South East Regional Economy: Final Report.* Institute of Employment Research, University of Warwick, and BMG Research. London: South East England Development Agency.

Harrison, Jill Lindsey. 2011. *Pesticide Drift and the Pursuit of Environmental Justice.* Cambridge, MA: MIT Press.

HRSDC (Human Resources and Skills Development Canada). 2012. "Temporary Foreign Worker Program Labour Market Opinion Statistics. Annual Statistics 2008-2011, Table 10a." Released 30 April.

Industrial Accident Victims Group of Ontario and Justice for Migrant Workers. 2009. "Intervenor Submission, *Attorney General (Ontario) v. Fraser.*"

Justicia/Justice for Migrant Workers. 2005. "The Seasonal Agricultural Workers Program." http://www.justicia4migrantworkers.org/bc/pdf/sawp.pdf.

Katz, Cindi. 2001. "Vagabond Capitalism and the Necessity of Social Reproduction." *Antipode* 33 (4): 709-28, doi:10.1111/1467-8330.00207.

Krimsky, Sheldon. 2000. *Hormonal Chaos: The Scientific and Social Origins of the Environmental Endocrine Hypothesis*. Baltimore: Johns Hopkins University Press.

Levenstein, Charles, and John Wooding. 2000. "Deconstructing Standards, Reconstructing Worker Health." In *Reclaiming the Environmental Debate: The Politics of Health in a Toxic Culture,* edited by Richard Hofrichter, 39-55. Cambridge, MA: MIT Press.

Lewchuk, Wayne, Alice De Wolff, Andy King, et al. 2006. "The Hidden Costs of Precarious Employment: Health and the Employment Relationship." In *Precarious Employment: Understanding Labour Market Insecurity in Canada,* edited by Leah F. Vosko, 141-62. Montreal and Kingston: McGill-Queen's University Press.

McLaughlin, Janet. 2009. "Trouble in Our Fields: Health and Human Rights among Mexican and Caribbean Migrant Farm Workers in Canada." PhD dissertation, University of Toronto.

Mintz, Sidney W. 1986. *Sweetness and Power: The Place of Sugar in Modern History.* New York: Penguin Books.

Mohapatra, Prabhu P. 2004. "Assam and the West Indies, 1860-1920: Immobilizing Plantation Labour." In *Masters, Servants, and Magistrates in Britain and the Empire, 1562-1955,* edited by Douglas Hay and Paul Craven, 455-80. Chapel Hill: University of North Carolina Press.

Panitch, Leo, and Donald Swartz. 2008. *From Consent to Coercion: The Assault on Trade Union Freedoms.* Toronto: University of Toronto Press.

Picchio, Antonella. 1992. *Social Reproduction: The Political Economy of the Labour Market.* Cambridge: Cambridge University Press.

Preibisch, K. 2007. "Foreign Workers in Canadian Agriculture: Not an All-Male Cast." *FOCALPoint: Canadian Foundation for the Americas* 6: 8-10. Special Edition: Labour Migration and Development. http://www.focal.ca/pdf/focalpoint_se_may-june2007.pdf.

–. 2010. "Pick-Your-Own Labor: Migrant Workers and Flexibility in Canadian Agriculture." *International Migration Review* 44 (2): 404-41, doi:10.1111/j.1747-7379.2010.00811.x.

Preibisch, K., and Evelyn Encalada. 2010. "The Other Side of El Otro Lado: Mexican Migrant Women and Labour Flexibility in Canadian Agriculture." *Signs: Journal of Women in Culture and Society* 35 (2): 289-316.

Salvatore, Alicia L., Jonathan Chevrier, Asa Bradman, et al. 2009. "A Community-Based Participatory Worksite Intervention to Reduce Pesticide Exposures to Farmworkers and Their Families." *American Journal of Public Health* 99 (S-3): S578-S581, doi:10.2105/AJPH.2008.149146.

Sanborn, Margaret, Donald Cole, Kathleen Kerr, et al. 2004. *Pesticides Literature Review.* Toronto: Ontario College of Family Physicians.

Sargeant, Malcolm, and Eric Tucker. 2009. "Layers of Vulnerability in Occupational Health and Safety for Migrant Workers: Case Studies from Canada and the UK." *Policy in Practice in Health and Safety* 7 (2): 51-73.

Satzewich, Vic. 1991. *Racism and the Incorporation of Foreign Labour: Farm Labour Migration to Canada since 1945.* New York: Routledge.

Scott, Dayna Nadine. 2008. "Confronting Chronic Pollution: A Socio-Legal Analysis of Risk and Precaution." *Osgoode Hall Law Journal* 46 (2): 293-343. http://www.ohlj.ca/english/documents/OHLJ46-2_Scott_ConfrontingChronicPollution.pdf.

Sever, L.E., T.E. Arbuckle, and A. Sweeney. 1997. "Reproductive and Developmental Effects of Occupational Pesticide Exposure: The Epidemiologic Evidence." *Occupational Medicine* 12 (2): 305-25.

Smith, Adrian A. 2005. "Legal Consciousness and Resistance in Caribbean Seasonal Agricultural Workers." *Canadian Journal of Law and Society* 20 (2): 95-122, doi: 10.1353/jls.2006.0027.

Smith, Doug. 2000. *Consulted to Death: How Canada's Workplace Health and Safety System Fails Workers*. Winnipeg: Arbeiter Ring Publishing.

Smith, Rick, and Bruce Lourie. 2009. *Slow Death by Rubber Duck: How the Toxic Chemistry of Everyday Life Affects Our Health*. Toronto: Knopf Canada.

Swinton, Katherine. 1983. "Enforcement of Occupational Health and Safety Legislation: The Role of the Internal Responsibility System." In *Studies in Labour Law,* edited by Ken Swan and Katherine Swinton, 145-60. Toronto: Butterworths.

Tucker, Eric. 1990. *Administering Danger in the Workplace: The Law and Politics of Occupational Health and Safety Regulation in Ontario, 1850-1914*. Toronto: University of Toronto Press.

–. 2006. "Will the Vicious Circle of Precariousness Be Unbroken? The Exclusion of Ontario Farmworkers from the Occupational Health and Safety Act." In *Precarious Employment: Understanding Labour Market Insecurity in Canada,* edited by Leah F. Vosko, 256-76. Montreal and Kingston: McGill-Queen's University Press.

Vaughan, Elaine. 1993a. "Chronic Exposure to an Environmental Hazard: Risk Perceptions and Self-Protective Behaviour." *Health Psychology* 12 (1): 74-85, doi:10.1037/0278-6133.12.1.74.

–. 1993b. "Individual and Cultural Differences in Adaptation to Environmental Risks." *American Psychology* 48 (6): 673-80, doi:10.1037/0003-066X.48.6.673.

Villarejo, Don, and Stephen A. McCurdy. 2008. "The California Agricultural Workers Health Survey." *Journal of Agricultural Safety and Health* 14 (2): 135-46.

Vosko, Leah. 2006. "Precarious Employment: Towards an Improved Understanding of Labour Market Insecurity." In *Precarious Employment: Understanding Labour Market Insecurity in Canada,* edited by Leah Vosko, 3-42. Montreal and Kingston: McGill-Queen's University Press.

Wilk, Valerie A. 1986. *The Occupational Health of Migrant and Seasonal Farmworkers in the United States*. 2nd ed. Washington, DC: Farmworker Justice Fund.

Woodruff, Tracey J., Amy D. Kyle, and Frederic Y. Bois. 1994. "Evaluating Health Risks from Occupational Exposure to Pesticides and the Regulatory Response." *Environmental Health Perspectives* 102 (12): 1088-96.

Conclusion Thinking about Thresholds, Literal and Figurative

Dayna Nadine Scott

As the chapters in this volume make clear, we are interested in critical, engaged theoretical work on the topic of gender and environmental health. At the same time, we strive continuously to put science, policy, and theory into a concrete social and political context. This includes asking what it means at the end of the day, and what specific steps might be taken, and by whom, to make the changes that would not only bring about better policy but also contribute to social transformation in thinking about toxics, gender, and environmental health. We are seizing the opportunity, with this volume, to mark a moment in thinking about thresholds: literal and figurative.

What is it about this moment that we find compelling, and why do we think it presents us with the opportunity to consider thresholds? As mentioned at the outset, research is starting to accumulate that links exposures to certain chemicals, at very low doses, at certain key times, to serious environmental health harms. These "critical windows of vulnerability" have a distinctly biological, developmental, and thus gendered nature. The fact that the exposures in question are very low doses translates into particular regulatory challenges – almost all of our current regulatory regimes are based on the notion of a threshold. This is demonstrated clearly by Jyoti Phartiyal with respect to the Guidelines for Canadian Drinking Water Quality; it is also true for the air pollution regulations that affect the Aamjiwnaang First Nation as described in the introduction (Scott 2008), and it is true for the risk assessment of toxics generally under the Canadian Environmental Protection Act,

1999 (CEPA), as explained by Dayna Nadine Scott and Sarah Lewis. Our legal and regulatory regimes have clearly not kept pace with the new toxicological (and endocrinological) realities.

The nature of contemporary pollution harms is changing. First, they are diffuse, body-altering, intergenerational, and very difficult to link to any particular chemical or source. Second, they are unevenly distributed, including along a gender axis. And third, they derive from continuous low-dose exposures that are largely within legally sanctioned limits. It is this third characteristic that demonstrates how deeply embedded the idea of "threshold" is in our policy and regulatory structures. New strategies of resistance and new innovative regulatory solutions are required. In order to make inroads, we – collectively – need to strive more than ever to strengthen our connections across disciplines. We, as social scientists (and lawyers!), need to truly understand the direction of the research emerging from the natural sciences. As natural scientists, we need to find a way to keep the social and the political in view. The consideration of gender in research on chemicals is a complex matter, and we hope that by working with critical, feminist, environmental justice activists, we can improve the design of studies and the knowledge we are able to take from them.

There is something else about the contemporary moment that we find compelling. We have witnessed, over the past several years, a willingness – if uneven and tentative – on the part of governments at all levels to begin rethinking regulatory approaches to the management of health risks from exposures to toxic chemicals. This is not to say that the recent spate of reforms has been sufficient, or even entirely positive, but simply that reforms are happening. There is space for us to engage. As explored in Chapter 3, the federal government announced a bold new initiative in 2006 called "the Challenge" as part of its Chemicals Management Plan. Under the Challenge, the government identified 200 high-priority chemicals for which the regulator indicated it was "predisposed" to a finding of toxicity. The presumption, it was thought, would apply unless the challenged stakeholders – namely, the industry that produces, imports, or uses the listed chemicals – submitted information sufficient to rebut it (Scott 2009). Notwithstanding the fact that the implementation of this predisposition under the Challenge proved to be less revolutionary than initially hoped (see the chart in Chapter 3), the significance of the nod in this direction should not be missed. The "presumption of toxicity" would have reversed a long-standing policy in Canada and the United States in which the burden of proof with respect to chemical safety has rested on those wishing to challenge the toxicity of a particular chemical, not with those wishing to profit from its use (Scott 2009; Collins and McLeod-Kilmurray 2011). It may be that the Challenge, in hindsight, will not be seen as the first move of many towards a more precautionary regulatory trajectory, or a "reverse onus" situation, but it marks the first significant movement on this file in some time. There is a sense of flux as regulatory reforms to toxic substances regimes are being implemented across the European Community with the controversial REACH initiative, and are being contemplated elsewhere in North America (Abelkop and Graham, 2014).

In Ontario, the government enacted the Toxics Reduction Act in 2009.[1] If there is an approach more progressive and precautionary than shifting the burden of proof in chemical risk assessment, it is definitely "toxics use reduction" or the "clean production approach" (Tickner and Geiser 2004; Geiser 2004). This is the movement that says: we don't need to spend any more time and resources endlessly engaged in contested and convoluted risk assessment processes, we can simply get on with the business of finding safer alternatives, finding better ways to achieve our social objectives, and reducing our reliance on toxic substances (see, for example, O'Brien 2000). Ontario's legislation, however, doesn't go as far as Massachusetts's forward-looking law that requires manufacturers to systematically reduce their use of chemicals over time, and includes modest penalties that kick in if they fail to comply.[2] Ontario's law merely requires manufacturers to track and quantify their use and production of listed substances, and to report annually on a *plan* to reduce their use/manufacture of those substances (Castrilli 2011). As Castrilli says, however, it is the first Ontario environmental law to approach toxics management by looking upstream, at pollution prevention, rather than at the end of the pipe, as in pollution control or abatement.

The thinking behind this scheme is the same logic that underlies the National Pollutant Release Inventory (NPRI), a federal database established under CEPA that requires large emitters of certain substances to report annually on the volumes of their emissions and makes that information publicly available on the web. The logic is that once managers are forced to turn their minds to the emissions and face the public scrutiny related to them, they will naturally work to find ways to reduce them over time. The evidence from years of experience with NPRI is mixed (see Antweiler and Harrison 2003), but it seems that in these times of "smart regulation" (Grabosky and Gunningham 1998) the best our governments can muster is a weak commitment, incorporating voluntary measures that are intended, apparently, to "incentivize" corporate actors to move in the direction we desire (Gorrie 2009).

This is complicated by the fact that these initiatives are often demanded and celebrated from the grassroots: they are framed in the language of the citizen's "right to know." Take as an example Toronto's Environmental Reporting and Disclosure by-law, which came into effect in 2010.[3] The program is aimed at small and medium-sized businesses throughout the city that use or emit toxic substances, and is geared towards empowering residents to act in their own neighbourhoods to put pressure on businesses to come clean. It was largely influenced by an inspired campaign by the advocacy group Toronto Environmental Alliance called "Secrecy Is Toxic," in which Torontonians were invited to contribute to a wiki-map revealing polluting activities throughout the city.

These new reforms are not perfect. What they have in common, of course, is that they demand a lot from non-state actors and not much from state actors. The trend has further penetrated into the campaigns of advocacy groups: from organizations seeking to protect Canadians from toxics in consumer products and cosmetics, such as Environmental Defence, to those seeking to implement breast cancer prevention, such as Breast Cancer Action Montreal, all have promoted labelling campaigns in recent years based on the idea of a citizen's right to know. We are conflicted about these campaigns: we too want to see carcinogens and endocrine disruptors taken out of consumer products and cosmetics, but we see the focus on information as misguided. Labelling of toxic and potentially toxic chemicals might be a feasible avenue for reducing exposures for some women, as MacKendrick aptly demonstrates in Chapter 2, but it is essential that we recognize that women are disproportionately responsible for putting into place and carrying out the practices needed to avoid exposures to toxic chemicals, *and* that women vary dramatically in their abilities to make effective use of labels. In a recent blog post with Robyn Lee, we argued that the work of "precautionary consumption" is undeniably women's work (Lee and Scott 2014). Since women assume primary responsibility for ensuring the health of family members, and since "smart shopping" takes large investments of time and energy, we argue that campaigns to reduce chemical exposures from consumer goods through labelling are misguided because they fail to take into account this unequal division of labour (see also MacKendrick 2014).

Further, while the most educated and motivated of us may be able to buy the right sofas and carpets and electronics for our homes, none of us will choose these for our schools, offices, hospitals, or libraries. Our control over these exposures will continue to be limited, even with labelling in place. There is also the risk that when faced with a huge number of toxic and potentially toxic chemicals to be avoided, we will make only a few symbolic changes in our consumption habits, potentially precluding further political engagement (MacGregor 2006). Individual changes in consumption, although important, cannot replace collective action aimed at regulatory reform. While education of consumers and pressure applied by consumers on manufacturers are important avenues for change, it is essential that government take primary responsibility for regulating toxic chemicals and getting them off store shelves, so that everyone is protected. Fundamentally, as the orientation of this volume seeks to make clear, we must keep in mind that while some consumers would be able to choose to protect themselves from exposures to carcinogens with labelling in place, the workers inside the manufacturing facilities and the communities surrounding them will continue to be exposed until the substances are eliminated from our economies.

The types of reforms we see emerging to deal with the complexity of contemporary pollution harms are those characterized by the "smart regulation" movement. They focus on the incentivizing potential of information disclosure, making space for innovation initiated by industry actors and tapping into the energy and enthusiasm of ordinary citizens (Wood, 2006). Predictably, as incentives move into the foreground, enforcement fades into the background. This is by design, not by accident. Disempowered communities, workers, and women will not be served by these reforms.

Yet, we began this discussion of the critical moment in thinking about gender and environmental health with more optimism than this. There does seem to be a willingness to begin rethinking regulatory approaches to the management of toxic chemicals. The regulatory reforms are not yet in the form we would like to see, but the fact that they have been brought forward indicates that there is space to talk about toxics and environmental health out in the open. We have come to a point at which it is possible (and perhaps even politically necessary) for the federal government to acknowledge its responsibility for ensuring that Canadians are not exposed to toxic substances in consumer goods; for provincial governments to understand that the long-term health of residents depends on fundamental changes to how we approach the use of toxic substances in our production processes; and for municipal governments to recognize that part of their mandate to protect public health includes regulating the use of toxics by businesses. Perhaps we have passed the threshold, figuratively speaking.

This place was reached, this threshold crossed, through the relentless efforts of environmental health activists, the women's environmental health movement, and environmental justice organizers. We have to continue this work, continue to push both the boundaries of our own understandings and the limits of our own perspectives. As our feminist political economy of pollution framework makes clear, we have to avoid the tendency to let the "chemical enemy" become the central focus of concern, rather than the system that produces the chemical and the social, political, and gender relations that enable it to continue being produced and consumed. Further, we have to seize these openings, these moments in contemporary thinking about chemicals and toxics in an effort to catalyze a process of social learning and transform our way of thinking about and governing our exposures to chemicals.

Notes

1 *Toxics Reduction Act,* S.O. 2009, c. 19 (http://www.e-laws.gov.on.ca/html/statutes/english/elaws_statutes_09t19_e.htm).
2 *Massachusetts Toxics Use Reduction Act,* 1989, General Laws, c. 21I (http://www.malegislature.gov/Laws/GeneralLaws/PartI/TitleII/Chapter21I).
3 *Environmental Reporting and Disclosure* (By-law 1293-2008), Toronto Municipal Code, c. 423 (http://www.toronto.ca/legdocs/municode/1184_423.pdf).

References

Antweiler, Werner, and Kathryn Harrison. 2003. "Toxic Release Inventories and Green Consumerism: Empirical Evidence from Canada." *Canadian Journal of Economics* 36 (2): 495-520.

Castrilli, Joseph. 2011. "Toxics Reduction: The New Paradigm in Ontario Environmental Law." Prepared for the Law Society of Upper Canada. http://s.cela.ca/files/811LSUC%20Six%20Minute%20Lawyer.pdf.

Collins, Lynda M., and Heather McLeod-Kilmurray. 2011. "Material Contribution to Justice? Toxic Causation after *Resurfice Corp. v. Hanke.*" *Osgoode Hall Law Journal* 48 (3 and 4): 411-56. http://ohlj.ca/english/documents/3_48_4_CollinMcleod_readyforprint_22-06-11.pdf.

Geiser, Ken. 2004. "Pollution Prevention." In *Environmental Governance Reconsidered: Challenges, Choices, and Opportunities,* edited by Robert F. Durant, Daniel J. Fiorino, and Rosemary O'Leary, 427-54. Cambridge, MA: MIT Press.

Gorrie, Peter. 2009. "McGuinty's Broken Promise on Toxic Chemicals." *Toronto Star,* 18 April. http://www.thestar.com/News/Insight/article/620441.

Grabosky, Peter, and Neil Gunningham. 1998. *Smart Regulation: Designing Environmental Policy.* Oxford: Oxford University Press.

Lee, Robyn, and Dayna Nadine Scott. 2014. "(Not) Shopping Our Way to Safety." Canadian Women's Health Network. http://www.cwhn.ca.

MacGregor, Sherilyn. 2006. *Beyond Mothering Earth: Ecological Citizenship and Politics of Care.* Vancouver: UBC Press.

MacKendrick, Norah. 2014. "More Work for Mother: Chemical Body Burdens as Maternal Responsibility." *Gender and Society* 28 (3).

O'Brien, Mary. 2000. *Making Better Environmental Decisions: An Alternative to Risk Assessment.* Cambridge, MA: MIT Press.

Scott, Dayna Nadine. 2008. "Confronting Chronic Pollution: A Socio-Legal Analysis of Risk and Precaution." *Osgoode Hall Law Journal* 46 (2): 293-343. http://www.ohlj.ca/english/documents/OHLJ46-2_Scott_ConfrontingChronicPollution.pdf.

–. 2009. "Testing Toxicity: Proof and Precaution in Canada's Chemicals Management Plan." *Review of European Community and International Environmental Law (RECIEL)* 18 (1): 59-76, doi:10.1111/j.1467-9388.2009.00621.x.

Tickner, Joel, and Ken Geiser. 2004. "The Precautionary Principle Stimulus for Solutions- and Alternatives-Based Environmental Policy." *Environmental Impact Assessment Review* 24: 801-24.

Wood, Stepan. 2006. "Voluntary Environmental Codes and Sustainability." In *Environmental Law for Sustainability,* edited by Benjamin J. Richardson and Stepan Wood, 229-76. Oxford and Portland: Hart Publishing.

Glossary

Key Technical and Scientific Concepts

biomonitoring: A method for analyzing tissues and fluids for biological markers of contaminant exposure.

bisphenol A (BPA): A substance commonly found in clear hard plastics, the linings of tin cans, and debit receipts. It has been found to have the potential to cause endocrine disruption.

body burden: The internal chemical load carried by organisms, including people.

brominated flame retardants (BFRs): Substances commonly added to consumer goods to reduce their flammability. This type contains bromine.

carcinogen: A substance that may cause or lead to cancer.

carcinogenicity: The ability of a substance to induce cancer.

chemical mode of action: A chemical's mechanism of toxicity, based on how it influences or interacts with the body.

chemical/health risk assessment: An evaluation carried out by governmental bodies estimating the risks associated with exposure to a given chemical and its potential harm to human health, based on chemical properties, available scientific and experimental research, biomonitoring data, and other relevant information.

chronic, low-dose exposures: Frequent or continuous exposure to a given chemical or contaminant that is present at low levels in the environment.

critical windows of vulnerability: Key developmental or reproductive life stages where the body can be more biologically vulnerable or influenced by exposure to chemicals in the environment.

cumulative/additive/aggregate/synergistic chemical impacts: The effects produced by the interaction of a combination of chemicals in the environment that may not be produced by a single chemical on its own.

developmental toxicity: The ability of a chemical to adversely affect a developing organism, either as a result of parental exposure or exposure in utero or after birth. The adverse effects may manifest later in life.

endocrine-disrupting chemicals (EDCs): Certain synthetic chemicals that are structurally similar to hormones in the body and mimic hormone action by binding with and activating available hormone receptors. These interactions can trigger changes in how cells and organs function, and have an impact on a diverse array of metabolic, epigenetic (gene imprinting), growth, and reproductive processes in the body. Disruptions can contribute to reproductive and developmental disorders and various cancers later on in life and within subsequent generations.

epidemiology: The study of the distribution of disease in populations, and the study of the factors influencing that distribution.

epigenetic changes: Changes to the imprinting of genes or gene expression, usually as a result of chemical exposure.

fibromyalgia: A medical disorder characterized by chronic widespread pain combined with symptoms of fatigue, sleep disturbance, joint stiffness, depression, anxiety, and stress, among others. Evidence has been growing regarding the connection between fibromyalgia and exposure to endocrine-disrupting chemicals in the environment.

gender: The culturally defined characteristics, differences, and roles that are socially constructed and assigned to women and men (see more detailed discussion in the Introduction).

genotoxicity: The degree to which an agent is known to cause damage to DNA, or cause genetic mutations that can result in cancer.

in vitro: Experiments or studies conducted using components of an organism, isolated from their usual biological context within a test tube or petri dish, in order to permit more detailed analysis at the molecular level.

in vivo: Experiments or studies that use a whole, living organism to carry out tests.

latency effect: The tendency for there to be a long time delay between a toxic exposure and the onset of a related illness.

lead: A heavy metal, once commonly used as an additive in gasoline and paint, for which absorption and/or ingestion is known to cause poisoning. It affects the brain, the nervous and digestive systems, and blood.

linear dose-response curve: An assumption of traditional toxicology dictating that the greater the dose of exposure to a chemical, the greater the harm to human health.

longitudinal chemical effects: The effects of chemical exposure on human health over an extended period of time.

margin of exposure: An evaluation of exposure carried out in government chemical risk assessments, which attempts to determine whether a chemical is safe by calculating the difference between the estimated threshold level and estimated exposure level. The estimated threshold is the level at which a chemical is considered harmful to human health, whereas estimated exposure accounts for the level of exposure that might be experienced by an individual.

maximum acceptable concentrations (MACs): Government-set guidelines for the maximum level of a substance or contaminant in drinking water.

microbiological contaminants: Micro-organisms such as viruses and bacteria that may enter water from sewage treatment plants, septic systems, or agricultural livestock operations, for example.

multiple chemical sensitivity (MCS): Also known as environmental sensitivity or intolerance, MCS describes an acquired environment-linked condition where an individual reacts to substances at exposure levels commonly tolerated by most people. MCS often stems from an initiating toxic exposure, which leads to a loss of tolerance or hypersensitivity to low levels of diverse and unrelated chemical triggers in the environment.

mutagenicity: The degree to which an agent can induce or increase the frequency of mutation in genes.

nitrate: An inorganic compound that enters water primarily through the use of agricultural fertilizers.

non-linear dose-response curve: A new understanding of chemical toxicology based on emerging environmental health research indicating that in certain cases, low doses of a substance can result in greater harm to human health than higher doses.

off-gassing: The release of chemicals into the air from products in the indoor environment under normal conditions of temperature and pressure. Commonly noted as the odor associated with a new carpet or a new car.

PBiT: Chemicals that are characterized as persistent, bioaccumulative, and inherently toxic.

PCBs (polychlorinated biphenyls): A broad family of synthetic chemicals that were widely manufactured for use in industrial and commercial applications, such as electrical equipment, because of their non-flammability, chemical stability, and insulating properties. They are now accepted as highly toxic and bioaccumulative, and their use and manufacture have been banned in most parts of the world.

phthalates: A group of synthetic chemicals used to soften plastics, especially polyvinyl chloride (PVC). They are present in an astonishing array of consumer and pharmaceutical products, and are beginning to be banned from children's toys and other applications over health concerns in many jurisdictions.

POPs (persistent organic pollutants): A broad class of substances that persist in the environment, bioaccumulate, and present a range of risks to human health and the environment. An international convention now prevents the use, manufacture, and import of the twelve worst POPs, known as the "dirty dozen," including the pesticide DDT and PCBs, and controls the release of unintentionally produced POPs such as dioxins and furans.

precautionary consumption (PC): A set of practices for "green" shopping thought to exercise precaution at the individual consumer level, carried out disproportionately by women.

reproductive toxicity: The toxic effects of chemical exposure on an individual's reproductive system, or on the development and reproductive system of their offspring.

sex: The biological characteristics that distinguish men's and women's bodies (for a more detailed discussion, see the Introduction).

sex-/gender-"disaggregated" data: Scientific or statistical data that are separated or divided based on sex or gender.

social determinants of health: Factors that influence an individual's achievement of optimal health status and well-being. Factors are based on a person's social location and may include: age, sex, biology, genetic constitution, gender, culture, race, ethnicity, sexuality, physical ability, income level, social status, social support networks, level of education, employment status, type or nature of employment, workloads and working conditions, levels of stress and violence in their lives, access to nutritious foods and clean air or water, reproductive functions and ability to practise safe sex, personal health practices, coping skills, support, healthy development as a child and access to health services and resources, physical environment, and the quality of the greater ecosystem. Individuals who experience intersecting disadvantages or discrimination in any of these areas may face greater challenges in living a safe and healthy life.

social location: A person's place in society, as determined by socioeconomic status, race, level of education, age, and so on.

sodium laureth sulfate: A chemical detergent and surfactant found in many personal care products, such as toothpaste and shampoo.

threshold: The exposure level for a chemical substance at which models predict that health effects are expected to occur. Below this level, adverse effects are expected to be rare. The concept of "threshold" depends upon the idea that "the dose makes the poison"; in other words, increasing doses are expected to correspond with a greater chance of health effects, and low doses are presumed to correspond with "no effect."

tolerable daily intake: A value that dictates the "safe" limit of chemical exposure in relation to various pathways, including air, food, and drinking water.

toxic chemicals/substances: Can be found naturally in the environment, or can be synthetic.

toxicity hazard endpoint: A biological event used to determine when a change in the normal function of the human body occurs as a result of toxic exposure. Such an event can include the growth of cancerous tumours or the development of reproductive irregularities.

toxicology: A scientific discipline that is concerned with the nature, effects, and detection of poisons.

trichloroethylene (TCE): A volatile and highly toxic organic compound often found as a solvent.

triclosan: An antiseptic that is active against a wide range of bacteria and fungi. It is widely present in hand soaps and gels encouraged from a public health perspective, but its effectiveness and toxicity are now the subject of intense controversy.

volatile organic compounds (VOCs): Organic chemicals with a high vapour pressure at room temperature; they often end up in indoor air. Chronic exposures to VOCs can be dangerous to human health or cause harm to the environment. Low doses and a latency effect mean that research into the health effects of VOC exposure is difficult.

Key Laws, Policies, and Regulations

Canadian Environmental Protection Act, 1999 (CEPA): The primary legislative tool for managing toxic chemical risks. Includes requirements for assessing and managing substances in use in Canada or being released into the Canadian environment.

"CEPA-toxic": Describes substances that meet the definition of "toxic" under s. 64 of the Canadian Environmental Protection Act, as assessed by the Minister of Environment and the Minister of Health.

Chemicals Management Plan (CMP): A joint initiative of Health Canada and Environment Canada launched in 2006 to improve the degree of protection against hazardous chemicals in Canada and ensure their proper management through a number of measures.

Cosmetic Ingredient Hotlist: A record of prohibited and restricted cosmetic ingredients published by Health Canada. Once a chemical is placed on the hotlist, the government can require industry to remove the ingredient from a formulation, reduce the concentration of the ingredient, provide evidence that the product is safe for its intended use, or confirm that the product is labelled as required.

Domestic Substances List (DSL): An inventory of substances that were already being used in Canada or being released into the Canadian environment when the requirement for pre-market approval was introduced in 1999, and that may have toxic risks.

Guidelines for Canadian Drinking Water Quality: Voluntary measures developed by Health Canada and the Federal-Provincial-Territorial Committee on Drinking Water that set maximum acceptable concentrations (MACs) as well as aesthetic and operational guidelines for microbiological, chemical, and radiological substances found in drinking water.

hazards identification approach: A form of chemicals evaluation that detects a chemical's effect on key events in biological processes known or suspected to raise the risk of development of a specific disease or disorder.

List of Toxic Substances (in Schedule 1 of the Canadian Environmental Protection Act, 1999): A list of chemicals that meet the criteria for toxicity under CEPA, based on the evaluation of PBiT (persistence, bioaccumulation, inherent toxicity), risk of exposure, and risk of carcinogenicity, mutagenicity, developmental and reproductive toxicity. Once a chemical is added to the list, the government may introduce risk management measures to restrict or control its use, manufacture, or importation.

Ministerial Challenge ("The Challenge"): An element of the federal government's Chemicals Management Plan that calls on industry to provide information about the properties, uses, releases, and management of over 200 high-priority chemicals that are PBiT (persistent, bioaccumulative, and inherently toxic) and have a high likelihood of exposure. These chemicals are designated as toxic under the Canadian Environmental Protection Act unless industry provides evidence to the contrary.

Pest Control Products Act: An act that controls the chemical ingredients in, and use of, pesticides in Canada.

Precautionary Principle: Allows for restrictions on human activities suspected to pose a threat to the environment or human health, by stating that proof of safety must be established before new technologies are introduced.

risk management measures: Restrictions placed on substances deemed toxic by the federal government under the Canadian Environmental Protection Act, 1999. Measures include regulations, pollution prevention plans, environmental emergency plans, guidelines, and voluntary codes of practice, among others.

Safe Drinking Water Act (2002): A law regarding the treatment and distribution of drinking water in Ontario, enacted in response to recommendations in the report of the Walkerton Inquiry. Among its directives, the act creates, through regulation, legally binding standards for contaminants in drinking water; makes it mandatory to use licensed and accredited laboratories for drinking water testing and to report adverse test results where contaminants in drinking water do not meet drinking water quality standards; requires that all operators of municipal drinking water systems be trained and certified; and establishes a licensing regime for drinking water systems.

Significant New Activity (SNAc) Notices: Requests by the federal government for submission of additional information on existing chemicals under the Chemicals Management Plan when the government suspects that new activities may increase exposure potential and contribute to a substance's becoming "CEPA-toxic."

toxics use reduction: A regulatory approach to pollution prevention that seeks to target and measure reductions in the upstream use of toxic materials, instead of the downstream assessment of risks to health and the environment.

Contributors

Bita Amani is an Associate Professor of Law at Queen's University, Faculty of Law, and is called to the Ontario bar.

Matthias Beck is a professor of public sector management at Queen's University Belfast. His main research interests are risk management and risk regulation, with a particular focus on the public sector, public/private partnerships, and state/business relationships in transitional and developed economies.

James T. Brophy worked for eighteen years as a director for the Occupational Health Clinics for Ontario Workers and now teaches ocasionally in the Department of Sociology, Anthropology and Criminology at the University of Windsor. His current research focus is on occupational and environmental risk factors for breast cancer.

Samantha Cukier is currently pursuing a PhD at Johns Hopkins Bloomberg School of Public Health in the Department of Health Behavior and Society, focusing on the role of alcohol marketing in the normalization of alcohol use.

Robert DeMatteo was the Director of Occupational Health and Safety for the Ontario Public Service Employees Union for twenty-six years, and is engaged in a comprehensive research and advocacy program on the health impact of worker exposures in the plastics industry.

Troy Dixon holds a BA in Political Science from York University and an MA from the University of Toronto. He is a former research assistant with the National Network on Environments and Women's Health.

Warren G. Foster is a professor in the Reproductive Biology Division in the Department of Obstetrics and Gynecology at McMaster University. He is an expert in reproductive/developmental toxicology and reproductive endocrinology, and currently investigates the role of environmental pollutants in human health.

William Fraser is a professor in the Department of Obstetrics and Gynaecology at the University of Montreal and an adjunct faculty member at the Department of Social and Preventive Medicine. He is the holder of a Canada Research Chair in Perinatal Epidemiology with a program of research broadly focused on maternal-fetal medicine.

Michael Gilbertson worked for many years as a biologist studying the effects of persistent toxic substances on the health of fish, wildlife, and humans in the context of the Great Lakes. He is currently involved in policy work on endocrine disruptors.

Laila Zahra Harris holds an MA in Anthropology from the University of Guelph. She is a former graduate fellow in Women's Health and the Environment with the National Network on Environments and Women's Health.

Margaret M. Keith has taught in the Department of Sociology, Anthropology and Criminology at the University of Windsor and is affiliated with the University of Stirling in Scotland. She has pioneered and developed participatory mapping exercises for understanding occupational and environmental harms tied to chemical exposures.

Sarah Lewis holds an MES from York University, with areas of interest in environmental justice, gender-environment relationships, local knowledge production, and grassroots activism.

Norah MacKendrick is an assistant professor in the Department of Sociology at Rutgers University, New Jersey. Her work brings together the study of gender, environment, and politics.

Josephine Mandamin is an Anishinawbe grandmother best known for leading the "Mother Earth Water Walk," an annual walk around the five Great Lakes to raise awareness of political and spiritual issues around water.

Patricia Monnier is an associate professor in the Department of Obstetrics and Gynaecology at McGill University. She is a specialist in reproductive endocrinology and infertility, with a PhD in cell biology and immunology and a research program focused on environmental impacts on reproductive health.

Jean Morrison is a women's services coordinator at Addiction Services, Annapolis Valley, Nova Scotia.

Jyoti Phartiyal is the projects coordinator for the National Network on Environments and Women's Health, and a candidate for an MES from York University.

M. Ann Phillips is an Ara'Guacu' Onile Wellness Dojo founder and creative director.

Lauren Rakowski holds a JD from Osgoode Hall Law School (2012) and has a Master's in Environment and Sustainability (MES) from the University of Western Ontario.

Nancy Ross is pursuing a PhD in the School of Social and International Studies, Bradford University, UK, while continuing to work as the Women Services Coordinator, Addiction Services, South Shore Health, Nova Scotia.

Annie Sasco is a team leader, epidemiology for Cancer Prevention Team, Victor Segalen Bordeaux 2 University.

Dayna Nadine Scott is the director of the National Network on Environments and Women's Health and an associate professor at Osgoode Hall Law School and the Faculty of Environmental Studies at York University.

Dugald Seely is the director of the Department of Research and Clinical Epidemiology at the Canadian College of Naturopathic Medicine, and executive director of the Ottawa Integrative Cancer Centre.

Adrian A. Smith is a socio-legal scholar studying the legal regulation of labour migration. He is an assistant professor in the Department of Law and Legal Studies at Carleton University.

Tasha Smith is a women's services coordinator at Addiction Services, South West Nova Scotia.

Alexandra Stiver has an MA in Anthropology and was a research fellow at the National Network on Environments and Women's Health in 2009.

Maria P. Velez is a fellow in reproductive endocrinology and infertility in the Department of Obstetrics and Gynaecology at the University of Montreal, and a PhD candidate in Public Health/Epidemiology at the Department of Social and Preventive Medicine.

Aimée L. Ward is an assistant research fellow currently coordinating the Adolescent Mobility Health Consortium (AMHC) within the Injury Prevention Research Unit, Department of Preventive and Social Medicine, Dunedin School of Medicine in New Zealand.

Andrew E. Watterson is a professor of health, the director of the Centre for Public Health and Population Health, and the head of the University of Stirling's interdepartmental Occupational and Environmental Health Research Group. His current research interests relate to the interface between science, policy, regulation, and civil society.

Sarah Young is a research associate for the Canadian College of Naturopathic Medicine, and program coordinator for the Ottawa Integrative Cancer Centre.

Index